Polymer Nanocomposite Materials

Polymer Nanocomposite Materials

Applications in Integrated Electronic Devices

Edited by

Ye Zhou
Guanglong Ding

Editors

Ye Zhou
Shenzhen University
Institute for Advanced Study
Room 358, Administration Building
518060 Shenzhen
China

Guanglong Ding
Shenzhen University
Institute for Advanced Study
Room 400-3, College of EST
518060 Shenzhen
China

Cover © Alexa_Space/Shutterstock

All books published by **Wiley-VCH** are carefully produced. Nevertheless, authors, editors, and publisher do not warrant the information contained in these books, including this book, to be free of errors. Readers are advised to keep in mind that statements, data, illustrations, procedural details or other items may inadvertently be inaccurate.

Library of Congress Card No.:
applied for

British Library Cataloguing-in-Publication Data
A catalogue record for this book is available from the British Library.

Bibliographic information published by the Deutsche Nationalbibliothek
The Deutsche Nationalbibliothek lists this publication in the Deutsche Nationalbibliografie; detailed bibliographic data are available on the Internet at <http://dnb.d-nb.de>.

© 2021 WILEY-VCH GmbH, Boschstr. 12, 69469 Weinheim, Germany

All rights reserved (including those of translation into other languages). No part of this book may be reproduced in any form – by photoprinting, microfilm, or any other means – nor transmitted or translated into a machine language without written permission from the publishers. Registered names, trademarks, etc. used in this book, even when not specifically marked as such, are not to be considered unprotected by law.

Print ISBN: 978-3-527-34744-5
ePDF ISBN: 978-3-527-82648-3
ePub ISBN: 978-3-527-82650-6
oBook ISBN: 978-3-527-82649-0

Typesetting   SPi Global, Chennai, India
Printing and binding CPI Group (UK) Ltd, Croydon, CR0 4YY

Printed on acid-free paper

C103159_040321

Contents

Preface *xi*

1 Introduction of Polymer Nanocomposites *1*
Teng Li, Guanglong Ding, Su-Ting Han, and Ye Zhou
1.1 Introduction *1*
1.2 The Advantage of Nanocomposites *3*
1.3 Classification of Nanoscale Fillers *5*
1.3.1 One-Dimensional Nanofillers *5*
1.3.2 Two-Dimensional Nanofillers *6*
1.3.3 Three-Dimensional Nanofillers *6*
1.4 The Properties of Polymer Nanocomposites *6*
1.5 Synthesis of Polymer Nanocomposites *7*
1.5.1 Ultrasonication-assisted Solution Mixing *8*
1.5.2 Shear Mixing *9*
1.5.3 Three Roll Milling *9*
1.5.4 Ball Milling *10*
1.5.5 Double-screw Extrusion *10*
1.5.6 In Situ Synthesis *10*
1.6 Conclusions and Future Outlook *11*
References *11*

2 Fabrication of Conductive Polymer Composites and Their Applications in Sensors *21*
Jiefeng Gao
2.1 Introduction *21*
2.2 Fabrication Methods for CPCs *22*
2.2.1 Melt Blending *23*
2.2.2 Solution Blending *25*
2.2.3 In Situ Polymerization *27*
2.3 Morphologies *27*
2.3.1 Random Dispersion of Nanofiller in the Polymer Matrix *27*
2.3.2 Selective Distribution of Nanofillers on the Interface *29*
2.3.2.1 Segregated Structure *29*

2.3.2.2	Surface Coating	31
2.4	Application in Sensors	32
2.4.1	Strain Sensor	33
2.4.2	Piezoresistive Sensor	33
2.4.3	Gas Sensor	35
2.4.4	Temperature Sensor	38
2.5	Conclusion	40
	References	41

3 Biodegradable Polymer Nanocomposites for Electronics 53
Wei Wu

3.1	Introduction	53
3.2	Biodegradable Polymer Nanocomposites in Electronics	55
3.2.1	Polylactide	55
3.2.2	PCL	58
3.2.3	PVA	59
3.2.4	PVP	61
3.2.5	Cellulose	62
3.2.6	Chitosan	64
3.2.7	Silk	65
3.3	Challenges and Prospects	66
	List of Abbreviations	67
	References	67

4 Polymer Nanocomposites for Photodetectors 77
Raj Wali Khan, Zheng Wen, and Zhenhua Sun

4.1	Introduction	77
4.2	Photodetector Brief	79
4.2.1	Photodiode	80
4.2.2	Photoconductor	80
4.3	Photodetectors Based on Novel Semiconductors	82
4.4	Photodetectors Based on Polymer Nanocomposites	87
4.4.1	Polymer–Polymer Nanocomposite	88
4.4.2	Polymer–Small Molecular Organic Nanocomposite	98
4.4.2.1	MEH-PPV–Small Molecular Organic Nanocomposite	98
4.4.2.2	P3HT-Small Molecular Organic Nanocomposite	99
4.4.3	Polymer–Polymer–Small Molecular Organic Nanocomposite	107
4.4.4	Polymer–Small Molecular Organic–Small Molecular Organic Nanocomposite	110
4.4.5	Polymer–Inorganic Nanocrystals Nanocomposite	112
4.4.5.1	MEH-PPV–Inorganic Nanocrystals Nanocomposite	112
4.4.5.2	P3HT–Inorganic Nanocrystals Nanocomposite	115
4.4.6	Polymer–Small Molecular Organic–Inorganic Nanocrystals Nanocomposite	120
4.5	Outlook	123

	List of Abbreviations *123*	
	References *124*	

5	**Polymer Nanocomposites for Pressure Sensors** *131*	
	Qi-Jun Sun and Xin-Hua Zhao	
5.1	Introduction *131*	
5.2	Parameters for Pressure Sensors *132*	
5.2.1	Pressure Sensitivity *132*	
5.2.2	Linear Sensing Range *134*	
5.2.3	LOD and Response Speed *134*	
5.2.4	Reliability *134*	
5.3	Working Principles and Examples of Polymer Nanocomposite Based Pressure Sensors *135*	
5.3.1	Capacitive Pressure Sensors *135*	
5.3.2	Piezoresistive Pressure Sensors *137*	
5.3.3	Piezoelectric and Triboelectric Tactile Sensors Based on Polymer Nanocomposites *143*	
5.4	Applications of the Polymer Nanocomposite Based Pressure Sensors *148*	
5.4.1	Human Wrist Pulse Detection *148*	
5.4.2	Subtle Human Motion Detection *149*	
5.4.3	Texture Roughness Detection *151*	
5.4.4	E-skin Application *152*	
5.5	Performance of Pressure Sensors with the Polymer Nanocomposites Reported Over the Past Decade *153*	
5.6	Conclusion *154*	
	References *154*	

6	**The Application of Polymer Nanocomposites in Energy Storage Devices** *157*	
	Ningyuan Nie, Mengmeng Hu, Jie Liu, Jiangqi Wang, Panpan Wang, Hua Wang, Zhenyuan Ji, Zhe Chen, and Yan Huang	
6.1	Introduction *157*	
6.2	Electrodes *158*	
6.2.1	For Battery *158*	
6.2.1.1	Polymer–Graphene/Carbon Nanotube *158*	
6.2.1.2	Polymer Inorganic *161*	
6.2.1.3	Polymer–Organic Salt Graphene *163*	
6.2.2	For Supercapacitor *164*	
6.2.2.1	Polymer–Metal Oxide *165*	
6.2.2.2	Polymer–Graphene/Carbon Nanotube *165*	
6.2.2.3	Polymer–Metal Oxide–Graphene/Carbon Nanotubes *169*	
6.3	Electrolytes *171*	
6.3.1	For Battery *171*	
6.3.2	For Supercapacitor *172*	

6.4	Separator	174
6.4.1	For Battery	174
6.4.2	For Supercapacitors	175
6.5	Conclusion	176
	References	177

7	**Functional Polymer Nanocomposite for Triboelectric Nanogenerators**	**189**
	Xingyi Dai, Jiancheng Han, Qiuqun Zheng, Cheng-Han Zhao, and Long-Biao Huang	
7.1	Introduction	189
7.2	Triboelectric Nanogenerators	190
7.3	Functional Polymer Nanocomposite	194
7.4	Self-healing Triboelectric Nanogenerators	197
7.5	Shape Memory Triboelectric Nanogenerators	201
7.6	Biodegradable Triboelectric Nanogenerators	204
7.7	Conclusion	208
	References	208

8	**Polymer Nanocomposites for Resistive Switching Memory**	**211**
	Qazi Muhammad Saqib, Muhammad Umair Khan, and Jinho Bae	
8.1	Introduction	211
8.2	Resistive Switching Memory for Polymer Nanocomposite	213
8.2.1	Resistive Switching	213
8.2.2	Resistive Switching Memory Operating Mechanism	214
8.2.2.1	Formation and Rupture of Conductive Filaments	214
8.2.2.2	Cations and Anions Migration	216
8.2.2.3	Electrons Trapping and De-tapping	216
8.2.2.4	Other Conduction Mechanisms	216
8.2.3	Fabrication Techniques	217
8.2.4	Polymer Nanocomposite Materials	218
8.3	Polymer Nanocomposite Based RSM Devices	218
8.3.1	Oxide Based Polymer Nanocomposite RSM	218
8.3.2	Metal Based Nanoparticles for Polymer Nanocomposite RSM	222
8.3.3	Graphene Based Polymer Nanocomposite RSM	224
8.3.4	Quantum Dot Based Polymer Nanocomposite RSM	227
8.3.5	Polymer Based Nanocomposites for RSM	229
8.3.6	2D Material Based Polymer Nanocomposites RSM	231
8.3.7	Other Materials Used for Polymer Nanocomposite Based RSM	232
8.4	Concluding Remarks	233
	Acknowledgments	234
	References	234
8.A	Performance Comparison According to Device Material and Structure	243

9	**Polymer Nanocomposites for Temperature Sensing and Self-regulating Heating Devices** *247*
	Yi Liu, Han Zhang, and Emiliano Bilotti
9.1	Introduction *247*
9.2	Conducting Mechanism and Percolation Theory *248*
9.3	PTC Theory *249*
9.4	Main Factors Influencing the PTC Effect *250*
9.4.1	Effect of Filler Size and Shape *250*
9.4.2	Effect of Filler Dispersion and Distribution *253*
9.4.3	Effect of Mixed Filler *254*
9.4.4	Effect of Polymer Thermal Expansion and Crystallinity *255*
9.4.5	Effect of Polymer Transition Temperature *257*
9.4.6	Effect of Polymer Blend *257*
9.5	Temperature Sensors *259*
9.6	Self-regulating Heating Devices *259*
9.7	Conclusions *262*
	References *263*
10	**Polymer Nanocomposites for EMI Shielding Application** *267*
	Ajitha A. Ramachandran and Sabu Thomas
10.1	Introduction *267*
10.2	Mechanism of EMI Shielding of Polymer Composites *268*
10.2.1	Materials for EMI Shielding *269*
10.3	Polymer Nanocomposites for EMI Shielding Application *270*
10.3.1	Nanofiller Incorporated Conducting Polymer Composites *270*
10.3.2	Polymer Blend Nanocomposites for Electromagnetic Interference (EMI) Shielding *271*
10.3.3	Conducting Polymers for EMI Shielding Application *272*
10.4	Characterization Techniques Used for the Electrical Studies of Polymer Composites *274*
10.4.1	Conductivity Studies of Polymer Composites *274*
10.4.2	Electromagnetic Interference (EMI) Shielding Studies *276*
10.5	Conclusion *278*
	References *279*

Index *285*

Preface

Polymer nanocomposites combining the merits of polymers (e.g. light weight, flexibility, low cost) and functional properties of nanomaterials caused by small size effect, quantum size effect, and surface/boundary effect show adjustable optical, electrical, biological, and mechanical characteristics and attract extensive researches. Various polymer nanocomposites with amazing performances have been prepared and utilized for developing integrated electronic devices in a number of emerging areas and exhibit huge commercial value, thanks to their simple preparation techniques and countless combinations. This book highlights the recent researches about the basic conceptions, preparation/characterization techniques, properties, device design strategies, and intriguing applications of polymer nanocomposites. The existing/potential application prospects and challenges for the polymer nanocomposites are also discussed. We expect that this book can offer a well-timed assistance to the academic researchers in the rapidly expanding applications including environment, sensor, energy conversion/storage, biology, and information storage as a simple and convenience instrument.

Herein, we would like to thank all the authors who have made contributions in this book. We want to express our sincerest appreciation and respect to Ms. Katherine Wong, Dr. Shaoyu Qian, Ms. Pinky Sathishkumar, Mathangi Balasubramanian and other editors at Wiley for all the help offered during the whole book editing process. We also want to thank all the readers interested in this book. In this book, we have introduced the concepts, properties, and mechanisms of polymer nanocomposites and summarized their recent applications in some hottest fields. The application challenges, commercial prospects, and potential research directions of polymer nanocomposites are also pointed out and discussed. We aim to provide a comprehensive, popular, and up-to-date book for the researchers. Although we have done our best to make this book better, there still inevitably are some omissions and mistakes. Please grant your criticisms and instructions.

We hope that this book can provide references and guides for researchers in polymer nanocomposites based electron devices, as well as promote the interests of the students to this field.

01 June 2020

Ye Zhou
Guanglong Ding

1

Introduction of Polymer Nanocomposites

Teng Li[1], Guanglong Ding[2], Su-Ting Han[1], and Ye Zhou[2]

[1] Shenzhen University, Institute of Microscale Optoelectronics, Room 909, Shenzhen, Guangdong 518060, China
[2] Shenzhen University, Institute for Advanced Study, Room 358, Shenzhen, Guangdong 518060, China

1.1 Introduction

Polymers have been one of the most important components in almost every area of human activity today. Nowadays, polymers as multifunctional materials gradually replace metals, glass, paper, and other traditional materials in various applications due to its lightweight, flexibility, and low cost [1]. In most of their applications, the applied materials are not composed of a single chemical component but mixture systems of multiple components with polymers and other additives. By incorporating different additives, such as metal, minerals, or even air, a wide variety of materials with unique physical properties and competitive production costs can be produced. For example, glass fiber-reinforced plastics are composite materials manufactured by laminating unsaturated polyester resin with glass fiber and filler, which can increase mechanical strength and heat resistance [2].

In addition, scientific research shows that the size of filling material in fiber reinforced composites has a great influence on the material properties, since the size of the filling particles largely determines the surface interactions of adhesion, particle movement, dispersion, and bonding between the surface and matrix [3]. With the particle size of the filler that gradually reduces to the nanoscale, some properties depending on the interface have undergone great changes, such as gas adsorption, chemical activity, electrical properties, and catalytic activity. Examples of different sizes of materials are shown in Figure 1.1, and a hydrogen atom is about 0.1 nm in size, while a human hair is 10^4 nm in diameter. Among them, nanomaterials are employed to describe the materials that have at least one dimension in the size range from approximately 1 to 100 nm [4]. Different from the bulk and microscale materials, nanomaterials are unique in that they have many unusual, useful, and interesting properties. For example, bulk gold is a very stable precious metal in golden color, which can be kept for a long time under atmospheric environment, so it is used as the initial currency by people. Unlike bulk gold, gold nanoparticles dispersed in water will show different colors according to the size of nanoparticles,

Polymer Nanocomposite Materials: Applications in Integrated Electronic Devices, First Edition.
Edited by Ye Zhou and Guanglong Ding.
© 2021 WILEY-VCH GmbH. Published 2021 by WILEY-VCH GmbH.

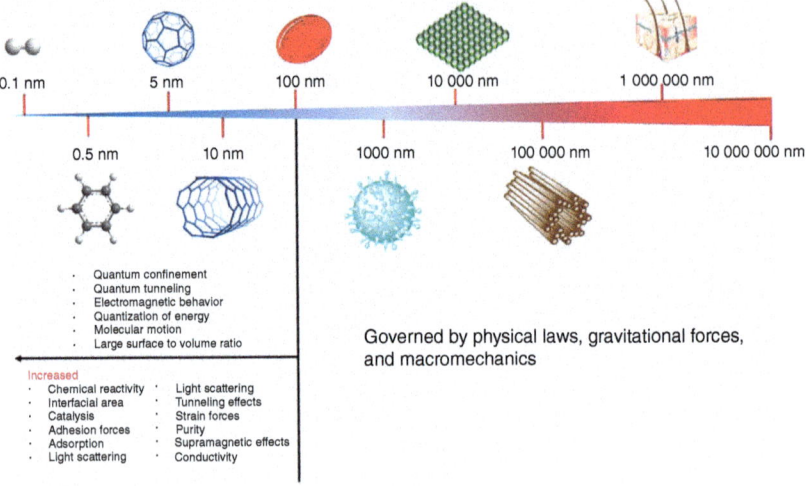

Figure 1.1 Nanomaterials peculiarities of size scale.

and they have high reactivity that can even be used as catalyst at low temperature [5]. Since most of the properties of nanomaterials depend on their size, shape, and surface structure, their ultrafine size always tends nanomaterials to aggregate into bulk materials, especially without proper stabilization in their formation and application [6]. This is because the agglomeration process makes the high surface energy and activity of nanomaterials decrease to a more stable state. Therefore, in order to preserve the properties of nanomaterials, it is necessary to distribute them uniformly in matrices to prevent from aggregating into bulk materials [7].

Polymer nanocomposites (PNCs) are the mixture of polymers and nanomaterials, having at least one-dimensional structure and one component material in the nanometer regime of less than 100 nm. Combining nanomaterials into the polymer matrix not only makes it possible to produce a new class of properties provided by uniformly dispersed nanomaterials, but also greatly improves most of expected properties of the original polymer, such as mechanical properties, heat resistance, biodegradability, and so on [8]. As early as 1970, the term "nanocomposites" was first proposed by Theng [9], and PNCs began to develop in commercial research institutions and academic laboratories in the late 1980s [10, 11]. Over the past decade, PNCs have made great progress in various fields, which is reflected by the exponential growth of publications from their inception (Figure 1.2). The existence of nanomaterials in polymer matrix changes the surface chemical and physicochemical properties of PNCs, where the geometry, surface chemistry, aspect ratio, and size of nanomaterials are the key parameters to regulate these performances. Therefore, PNCs are a new class of materials with unique properties, which are far superior to traditional doped and composite polymer systems. The large interface interaction between nanomaterials and polymer matrix surfaces and the difference of nanoscale fundamentally distinguish PNCs from the traditional system. The development of nanomaterials and polymer science and technology

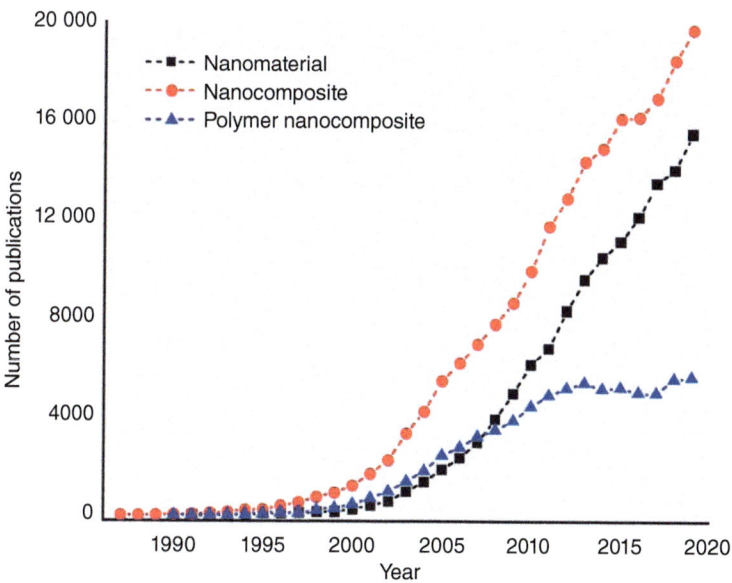

Figure 1.2 Number of publications per year on "nanomaterials," "nanocomposites," and "polymer nanocomposites," according to SciFinder Scholar on 30 April 2020.

has promoted many applications of PNCs, which cover almost all fields of polymer material application fields, such as microelectronics, magnetic electronics, biological materials, sensor, energy storage, and so on [12]. Therefore, the chapters include unique perspectives of different experts with their knowledge and understanding of PNCs in this book.

1.2 The Advantage of Nanocomposites

Since the fillers of nanocomposites are nanoscale, the performances of nanocomposites can be improved by the advantages of the reduction of filler size and the increased surface area. In terms of size, the filler is 3 orders of magnitude smaller than the traditional substitute. In addition, the quantum confinement effects caused by the nanomaterials will lead to new physical phenomena, which can be applied in electrical and optical research. Many of these properties are related to the size of the polymer chain, and the polymer chain close to the fillers is affected by the interaction between the packing surface and the polymer matrix, which is different from the polymer chain far away from the interface. The size of polymer chain can be reflected the radius of gyration R_g, and the thickness of the interface regions (t) around the particle is independent of the particle size. Therefore, the volume of interface material ($V_{interface}$) relative to the volume of particle ($V_{particle}$) will increase with the decrease of particle size.

Figure 1.3 shows the functional relationship between the $V_{interface}/V_{particle}$ varies and the aspect ratio of particles [13]. The aspect ratio reflects the shape of the

1 Introduction of Polymer Nanocomposites

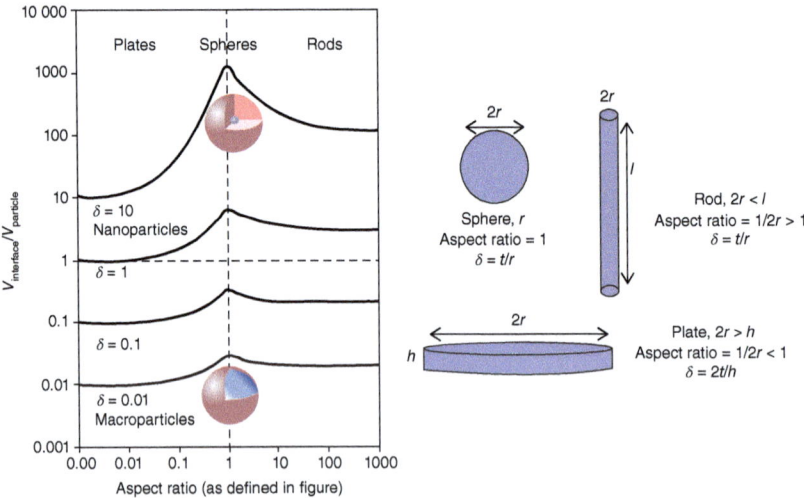

Figure 1.3 The graph on the left shows the function relationship between the ratio of interfacial volume to the particle volume ($V_{interface}/V_{particle}$) and the particle aspect ratio. The red shell represents the interface of particle, where the blue nucleus represents the particle. The graph on the right defines the particle aspect ratio and the ratio of the interfacial thickness to the particle size (δ) with different shapes (r is radius, l is length, h is height). The interface thickness (t) is considered to be independent of particle size. When the particle size is reduced to less than 100 nm, the physical properties can be controlled by the volume of the interface around the particle, which is especially obvious for the sphere and rod. Source: Winey and Vaia [13].

particles, which can be divided to plate (aspect ratio <1), sphere (aspect ratio = 1), and rod (aspect ratio >1). δ represents the size of the filler, that is, the ratio of the interface thickness t to the minimum dimension size of the particle. For spherical and rod-shaped particles, δ is equal to the t/r, but δ is $2t/h$ in plate-shaped particles. When the particle is microscale, δ is approximately equal to 0.01, and the particle volume exceeds the volume of the interface region in all shapes. As the particle size decreases, $V_{interface}/V_{particle}$ values gradually increases. When δ goes above one, $V_{interface}$ is going to exceed $V_{particle}$. When the particles reach the nanoscale ($\delta = 10$), the interface volume is more than 10 times that of the particle. Moreover, particles with different shapes have different $V_{interface}/V_{particle}$ in the same δ. The three-dimension sphere has the highest value, followed by the two-dimensional rod and the one-dimensional plate. With the decrease of particle size, the gap becomes more obvious, and even the $V_{interface}/V_{particle}$ of spherical particles is 2 orders of magnitude larger than that of plate-shaped particles. Therefore, the addition of nanoscale fillers has a great impact on the performance of polymer in PNCs. Even if the volume fraction of fillers is very small, the resulting interface region volume will be very large.

As the interaction between polymer and particle is strengthened in PNCs, the interparticle interface and coordination will be reflected in the macroscopic properties. Due to the nanoscale of particles, the secondary forming constituents have a very high aspect ratio of over 100. When the volume fraction is 1–5%, these fillers can reach the percolation thresholds, which refer to the critical value of the volume fraction of the packed particles that can mutate a certain physical property of the composite material system. Therefore, the mechanical and transport performances of PNCs can be greatly improved under the condition of low load nanoparticles. Especially for the conductive particles, when the volume fraction of these particles increases to a certain critical value in polymer, conductivity of the polymer suddenly increases sharply from insulator to conductor, and the change range is up to 10 orders of magnitude.

1.3 Classification of Nanoscale Fillers

So far, various types of nanomaterials have been found to be able to form PNCs with polymers. According to different applications, nanoparticles with corresponding properties can be selected into the polymer system to achieve the expected performance. In general, these nanofillers suitable for PNC applications can be mainly divided into one-, two-, and three-dimensional materials according to their different dimensions (Table 1.1).

1.3.1 One-Dimensional Nanofillers

One-dimensional nanofillers are plate-like materials with one-dimensional dimensions less than 100 nm, which are usually a few nanometers thick and relatively long sheets [14]. Most one-dimensional nanofillers have unique morphology

Table 1.1 Overview of nanomaterials classified by their nanoscale dimensions.

Plate	Rod	Sphere
• Montmorillonite clays (MMT)	• Carbon nanofibers (CNFs)	• Nano-silica (n-silica)
• Nanographene platelets (NGPs)	• Carbon nanotubes (CNTs)	• Nano-alumina (n-Al$_2$O$_3$)
• Layered double hydroxide (LDHs)	• Halloysite nanotubes (HNTs)	• Nano-silver (n-Ag)
	• Nickel nanostrands (NiNs)	• Nano-titanium dioxide (n-TiO$_2$)
	• Aluminum oxide nanofibers (Nafen)	• Nano-silicon carbide (n-SiC)
		• Nano-zinc oxide (n-ZnO)
		• POSS

characteristics, such as nanoplate [15], nano-disk [16], nano-wall [17–23], etc., which play an important role in functional nano-devices [24, 25]. Recently, the widely studied materials are montmorillonite clays (MMT) [26], nanographene platelets (NGPs) [27, 28], ZnO nanosheets [29–31], Fe_3O_4 nanosheets [30], and so on, which have excellent electrical, optical, and magnetic properties [32], and are widely used in the fields of micro–nano electronics, biosensors, and chemical engineering [33]. The one-dimensional nanofillers are common nanomaterials in electronic and thermal devices due to their shape characteristics.

1.3.2 Two-Dimensional Nanofillers

Two-dimensional fillers are the materials with two dimensions less than 100 nm, and they are mostly in the form of rods [14]. The typical two-dimensional nanomaterials are carbon nanofibers (CNFs), carbon nanotubes (CNTs), halloysite nanotubes (HNTs), nickel nanostrands (NiN_S), and aluminum oxide nanofibers (Nafen). In addition, the most common two-dimensional nanofillers in PNCs are nanotubes [34], plant fibers [35–39], nanowires [40], carbon fibers [41–44], oxides [45–55], graphene [56, 57], molybdenum disulfide (MoS_2) [58], and hexagon boron nitride (*h*-BN) [59]. Compared with one- and three-dimensional fillers, two-dimensional fillers have better flame retardancy and striped characteristic, resulting in wide applications in the fields of catalysis, electronics, optics, sensing, and energy [3, 26, 60–62].

1.3.3 Three-Dimensional Nanofillers

Three-dimensional nanofillers are nanomaterials with three dimensions on the nanometer scale, so they are mostly spherical or cube-shaped [63], which is also commonly referred to zero-dimensional particles. The most common three-dimensional fillers are polyhedral oligomeric silsesquioxane (POSS), nanosilicon, nanometal particles, nanometal oxides, and quantum dots (QDs) [33, 64]. Among them, metals and metal oxide nanoparticles have the advantages of high stability, catalytic activity, and easy preparation, and they are often used in the fields of catalysis [65], purification [66–69], coatings [70–74], and biological fields [75, 76], together with various polymers. One-, two-, and three-dimensional nanofillers all have various special properties, and will ultimately promote the remarkable performance of PNCs by loading in compatible polymers.

1.4 The Properties of Polymer Nanocomposites

In PNCs, many properties of the original polymer can be greatly improved, as well as new properties resulting from the addition of nanoparticles. As shown in Figure 1.4, the main properties of PNCs are listed, covering physical, chemical, and biological areas. In general, the improvement level of properties is determined by the size,

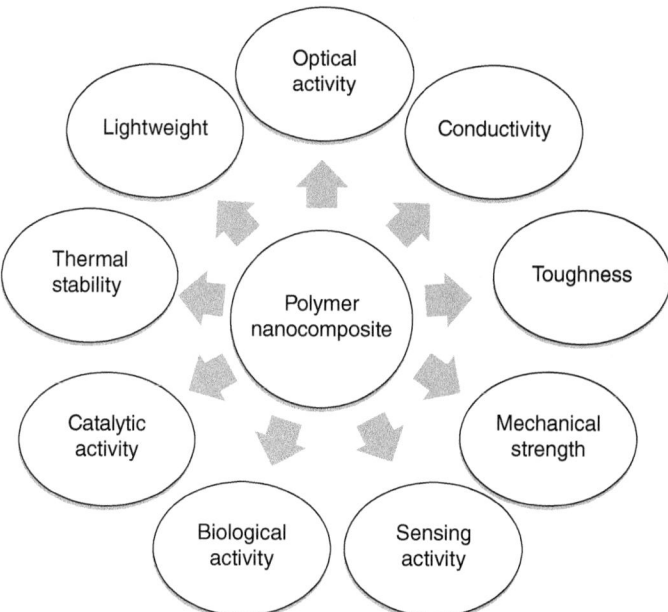

Figure 1.4 Significant properties of polymer nanocomposites.

loading capacity, aspect ratio, dispersion uniformity, and interface interactions of the nanofillers with polymer matrix [4].

For example, most polymers don't possess conductivity except some conducting polymers, which is due to the covalent bonding of polymers and the lack of electron channels or ion migration. Interestingly, new PNCs formed by adding conductive nanofillers to insulating polymers exhibit many electrical properties. As early as 1994, Ajayan et al. used CNTs as reinforcement materials to prepare PNCs [77]. Since then, there have been a lot of researches on using CNTs as fillers to improve the electrical properties of PNCs. Only a small volume fraction of such fillers is needed to improve the electrical properties of polymers by several orders of magnitude effectively [78].

1.5 Synthesis of Polymer Nanocomposites

In the synthesis of PNCs, it is necessary to uniformly distribute the fillers into matrix in order to realize the functions of fillers. However, due to the fact that the fillers are nanoscale, the uniform dispersion is much different from that of the microscale fillers, which is mainly manifested in the following aspects. First, if the filling operation is carried out according to the volume fraction, much more nanometer fillers than the microfillers are required at the same volume fraction. Therefore, the nanoparticles in matrix are very crowded with greater van der Waals and electrostatic interactions between the particles, making it difficult to distribute evenly. Second, the anisotropic nanofillers have a very high aspect ratio, which makes them

Table 1.2 Summary of common methods for synthesis of polymer nanocomposites.

Technique	Suitable filler	Suitable matrix	Solvent	Controlling factors
Ultrasonication-assisted solution mixing	All types	Liquid or viscous monomers or oligomers of thermosets	Required	Sonication power and time
Shear mixing	Nanosheets	Liquid or viscous monomers or oligomers of thermosets	Required	Shapes of the rotor blades, rotating speed and time
Three roll milling	Nanosheets and nanotubes	Liquid or viscous monomers or oligomers of thermosets	Not required	Speed of roller, gap between adjacent roller
Ball milling	All types	Liquid or solid thermoplastics and thermosets	Not required	Time of milling, ball size, rotating speed, ball/nanofiller ratio
Double-screw extrusion	All types	Solid thermoplastics	Not required	Processing temperature, screw configuration, rotation speed
In situ synthesis	All types	Liquid or viscous monomers or oligomers of thermosets	Required	Chemical reaction conditions, temperature, condensation rate

more prone to agglomerate. For example, monolayer graphene has aspect ratio of about 10^4, so they tend to reduce their surface energy by π–π stacking. Third, a large amount of nanofillers with huge surface area is loaded in the polymer matrix, which will produce a large interface area and change the overall performances of the PNC. Therefore, the decisive step in the synthesis of PNCs is the uniform dispersion of nanofillers in polymer matrix. As shown in Table 1.2, the common methods to disperse nanofillers and prevent the aggregation of nanoparticles by using external energy are summarized.

1.5.1 Ultrasonication-assisted Solution Mixing

The most widely used approach to produce PNCs is ultrasonication-assisted solution mixing [79–83]. In this method, the nanofillers and polymer are initially dissolved in a solution. Then the nanofillers are evenly distributed in the matrix in assistant of the ultrasound. Afterwards, the PNCs are obtained by evaporation of the solvent.

The nanoparticles are separated from the agglomeration state to the smaller units by the ultrasonic energy, which is higher than the energy of interaction between the nanomaterials in the aggregates. With the increase of ultrasonic time, the aggregates of nanofillers are broken down into smaller ones, and even become individual nanoparticles independent of other nanoparticles in the polymer. In addition, this process often occurs at a high temperature, which can initiate in situ polymerization of reactive monomers or their soluble prepolymers with nanomaterials to enhance interfacial interactions [84].

Due to the simple operation and stable performance, the ultrasonic-assisted solution mixing method has been widely used in the researches of new nanocomposites. However, due to the poor effect of ultrasound in high viscosity solution, most of the polymers need to be dissolved in a high boiling point solvent and maintain a low concentration, which will affect the process of solvent removal and ultimately reduce the quality of the nanocomposites. Therefore, when using this method, it is important to pay attention to the choice of solvent.

1.5.2 Shear Mixing

Compared with the ultrasonic-assisted method, shear mixing is a much more common and simple method, which only requires the stirring process and has the potential for industrial mass production [85]. In the process of stirring, the shear force generated by stirrer rotating is used to separate the aggregates of nanofillers. Due to the low strength of the shear force, the nanoparticles will be separated under stirring and then aggregated again, so it is generally necessary to increase the speed of the agitator to complete the separation. This method generally does not destroy the structure of nanofillers; therefore it is suitable not only for separating loosely bound nanoaggregates, but also for stripping off some layered nanosheets. In addition, this method needs to be carried out in low viscosity solvent just like ultrasonic assisted method.

1.5.3 Three Roll Milling

Three roller milling is a method of dispersing nanofillers by the shearing force between rolls in high viscosity matrix, such as ink, paste material, coating, etc. The machine of three roll milling is composed of three cylindrical rollers with different rotating speeds, and the adjacent rollers rotate in the opposite direction. The particle size distribution and uniformity of the packing can be well controlled as the speeds of the rollers and the gap between them are adjustable. In addition, the shearing force generated between the rollers is higher than that generated by stirring, so the method can be applied to high viscosity materials, and carried out under the condition of little or no solvent. Therefore, this method is often used to disperse some anisotropic nanofillers, such as CNTs [86–89], graphene nanosheets [90–92], nanoclays [93–95], and so on.

However, it should be noted that the distance between adjacent rollers should be at least 1 µm, so the dispersion effect of nanospheres with three-dimensional direction less than 100 nm will not be good. The aggregates of nanospheres can only be turned into smaller units, not broken into individual particle. On the other hand, the rotation of the roller requires the addition of viscous materials, and nanofillers can only be dispersed in the thermosetting matrix but not in the thermoplastic matrix.

1.5.4 Ball Milling

Ball milling is widely used in metallurgy and mineral processing industry [96]. The principle of ball milling is to grind and mix powders in a closed space by using the huge shear force and compression force produced by hard ball collision. In the synthesis of PNCs, this method can disperse CNTs [97], graphene nanoparticles [98–101], silica nanoparticles [102], and BNs [103, 104] into thermoplastic and thermosetting polymers. The high shear force produced by ball milling can peel off some two-dimensional nanostructures, such as graphene, MoS_2, and BNs, but may not separate the interlayer structure connected by ionic bonding [105–110]. In addition, ball milling is not only suitable for solvent-free conditions but also solvent-free conditions, so nanofillers can be directly dispersed in some solid thermoplastic matrix, such as polyethylene (PE) [101, 111], polyphenylene sulfide [104, 112], and polymethyl methacrylate (PMMA) [102].

1.5.5 Double-screw Extrusion

Double-screw extrusion disperses nanofillers in thermoplastic matrix by huge shear force generated by high speed rotation of double-screw at high temperature [113, 114]. This method has been widely used in industry due to the advantages of solvent-free and environment-friendly technology. With this method, the fillers can be dispersed into the polymer in a high content way to achieve the well-controlled performance, and applied to different sizes of nanoparticles, such as graphene sheets [115], CNTs [116], and silicon dioxide [117]. This method needs higher temperature, which is helpful to reduce the viscosity of polymer and load more nanofillers, but also has the risk of decomposing polymers and nanofillers. The reason is owing to the existence of low thermal stability functional groups in the materials. When the temperature is too high, the fracture will occur, resulting in the deterioration of the performance of PNCs [118]. Moreover, the gap between the screws is too large to keep some aggregates of nanofillers evenly, which will not achieve the uniform monodispersing of nanofillers. So, it is necessary to combine other technologies to further improve the performance [119, 120].

1.5.6 In Situ Synthesis

In addition to the aforementioned methods of dispersing prepared nanofillers into polymers, another important synthesis strategy is in situ synthesis, which directly generates nanoparticles in polymers through molecular precursors [121]. This method can be divided into chemical and physical in situ synthesis [122]. Chemical in situ synthesis is used to synthesize nanoscale fillers by chemical reaction, such as the hydrothermal method and sol–gel method [123, 124]. The physical in situ synthesis is transforming the precursor of gas phase into inorganic nanoparticles through plasma action, and then condensing the organic compounds on the surface of inorganic particles to cover the polymer shell to form PNCs [125].

1.6 Conclusions and Future Outlook

In this chapter, the basic principles, properties, and synthesis methods of PNCs are clearly described. The composite material has unique structure and performance, and has a wide range of applications in many fields. The particle size, orientation, shape, dispersion, and volume dispersion of nanofillers affect the properties of PNCs. Most of the physical, chemical, and mechanical properties of PNCs depend on the interface interaction between the filler and the matrix. Therefore, the uniform dispersion of nanofillers is the most important consideration in the synthesis of PNCs. PNCs have recently become part of modern technology, but these areas are still in the early stage of development. With more and more scientists and engineers contributing to the understanding of PNCs, these functional materials will be applied in more and more fields.

References

1 Karak, N. (2019). *Fundamentals of Nanomaterials and Polymer Nanocomposites*. Elsevier.
2 Brunner, G. (2014). *Supercritical Fluid Science and Technology*, vol. 5 (ed. G. Brunner). Elsevier.
3 Pradhan, S., Lach, R., Le, H.H. et al. (2013). Effect of filler dimensionality on mechanical properties of nanofiller reinforced polyolefin elastomers. *ISRN Polym. Sci.* 2013: 1–9.
4 Jordan, J., Jacob, K.I., Tannenbaum, R. et al. (2005). Experimental trends in polymer nanocomposites-a review. *Mater. Sci. Eng., A* 393: 1–11.
5 Xiao, J. and Qi, L. (2011). Surfactant-assisted, shape-controlled synthesis of gold nanocrystals. *Nanoscale* 3: 1383–1396.
6 Koo, J. (2015). *An Overview of Nanomaterials*. Cambridge University Press.
7 Vigneshwaran, N., Ammayappan, L., and Huang, Q. (2011). Effect of Gum arabic on distribution behavior of nanocellulose fillers in starch film. *Appl. Nanosci.* 1: 137–142.
8 Hussain, F., Hojjati, M., Okamoto, M., and Gorga, R.E. (2006). Review article: polymer-matrix nanocomposites, processing, manufacturing, and application: an overview. *J. Compos. Mater.* 40: 1511–1575.
9 Theng, B.K.G. (1970). Interactions of clay minerals with organic polymers. some practical applications. *Clays Clay Miner.* 18: 357–362.
10 Usuki, A., Kawasumi, M., Kojima, Y. et al. (1993). Swelling behavior of montmorillonite cation exchanged for ω-amino acids by ∈-caprolactam. *J. Mater. Res.* 8: 1174–1178.

1 Introduction of Polymer Nanocomposites

11 Kojima, Y., Usuki, A., Kawasumi, M. et al. (2011). Mechanical properties of nylon 6-clay hybrid. *J. Mater. Res.* 8: 1185–1189.

12 Ray, S.S. and Bousmina, M. (2007). *Polymer Nanocomposites and Their Applications*. American Scientific Publishers.

13 Winey, K.I. and Vaia, R.A. (2011). Polymer nanocomposites. *MRS Bull.* 32: 314–322.

14 Verdejo, R., Bernal, M.M., Romasanta, L.J. et al. (2018). Reactive nanocomposite foams. *Cell. Polym.* 30: 45–62.

15 Nieto, A., Lahiri, D., and Agarwal, A. (2012). Synthesis and properties of bulk graphene nanoplatelets consolidated by spark plasma sintering. *Carbon* 50: 4068–4077.

16 Schmidt, F.P., Ditlbacher, H., Hohenester, U. et al. (2012). Dark plasmonic breathing modes in silver nanodisks. *Nano Lett.* 12: 5780–5783.

17 Jung, S.-H., Oh, E., Lee, K.-H. et al. (2008). Sonochemical preparation of shape-selective ZnO nanostructures. *Cryst. Growth Des.* 8: 265–269.

18 Siril, P.F., Ramos, L., Beaunier, P. et al. (2009). Synthesis of ultrathin hexagonal palladium nanosheets. *Chem. Mater.* 21: 5170–5175.

19 Dong, X., Ji, X., Jing, J. et al. (2010). Synthesis of triangular silver nanoprisms by stepwise reduction of sodium borohydride and trisodium citrate. *J. Phys. Chem. C* 114: 2070–2074.

20 Nayak, B.B., Behera, D., and Mishra, B.K. (2010). Synthesis of silicon carbide dendrite by the arc plasma process and observation of nanorod bundles in the dendrite arm. *J. Am. Ceram. Soc.* 93: 3080–3083.

21 Vizireanu, S., Stoica, S.D., Luculescu, C. et al. (2010). Plasma techniques for nanostructured carbon materials synthesis. a case study: carbon nanowall growth by low pressure expanding RF plasma. *Plasma Sources Sci. Technol.* 19: 34016.

22 Mann, A.K.P. and Skrabalak, S.E. (2011). Synthesis of single-crystalline nanoplates by spray pyrolysis: a metathesis route to Bi_2WO_6. *Chem. Mater.* 23: 1017–1022.

23 Tiwari, J.N., Tiwari, R.N., and Kim, K.S. (2012). Zero-dimensional, one-dimensional, two-dimensional and three-dimensional nanostructured materials for advanced electrochemical energy devices. *Prog. Mater Sci.* 57: 724–803.

24 Kim, K.S., Zhao, Y., Jang, H. et al. (2009). Large-scale pattern growth of graphene films for stretchable transparent electrodes. *Nature* 457: 706–710.

25 Bae, S., Kim, H., Lee, Y. et al. (2010). Roll-to-roll production of 30-inch graphene films for transparent electrodes. *Nat. Nanotechnol.* 5: 574–578.

26 Isitman, N.A., Dogan, M., Bayramli, E., and Kaynak, C. (2012). The role of nanoparticle geometry in flame retardancy of polylactide nanocomposites containing aluminium phosphinate. *Polym. Degrad. Stab.* 97: 1285–1296.

27 Shen, J., Hu, Y., Li, C. et al. (2009). Synthesis of amphiphilic graphene nanoplatelets. *Small* 5: 82–85.

28 Li, B. and Zhong, W.-H. (2011). Review on polymer/graphite nanoplatelet nanocomposites. *J. Mater. Sci.* 46: 5595–5614.

29 Umar, A. and Hahn, Y.B. (2006). ZnO nanosheet networks and hexagonal nanodiscs grown on silicon substrate: growth mechanism and structural and optical properties. *Nanotechnology* 17: 2174–2180.
30 Bai, W., Zhu, X., Zhu, Z., and Chu, J. (2008). Synthesis of zinc oxide nanosheet thin films and their improved field emission and photoluminescence properties by annealing processing. *Appl. Surf. Sci.* 254: 6483–6488.
31 Mani, G.K. and Rayappan, J.B.B. (2014). A simple and template free synthesis of branched ZnO nanoarchitectures for sensor applications. *RSC Adv.* 4: 64075–64084.
32 Li, B.L., Setyawati, M.I., Chen, L. et al. (2017). Directing assembly and disassembly of 2D MoS_2 nanosheets with DNA for drug delivery. *ACS Appl. Mater. Interfaces* 9: 15286–15296.
33 Vengatesan, M.R. and Mittal, V. (2016). *Nanoparticle- and Nanofiber-Based Polymer Nanocomposites: An Overview*. Wiley-VCH.
34 Yang, J., Zhang, Z., Friedrich, K., and Schlarb, A.K. (2007). Creep resistant polymer nanocomposites reinforced with multiwalled carbon nanotubes. *Macromol. Rapid Commun.* 28: 955–961.
35 Fahmy, T.Y.A., Mobarak, F., Fahmy, Y. et al. (2005). Nanocomposites from natural cellulose fibers incorporated with sucrose. *Wood Sci. Technol.* 40: 77–86.
36 Garcia de Rodriguez, N.L., Thielemans, W., and Dufresne, A. (2006). Sisal cellulose whiskers reinforced polyvinyl acetate nanocomposites. *Cellulose* 13: 261–270.
37 Fahmy, T.Y.A. and Mobarak, F. (2008). Nanocomposites from natural cellulose fibers filled with kaolin in presence of sucrose. *Carbohydr. Polym.* 72: 751–755.
38 Lee, K.-Y., Bharadia, P., Blaker, J.J., and Bismarck, A. (2012). Short sisal fibre reinforced bacterial cellulose polylactide nanocomposites using hairy sisal fibres as reinforcement. *Compos. Part A: Appl. Sci. Manuf.* 43: 2065–2074.
39 Ibrahim, I.D., Jamiru, T., Sadiku, E.R. et al. (2016). Impact of surface modification and nanoparticle on sisal fiber reinforced polypropylene nanocomposites. *J. Nanotechnol.* 2016: 1–9.
40 Lonjon, A., Laffont, L., Demont, P. et al. (2010). Structural and electrical properties of gold nanowires/P(VDF-TrFE) nanocomposites. *J. Phys. D* 43: 345401.
41 Xu, Y. and Hoa, S.V. (2008). Mechanical properties of carbon fiber reinforced epoxy/clay nanocomposites. *Compos. Sci. Technol.* 68: 854–861.
42 Pozegic, T.R., Anguita, J.V., Hamerton, I. et al. (2016). Multi-functional carbon fibre composites using carbon nanotubes as an alternative to polymer sizing. *Sci. Rep.* 6: 37334.
43 Ulus, H., Şahin, Ö.S., and Avcı, A. (2016). Enhancement of flexural and shear properties of carbon fiber/epoxy hybrid nanocomposites by boron nitride nano particles and carbon nano tube modification. *Fibers Polym.* 16: 2627–2635.
44 Ye, G. (2017). Preparation of poly(7-formylindole)/carbon fibers nanocomposites and their high capacitance behaviors. *Int. J. Electrochem. Sci.* 12: 8467–8476.

45 Lu, X., Chao, D., Chen, J. et al. (2006). Preparation and characterization of inorganic/organic hybrid nanocomposites based on Au nanoparticles and polypyrrole. *Mater. Lett.* 60: 2851–2854.

46 Subedi, D.P., Madhup, D.K., Sharma, A. et al. (2012). Retracted: study of the wettability of ZnO nanofilms. *Int. Nano Lett.* 2: 1.

47 Ślosarczyk, A., Barełkowski, M., Niemier, S., and Jakubowska, P. (2015). Synthesis and characterisation of silica aerogel/carbon microfibers nanocomposites dried in supercritical and ambient pressure conditions. *J. Sol–Gel Sci. Technol.* 76: 227–232.

48 Dhandapani, S., Nayak, S.K., and Mohanty, S. (2016). Compatibility effect of titanium dioxide nanofiber on reinforced biobased nanocomposites: thermal, mechanical, and morphology characterization. *J. Vinyl Add. Technol.* 22: 529–538.

49 Ma, J.-L., Chan, T.-M., and Young, B. (2016). Experimental investigation of cold-formed high strength steel tubular beams. *Eng. Struct.* 126: 200–209.

50 Saranya, M., Ramachandran, R., and Wang, F. (2016). Graphene-zinc oxide (G-ZnO) nanocomposite for electrochemical supercapacitor applications. *J. Sci. Adv. Mater. Devices* 1: 454–460.

51 Shehata, N., Gaballah, S., Samir, E. et al. (2016). Fluorescent nanocomposite of embedded ceria nanoparticles in crosslinked PVA electrospun nanofibers. *Nanomaterials* 6: 102.

52 Shehata, N., Samir, E., Gaballah, S. et al. (2016). Embedded ceria nanoparticles in crosslinked PVA electrospun nanofibers as optical sensors for radicals. *Sensors* 16: 1371.

53 Sunny, A.T., Vijayan, P.P., Adhikari, R. et al. (2016). Copper oxide nanoparticles in an epoxy network: microstructure, chain confinement and mechanical behaviour. *Phys. Chem. Chem. Phys.* 18: 19655–19667.

54 Alswata, A.A., Ahmad, M.B., Al-Hada, N.M. et al. (2017). Preparation of zeolite/zinc oxide nanocomposites for toxic metals removal from water. *Results Phys.* 7: 723–731.

55 Fambri, L., Dabrowska, I., Ceccato, R., and Pegoretti, A. (2017). Effects of fumed silica and draw ratio on nanocomposite polypropylene fibers. *Polymers* 9: 41.

56 Wang, X. and Song, M. (2013). Toughening of polymers by graphene. *Nanomater. Energy* 2: 265–278.

57 Paszkiewicz, S., Pawelec, I., Szymczyk, A., and Rosłaniec, Z. (2015). Thermoplastic elastomers containing 2D nanofillers: montmorillonite, graphene nanoplatelets and oxidized graphene platelets. *Polish J. Chem. Technol.* 17: 74–81.

58 Wang, X., Xing, W., Feng, X. et al. (2017). MoS_2/polymer nanocomposites: preparation, properties, and applications. *Polym. Rev.* 57: 440–466.

59 Ribeiro, H., Trigueiro, J.P.C., Silva, W.M. et al. (2019). Hybrid MoS_2/h-BN nanofillers as synergic heat dissipation and reinforcement additives in epoxy nanocomposites. *ACS Appl. Mater. Interfaces* 11: 24485–24492.

60 Rao, K.S., Senthilnathan, J., Ting, J.M., and Yoshimura, M. (2014). Continuous production of nitrogen-functionalized graphene nanosheets for catalysis applications. *Nanoscale* 6: 12758–12768.

61 Shahjamali, M.M., Salvador, M., Bosman, M. et al. (2014). Edge-gold-coated silver nanoprisms: enhanced stability and applications in organic photovoltaics and chemical sensing. *J. Phys. Chem. C* 118: 12459–12468.

62 Wan, J., Kaplan, A.F., Zheng, J. et al. (2014). Two dimensional silicon nanowalls for lithium ion batteries. *J. Mater. Chem. A* 2: 6051–6057.

63 Bhattacharya, M. (2016). Polymer nanocomposites-a comparison between carbon nanotubes, graphene, and clay as nanofillers. *Materials* 9: 262.

64 Kumar, A.P., Depan, D., Singh Tomer, N., and Singh, R.P. (2009). Nanoscale particles for polymer degradation and stabilization-Trends and future perspectives. *Prog. Polym. Sci.* 34: 479–515.

65 Shifrina, Z.B., Matveeva, V.G., and Bronstein, L.M. (2020). Role of polymer structures in catalysis by transition metal and metal oxide nanoparticle composites. *Chem. Rev.* 120: 1350–1396.

66 Sotto, A., Boromand, A., Balta, S. et al. (2011). Doping of polyethersulfone nanofiltration membranes: antifouling effect observed at ultralow concentrations of TiO_2 nanoparticles. *J. Mater. Chem.* 21: 10311–10320.

67 Huang, J., Zhang, K., Wang, K. et al. (2012). Fabrication of polyethersulfone-mesoporous silica nanocomposite ultrafiltration membranes with antifouling properties. *J. Membr. Sci.* 423–424: 362–370.

68 María Arsuaga, J., Sotto, A., del Rosario, G. et al. (2013). Influence of the type, size, and distribution of metal oxide particles on the properties of nanocomposite ultrafiltration membranes. *J. Membr. Sci.* 428: 131–141.

69 Zhao, S., Yan, W., Shi, M. et al. (2015). Improving permeability and antifouling performance of polyethersulfone ultrafiltration membrane by incorporation of ZnO-DMF dispersion containing nano-ZnO and polyvinylpyrrolidone. *J. Membr. Sci.* 478: 105–116.

70 Macyk, W., Szaciłowski, K., Stochel, G. et al. (2010). Titanium(IV) complexes as direct TiO_2 photosensitizers. *Coord. Chem. Rev.* 254: 2687–2701.

71 Paz, Y. (2010). Application of TiO_2 photocatalysis for air treatment: patents' overview. *Appl. Catal., B* 99: 448–460.

72 Su, W., Wang, S., Wang, X. et al. (2010). Plasma pre-treatment and TiO_2 coating of PMMA for the improvement of antibacterial properties. *Surf. Coat. Technol.* 205: 465–469.

73 Olad, A. and Nosrati, R. (2013). Preparation and corrosion resistance of nanostructured PVC/ZnO–polyaniline hybrid coating. *Prog. Org. Coat.* 76: 113–118.

74 Wang, N., Fu, W., Zhang, J. et al. (2015). Corrosion performance of waterborne epoxy coatings containing polyethylenimine treated mesoporous-TiO_2 nanoparticles on mild steel. *Prog. Org. Coat.* 89: 114–122.

75 Di Carlo, G., Curulli, A., Toro, R.G. et al. (2012). Green synthesis of gold-chitosan nanocomposites for caffeic acid sensing. *Langmuir* 28: 5471–5479.

76 Matos, A.C., Marques, C.F., Pinto, R.V. et al. (2015). Novel doped calcium phosphate-PMMA bone cement composites as levofloxacin delivery systems. *Int. J. Pharm.* 490: 200–208.

77 Ajayan, P.M., Stephan, O., Colliex, C., and Trauth, D. (1994). Aligned carbon nanotube arrays formed by cutting a polymer resin-nanotube composite. *Science* 265: 1212–1214.

78 Mao, C., Zhu, Y., and Jiang, W. (2012). Design of electrical conductive composites: tuning the morphology to improve the electrical properties of graphene filled immiscible polymer blends. *ACS Appl. Mater. Interfaces* 4: 5281–5286.

79 Jang, J., Bae, J., and Yoon, S.-H. (2003). A study on the effect of surface treatment of carbon nanotubes for liquid crystalline epoxide–carbon nanotube composites. *J. Mater. Chem.* 13: 676–681.

80 Stankovich, S., Dikin, D.A., Dommett, G.H. et al. (2006). Graphene-based composite materials. *Nature* 442: 282–286.

81 Yousefi, N., Gudarzi, M.M., Zheng, Q. et al. (2013). Highly aligned, ultralarge-size reduced graphene oxide/polyurethane nanocomposites: mechanical properties and moisture permeability. *Compos. Part A: Appl. Sci. Manuf.* 49: 42–50.

82 Yousefi, N., Sun, X., Lin, X. et al. (2014). Highly aligned graphene/polymer nanocomposites with excellent dielectric properties for high-performance electromagnetic interference shielding. *Adv. Mater.* 26: 5480–5487.

83 Shen, X., Wang, Z., Wu, Y. et al. (2016). Multilayer graphene enables higher efficiency in improving thermal conductivities of graphene/epoxy composites. *Nano Lett.* 16: 3585–3593.

84 Yousefi, N., Lin, X., Zheng, Q. et al. (2013). Simultaneous in situ reduction, self-alignment and covalent bonding in graphene oxide/epoxy composites. *Carbon* 59: 406–417.

85 Paton, K.R., Varrla, E., Backes, C. et al. (2014). Scalable production of large quantities of defect-free few-layer graphene by shear exfoliation in liquids. *Nat. Mater.* 13: 624–630.

86 Gojny, F.H., Wichmann, M.H.G., Köpke, U. et al. (2004). Carbon nanotube-reinforced epoxy-composites: enhanced stiffness and fracture toughness at low nanotube content. *Compos. Sci. Technol.* 64: 2363–2371.

87 Thostenson, E.T. and Chou, T.-W. (2006). Processing-structure-multi-functional property relationship in carbon nanotube/epoxy composites. *Carbon* 44: 3022–3029.

88 Viets, C., Kaysser, S., and Schulte, K. (2014). Damage mapping of GFRP via electrical resistance measurements using nanocomposite epoxy matrix systems. *Composites Part B* 65: 80–88.

89 Souri, H., Nam, I.W., and Lee, H.K. (2015). Electrical properties and piezoresistive evaluation of polyurethane-based composites with carbon nano-materials. *Compos. Sci. Technol.* 121: 41–48.

90 Ahmadi-Moghadam, B. and Taheri, F. (2014). Effect of processing parameters on the structure and multi-functional performance of epoxy/GNP-nanocomposites. *J. Mater. Sci.* 49: 6180–6190.

91 Chandrasekaran, S., Sato, N., Tölle, F. et al. (2014). Fracture toughness and failure mechanism of graphene based epoxy composites. *Compos. Sci. Technol.* 97: 90–99.

92 Li, Y., Zhang, H., Bilotti, E., and Peijs, T. (2016). Optimization of three-roll mill parameters for in-situ exfoliation of graphene. *MRS Adv.* 1: 1389–1394.

93 Dalir, H., Farahani, R.D., Nhim, V. et al. (2012). Preparation of highly exfoliated polyester-clay nanocomposites: process-property correlations. *Langmuir* 28: 791–803.

94 Park, J.-J. and Lee, J.-Y. (2013). Effect of nano-sized layered silicate on AC electrical treeing behavior of epoxy/layered silicate nanocomposite in needle-plate electrodes. *Mater. Chem. Phys.* 141: 776–780.

95 Kothmann, M.H., Ziadeh, M., Bakis, G. et al. (2015). Analyzing the influence of particle size and stiffness state of the nanofiller on the mechanical properties of epoxy/clay nanocomposites using a novel shear-stiff nano-mica. *J. Mater. Sci.* 50: 4845–4859.

96 Zhang, D.L. (2004). Processing of advanced materials using high-energy mechanical milling. *Prog. Mater Sci.* 49: 537–560.

97 Gupta, T.K., Singh, B.P., Mathur, R.B., and Dhakate, S.R. (2014). Multi-walled carbon nanotube-graphene-polyaniline multiphase nanocomposite with superior electromagnetic shielding effectiveness. *Nanoscale* 6: 842–851.

98 Wu, H., Zhao, W., Hu, H., and Chen, G. (2011). One-step in situ ball milling synthesis of polymer-functionalized graphene nanocomposites. *J. Mater. Chem.* 21: 8626–8632.

99 Jiang, X. and Drzal, L.T. (2012). Reduction in percolation threshold of injection molded high-density polyethylene/exfoliated graphene nanoplatelets composites by solid state ball milling and solid state shear pulverization. *J. Appl. Polym. Sci.* 124: 525–535.

100 Tang, L.-C., Wan, Y.-J., Yan, D. et al. (2013). The effect of graphene dispersion on the mechanical properties of graphene/epoxy composites. *Carbon* 60: 16–27.

101 Gu, J., Li, N., Tian, L. et al. (2015). High thermal conductivity graphite nanoplatelet/UHMWPE nanocomposites. *RSC Adv.* 5: 36334–36339.

102 Castrillo, P.D., Olmos, D., Amador, D.R., and Gonzalez-Benito, J. (2007). Real dispersion of isolated fumed silica nanoparticles in highly filled PMMA prepared by high energy ball milling. *J. Colloid Interface Sci.* 308: 318–324.

103 Donnay, M., Tzavalas, S., and Logakis, E. (2015). Boron nitride filled epoxy with improved thermal conductivity and dielectric breakdown strength. *Compos. Sci. Technol.* 110: 152–158.

104 Gu, J., Guo, Y., Yang, X. et al. (2017). Synergistic improvement of thermal conductivities of polyphenylene sulfide composites filled with boron nitride hybrid fillers. *Compos. Part A: Appl. Sci. Manuf.* 95: 267–273.

105 Lin, Y. and Connell, J.W. (2012). Advances in 2D boron nitride nanostructures: nanosheets, nanoribbons, nanomeshes, and hybrids with graphene. *Nanoscale* 4: 6908–6939.

106 Yao, Y., Lin, Z., Li, Z. et al. (2012). Large-scale production of two-dimensional nanosheets. *J. Mater. Chem.* 22: 13494–13499.

107 Lee, D., Lee, B., Park, K.H. et al. (2015). Scalable exfoliation process for highly soluble boron nitride nanoplatelets by hydroxide-assisted ball milling. *Nano Lett.* 15: 1238–1244.

108 Brent, J.R., Savjani, N., and O'Brien, P. (2017). Synthetic approaches to two-dimensional transition metal dichalcogenide nanosheets. *Prog. Mater Sci.* 89: 411–478.

109 Buzaglo, M., Bar, I.P., Varenik, M. et al. (2017). Graphite-to-graphene: total conversion. *Adv. Mater.* 29: 1603528.

110 Teng, C., Xie, D., Wang, J. et al. (2017). Ultrahigh conductive graphene raper based on ball-milling exfoliated graphene. *Adv. Funct. Mater.* 27: 1700240.

111 Gu, J., Guo, Y., Lv, Z. et al. (2015). Highly thermally conductive POSS-g-SiCp/UHMWPE composites with excellent dielectric properties and thermal stabilities. *Compos. Part A: Appl. Sci. Manuf.* 78: 95–101.

112 Gu, J., Xie, C., Li, H. et al. (2013). Thermal percolation behavior of graphene nanoplatelets/polyphenylene sulfide thermal conductivity composites. *Polym. Compos.* 35: 1087–1092.

113 Wu, C.L., Zhang, M.Q., Rong, M.Z., and Friedrich, K. (2002). Tensile performance improvement of low nanoparticles filled-polypropylene composites. *Compos. Sci. Technol.* 62: 1327–1340.

114 Fawaz, J. and Mittal, V. (2014). *Synthesis of Polymer Nanocomposites: Review of Various Techniques*. Wiley-VCH.

115 Kim, I.-H. and Jeong, Y.G. (2010). Polylactide/exfoliated graphite nanocomposites with enhanced thermal stability, mechanical modulus, and electrical conductivity. *J. Polym. Sci., Part B: Polym. Phys.* 48: 850–858.

116 Villmow, T., Pötschke, P., Pegel, S. et al. (2008). Influence of twin-screw extrusion conditions on the dispersion of multi-walled carbon nanotubes in a poly(lactic acid) matrix. *Polymer* 49: 3500–3509.

117 Zou, H., Wu, S., and Shen, J. (2008). Polymer/silica nanocomposites: preparation, characterization, properties, and applications. *Chem. Rev.* 108: 3893–3957.

118 Venugopal, G., Veetil, J.C., Raghavan, N. et al. (2016). Nano-dynamic mechanical and thermal responses of single-walled carbon nanotubes reinforced polymer nanocomposite thinfilms. *J. Alloys Compd.* 688: 454–459.

119 Moniruzzaman, M., Du, F., Romero, N., and Winey, K.I. (2006). Increased flexural modulus and strength in SWNT/epoxy composites by a new fabrication method. *Polymer* 47: 293–298.

120 Isayev, A.I., Kumar, R., and Lewis, T.M. (2009). Ultrasound assisted twin screw extrusion of polymer–nanocomposites containing carbon nanotubes. *Polymer* 50: 250–260.

121 Hanemann, T. and Szabó, D.V. (2010). Polymer-nanoparticle composites: from synthesis to modern applications. *Materials* 3: 3468–3517.

122 Caseri, W.R. (2013). Nanocomposites of polymers and inorganic particles: preparation, structure and properties. *Mater. Sci. Technol.* 22: 807–817.

123 Xiong, M., Zhou, S., Wu, L. et al. (2004). Sol–gel derived organic–inorganic hybrid from trialkoxysilane-capped acrylic resin and titania: effects of preparation conditions on the structure and properties. *Polymer* 45: 8127–8138.

124 Cao, Z., Jiang, W., Ye, X., and Gong, X. (2008). Preparation of superparamagnetic Fe_3O_4/PMMA nano composites and their magnetorheological characteristics. *J. Magn. Magn. Mater.* 320: 1499–1502.

125 Vollath, D. and Szabó, D.V. (1999). Coated nanoparticles: a new way to improved nanocomposites. *J. Nanopart. Res.* 1: 235–242.

2

Fabrication of Conductive Polymer Composites and Their Applications in Sensors

Jiefeng Gao

Yangzhou University, School of Chemistry and Chemical Engineering, No 180, Road Siwangting, Yangzhou, Jiangsu 225002, China

2.1 Introduction

Compared with traditional conductive materials like metals, conductive polymers (CPs) exhibit excellent flexibility, light weight, low cost, and easy processing characteristics. Therefore, CPs have become the research hotspot in recent decade and been widely used in many fields such as electromagnetic interference (EMI) shielding, sensors, and wearable electronics [1–3]. Generally, CPs can be divided into two categories: (i) structural (intrinsic) structural conductive polymers (SCPs) and (ii) conductive polymers composites (CPCs).

SCPs refer to a material possessing inherent conductivity provided by the structure carrier. The SCPs were firstly discovered in the 1970s, which aroused the attention of researchers on the chemical structure and the electrical properties of polymers [4, 5]. The common types of SCPs possessing conjugated bonds are polyacetylene, polypyrrole (PPy), polyaniline (PANI), polyphenylacetylene (PPA), and their derivatives. However, the value of electrical conductivity for the conjugated polymer materials is relatively small, usually ranging from 10^{-10} to 10^{-5} S cm^{-1}. The electrical conductivity of polymer materials can be enhanced by 10^4 S cm^{-1} by doping various molecules like salt ions, hyaluronic acid, and peptides [6, 7]. The doped SCPs show both electrical conducting property from metallic material and many merits like flexibility, lightweight, etc., from polymer material. Therefore, The SCPs have demonstrated their potential applications in flexible electronics, sensors, and EMI shielding [8–10]. However, the SCPs have their inherent deficiencies including poor solubility, mechanical properties, and high cost. Usually, further modification or compounding with other polymers is needed to improve their overall performance [11].

CPCs refer to one kind of electrically conductive polymer composite materials fabricated by incorporating the conductive fillers [carbon black (CB), carbon nanotubes (CNTs), graphene, etc.] into insulating polymer matrix [12–14]. The electrical conductivity of CPCs is highly related with the percolated conductive

Polymer Nanocomposite Materials: Applications in Integrated Electronic Devices, First Edition.
Edited by Ye Zhou and Guanglong Ding.
© 2021 WILEY-VCH GmbH. Published 2021 by WILEY-VCH GmbH.

network constructed by different conductive nanofillers. It is believed that most of the resistance comes from tunneling between adjacent conductive fillers dispersed in polymer matrix [15]. CPCs based on non-conductive polymer elastomers and conductive fillers have caused wide concern of both academic and industrial group and become a hot topic of polymer composites due to its easy preparation, controllable electrical property, excellent flexibility, and stretchability. In this chapter, we systematically introduce the development and prospect of CPCs through several parts including the fabrication method of CPCs, morphology, and microstructure of CPCs and their application as sensors.

2.2 Fabrication Methods for CPCs

As known, conductive fillers, especially nano-sized conductive particles, are easy to aggregate in the polymer due to their high-aspect-ratio, resulting in uneven distribution, which may deteriorate the comprehensive performance of CPCs [16]. Therefore, surface modification of nanofillers and special processing technique are required to enhance the dispersion of conductive nanofillers in the polymer, which mainly includes the following methods: (i) The physical blending [17]. The conductive particles are uniformly dispersed into the polymer melt matrix or polymer solution under the ultra-strong external field forces (shear, tensile, etc.). (ii) In situ polymerization [18]. The conductive particles were firstly dispersed in the organic solvent containing the polymer monomer. After the polymerization reaction, the previously evenly dispersed conductive particles were anchored in the polymer matrix in situ. (iii) Chemical modification of conductive filler [19]. After the chemical reaction, the surface of conductive particles is grafted with functional groups such as hydroxyl group, carboxyl group, amino group, etc. These groups have good interactions with the polymer, such as covalent bond and hydrogen bond, which can effectively avoid the nanofillers aggregation and hence improve their dispersion in the polymer matrix. (iv) Introduction of surfactant [20]. The surfactant will wrap around the conductive particles, increasing their compatibility in polymer solution or melt, thus improving the dispersion of the fillers.

For CPCs, the electrical conductivity is depended on the transportation of charge carriers (current) along the conductive network constructed by conductive fillers in the polymer matrix. Generally, a sudden increase of several orders of magnitude in conductivity (transition from insulator to conductor) can be found as the concentration of conducting phase reaches a critical value in the polymer matrix, which is defined as the percolation threshold. Above this threshold, the concentration dependence of the conductivity of the CPCs (σ) can be described by a scaling law

$$\sigma = \sigma_0[(\varphi - \varphi_c)]^t$$

where σ_0 represents the intrinsic conductivity of the fillers, t represents the electrical conductivity exponent, φ represents the volume fraction of conductive particles, and φ_c represents the volume fraction at percolation transition [21]. In a random conducting network, the exponent t only depends on the dimensionality of the network [22], which is called universal percolation behavior. Theoretical calculation suggests that $t = 1$–1.3 corresponds to two-dimensional networks, while $t = 2.0$ represents a three-dimensional network. The percolation threshold and electrically conductivity are affected by many factors such as nanofiller dimension and microstructure of the CPCs. For example, conductive nanofillers with a large aspect ratio (CNTs, graphene nanosheet) could form well connected conductive network, which is beneficial to a low percolation threshold in CPCs [23–25]. It is also reported that nanofiller filled CPCs exhibit better electrical and mechanical properties than CPCs incorporated with microsized fillers [26, 27]. Also, the electrically conductive behavior of CPCs has strong relationship with the structure of conductive network in the polymer matrix. The perfect interconnection between conductive nanofillers and formation of 3D continuous pathway in the polymer matrix are regarded as the key factor for construction of the conductive network [28]. As a result, the morphology and distribution of nanofiller in the polymer matrix play an important role determining the electrical conductivity of CPCs. The fabrication process and parameters can greatly influence the morphology of conductive network in polymer matrix and thus their electrical properties. There are many methods for fabrication of CPCs have been reported, among which melt blending, solution mixing and in situ polymerization are regarded as three major methods. However, it still remains challenge to prepare high performance CPCs with low percolation threshold while high conductivity [29, 30].

2.2.1 Melt Blending

Melt blending is processed by using kneading machine, molding machine, internal mixer or double screw extruder, etc. [31, 32] to evenly mix the polymer matrix and conductive fillers with the processing temperature above the melting point of polymer. Thus, high temperature and high shear force are needed in melt blending process to ensure the homogeneous dispersion of the conductive nanofiller in the melting polymer matrix. During the melting blending, the nanofillers are forced to disperse by the mechanical shear force and at the same time prevented from re-aggregation by the viscous polymer matrix. After the masterbatch is obtained, the final CPCs can be prepared by using different polymer processing technologies like spinning, hot press, and injection molding [33–35]. Melt blending is an environmental-friendly process method and is feasible for large-scale industrial production of the CPCs. Many recent studies have investigated the effect of fillers introduction, dispersion state, and processing factors on the physical and electrical properties of the CPCs prepared by the melt blending [36].

Kim [37] studied the influence of the CNTs concentration and dispersion state on thermal, rheological, and mechanical properties of the polybutylene terephthalate (PBT) nanocomposites. It was found that the storage and loss moduli of the composite increased with the increase of the CNTs content. The interaction between nanotube–nanotube and polymer–nanotube was regarded as the main cause for the nonterminal behavior of the PBT nanocomposites. Besides, it can be found that the heat distortion temperature and the thermal stability of the composite were

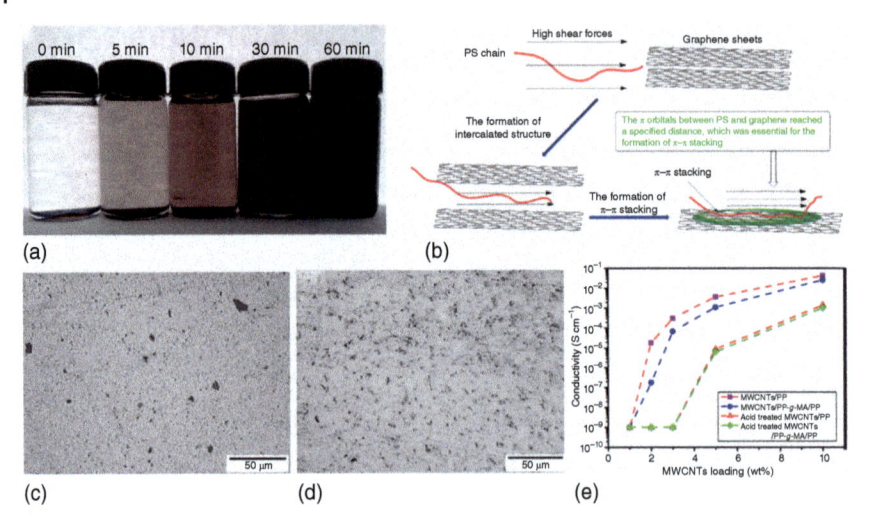

Figure 2.1 (a) Photographs of the PS/graphene/toluene suspension prepared by centrifuged at 8000 rpm for 30 minutes. (b) Schematic illustration for the formation of π–π stacking between graphene and PS in melt blending process. Source: (a)–(b) Reproduced with permission. [38] Copyright 2011, American Chemical Society. Microscopic morphology of (c) 1% pristine MWCNTs/PP; (d) 1% pristine MWCNTs/PP-*g*-MA/PP. (e) Electrical conductivity of various PP composites. Source: (c)–(e) Reproduced with permission. [39] Copyright 2009, Elsevier Ltd.

substantially enhanced at a low CNTs concentration. During the melt blending, the interaction between conductive fillers and polymer matrix greatly influences the dispersion state of the filler. For example, due to the interaction of π–π stacking between graphene and polystyrene (PS), the graphene could be easily dispersed in PS matrix during the melt blending [38]. It can be seen in Figure 2.1a that the suspension of PS/graphene without melting blending is transparent, suggesting the absence of graphene in the suspension. However, the solubility of graphene in toluene is greatly enhanced by prolong the melt blending time (5–60 minutes). This dark-colored suspension keeps stable and homogeneous even after three months or longer. The forming of π–π stacking between PS and graphene sheet in melt blending process is schematically demonstrated in Figure 2.1b. In melt blending, the PS chains were stretched by shear forces, forming some closely aligned aromatic rings parallel to the graphene sheet. Therefore, the interaction between PS and graphene was enhanced.

But mostly graphene and CNTs tend to aggregate in the polymer matrix [polymethyl methacrylate (PMMA), polystyrene-*block*-poly(ethylene-*co*-butylene)-*block*-polystyrene (SEBS), polypropylene (PP)] [40, 41]. To solve this problem, compatibilizers, chemical modification, or surfactants have often been used to improve the interaction between the filler and the nonpolar polymer matrix [42–44]. Pan et al. [39] studied the correlation between the dispersion state of multiwalled carbon nanotubes (MWCNTs) in PP and electrical conductivity of its composite. They adopted two methods to achieve the uniform dispersion of MWCNTs in PP matrix. One is chemical modification of MWCNTs and another one is incorporation of a master batch [polypropylene-*grafted*-maleic anhydride (PP-g-MA)] as a compatibilizer followed by simple melt blending. Through comparing the optical microscopic images of MWCNTs dispersion in PP in Figure 2.1c,d, we can get that the addition of compatibilizer could largely enhance the uniform dispersion of MWCNTs in PP matrix. Although the interfacial interaction between MWCNTs and PP could enhance the dispersion of MWCNTs, it may cause the reduction of the electrical conductivity of the composites (Figure 2.1e). Meanwhile, post-heat treatment can improve the connection of MWCNTs in the composite, leading to the increase of the electrical conductivity. Herein, a balance between uniform dispersion of CNTs and construction of conductive networks is vital to the enhancement of electrical conductivity for composites.

In short, melt blending is a simple method to fabricate CPCs. The electrical properties of the CPCs are strongly dependent on the processing parameters like mixing time, shear stress, and temperature as well as the surface modification and introduction of compatibilizers.

2.2.2 Solution Blending

Although melt blending is a useful and feasible processing method for the large-scale industrial production of the CPCs, the volume concentration of conductive particles is usually higher than 5% to obtain the composite with a high conductivity [45]. Such a high content of fillers would largely increase the melt viscosity [46, 47] and thus make the processing less smooth and also increase the cost of the material fabrication. As an alternative, solution mixing method can tackle this issue, because the nanofillers can be diluted in solvent to achieve relatively good dispersion [48, 49].

Solution mixing refers to a process for preparation of CPCs through mixing conductive elements and polymer matrix in a solvent, followed by cooling and solvent removal. Generally, this fabrication process contains three key steps: (i) preparation of filler suspension in a suitable solvent, (ii) mixing filler suspension with polymer, and (iii) precipitation or solvent evaporation of the mixed solution. While the majority of conductive fillers, especially carbon-based fillers (e.g. CNTs, graphene, and carbon black) show undesired dispersion in organic solvent due to their large specific surface area and high degree of graphitization thus low surface energy. Herein, it is difficult to get a uniformly dispersed suspension of carbon-based fillers just by mechanical stirring. Thus, powerful ultrasonication is often adopted to assist the nanofiller dispersion in a polymer solution [50–52]. In many cases, chemical modification or the dispersant is required. For instance, acid (such as sulfuric and/or nitric) is often used to modify the CNTs, and the graphitized structure of the CNTs is partially damaged and oxygen containing functional groups can be grafted onto the filler surface through covalent bonding [53, 54]. These functional groups can effectively promote the nanofiller dispersion in the solution due to the interaction (e.g. hydrogen bond) between fillers and solvent [16, 55].

Although covalent bonding is permanent and mechanically stable, chemical reaction can also, to a large degree, damage the sp^2 conformation of the carbon atom, thereby disrupting the conjugation of the CNT wall. As a result, thermal,

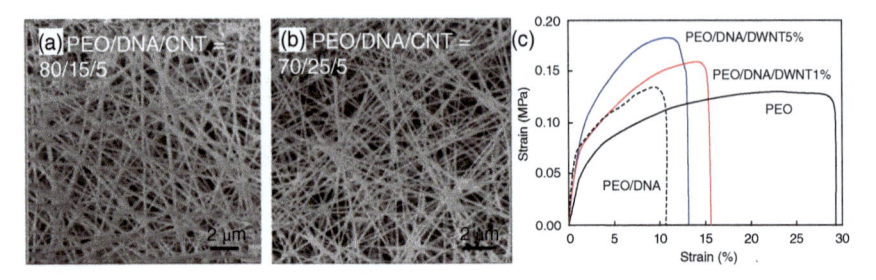

Figure 2.2 (a, b) The PEO/DNA nanofiber webs containing 5% DNA-dispersed DWNTs. (c) Stress–strain curves for nanofiber webs of pure PEO, PEO/DNA, and PEO/DNA/DWNT. Source: (a)–(c) Reproduced with permission. [57] Copyright 2013, American Chemical Society.

electrical, and mechanical properties of the chemically modified CNTs decreased dramatically, as compared with the pristine nanotubes [56]. To preserve the integrity of CNT microstructure while at the same time improve CNT dispersity, the noncovalent method to functionalize CNTs has received increasing attention, because the surface structure of the nanofillers is preserved and thus the physical and electrical properties of the nanofillers are preserved [19, 57]. Generally, the surfactants with perfect solubility with the solvent are usually used in noncovalent method, and these molecules are able to wrap CNTs, and the affinity between CNTs and the solvent is thus enhanced [58].

In recent years, amphiphilic molecules like polymers are non-covalently attached on the surface of nanofillers to overcome their aggregation [59, 60]. For example, poly(vinylpyrrolidone) (PVP) and polystyrene sulfonate (PSS) were used to improve the solubility of CNT in water and thus the CNT dispersion in polymer matrix [e.g. polyvinyl alcohol (PVA) and polyethylene oxide (PEO)] [61, 62]. Lee et al. [63] successfully dispersed MWCNTs with the concentration as high as 6.7 wt% in a PEO solution by attaching the amphiphilic poly(styrenesulfonic acid-*graft*-aniline) (PSS-*g*-ANI) to the surface of MWCNTs. The attachment of PSS-*g*-ANI has dramatically enhanced the dispersion of MWCNTs and dispersion of 1%, 10%, 20%, 30%, and 40% MWCNT/PEO solutions still remains homogeneous even after three days. Deoxyribonucleic acid (DNA) as a representative biopolymer has now also been considered as a powerful dispersing agent to disentangle strongly bundled CNTs in an aqueous solution, resulting from DNA's strong ability to wrap around the sidewalls of CNTs [64, 65]. It was reported that CNTs were individually dispersed in the DNA solution, and the uniform dispersion could be preserved in the polymer/CNTs/solvent ternary solution [65]. In PEO nanofiber composite, under the DNA assistant dispersion, double-walled carbon nanotubes (DCNTs) were finally trapped along the axis of the PEO nanofiber and acted as mechanically reinforcing filler and an electrical conductor [57]. The morphology of PEO nanofiber web consisting of long, homogeneous nanofibers with a diameter of <100 nm was obtained even with the content of 5% DNA-dispersed double-walled carbon nanotubes (DWNTs) shown in Figure 2.2a,b. Also, the modulus and tensile strength were largely improved by adding individually DNA dispersed DWNTs (Figure 2.2c).

2.2.3 In Situ Polymerization

In situ polymerization was firstly proposed by Imai et al. where polyimide (PI)/CB composite was obtained by dispersing the CB into the polymer salt monomer [66]. In fact, this is a unique solution based processing technology for preparing the CPCs, and chemical reaction is usually involved during the polymerization [67, 68]. The efficient polymer-chain graft onto the filler surface could form a perfect interface interaction between the filler and polymer matrices, which also improves the homogeneous dispersion of the fillers in the polymer matrix and influences the crystallization of polymer chains to some extent [69–71]. As a result, CPCs with a high weight content of the nanofillers can be obtained by this method [69, 72].

Zhu and coworkers prepared reduced graphene oxide (r-GO)/PI composites with different loadings of GO by in situ polymerization and the maximum content of GO can reach 30 wt%. During polymerization, a relatively high temperature was set to reduce GO into conductive r-GO in the polymer matrix. The electrical property of the obtained r-GO/PI was greatly enhanced, because of the conductive network formed by r-GO in the composite film and the conductivity could reach as high as 1.1×10^1 S m^{-1}, which is about 10^{14} times that of pure PI film [73]. The mechanical properties of GO/PMMA composites fabricated by in situ polymerization were also tested by Potts et al. [74]. It is found that the elastic modulus and tensile strength of the GO/PMMA composites could be improved even with 1 wt% loading of fillers.

2.3 Morphologies

Different fabrication techniques may lead to different morphologies of conductive networks in CPCs, which significantly influence the electrical properties of these composites [30]. Various morphologies including uniform dispersion of nanofiller in the polymer matrix, segregated structure, and selective decoration of the nanofiller on the skeleton of porous polymer materials are reported [75].

2.3.1 Random Dispersion of Nanofiller in the Polymer Matrix

The most straightforward and popular strategy to fabricate CPCs is the incorporation of conductive fillers (e.g. graphene, CNTs) in the polymer matrix by either melting or solution method. Naturally, the nanofillers are randomly located in the polymer matrix, and some of them may be squeezed out to polymer surface in particular at a high filler concentration [76]. The nanofillers can be uniformly distributed in the fibers, film, and three-dimensional polymer materials, depending on the processing technology used. Wet-spinning [77, 78] and electrospinning [75, 79] are two most common means to prepare one dimensional CPCs. Yu et al. [80] have successfully prepared conductive poly(styrenebutadiene-styrene) (SBS)/CNT fiber through wet-spinning the mixed solution of CNT and SBS elastomer. The schematic diagram of the fabrication procedure of SBS/CNT and the pictures of flexible SBS composite

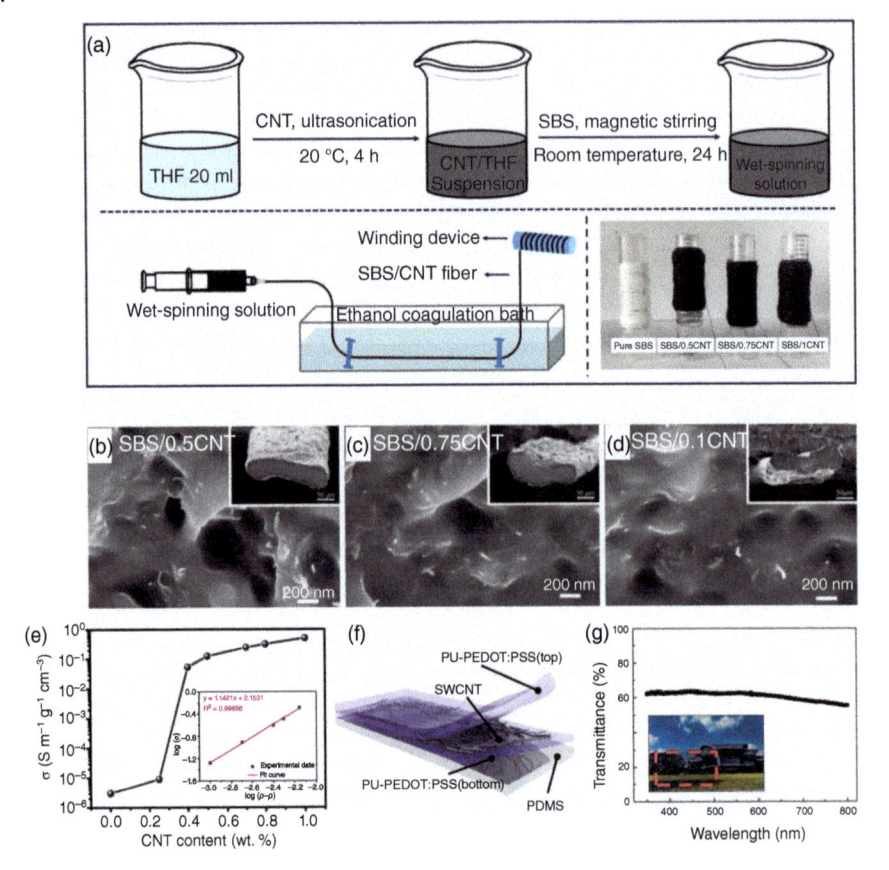

Figure 2.3 (a) Schematic diagram illustrating the preparation procedure for SBS/CNT fibers (SCFs) and photograph of SCFs with the various content of CNT. (b–d) The cross-sectional morphologies of SBS/CNT fibers containing different content of CNT. (e) The specific conductivities of SFCs as a function of different loading content of CNT. Source: (a)–(e) Reproduced with permission. [80] Copyright 2018, Elsevier Ltd. (f) Schematic illustration of the PU-PEDOT:PSS/SWCNT/PU-PEDOT:PSS with sandwiched structure on a polydimethylsiloxane (PDMS) substrate. (g) Transmittance of the integrated composite in the visible wavelength range from 350 to 700 nm. Source: (f)–(g) Reproduced with permission. [81] Copyright 2015, American Chemical Society.

with different content of CNTs are shown in Figure 2.3a. Figure 2.3b–d show the SEM micrographs of fracture SBS/CNT fibers with a content of 0.5%, 0.75%, and 1%, respectively. Uniformly dispersed CNTs in SBS matrix can be observed even at a CNT content of 1%. Also, the conductivity of the obtained 1-D conductive fiber composite improves with the increase of CNT content (Figure 2.3e).

Thin film-like CPCs can be fabricated by facile casting or template molding [82, 83] or polymer encapsulation [81, 84–86]. A stretchable, transparent, patchable nanohybrid conductive polymer film was reported by Roh et al. [81]. The CNTs were encapsulated in the interlayer of polyurethane (PU)-poly(3,4-ethylenedioxythiophene) polystyrenesulfonate (PEDOT:PSS) forming a sandwich-like structure, which is schematically illustrated in Figure 2.3f, where the CNTs functioned like a bridge that connects the conductive PEDOT to the PEDOT phases. The obtained sandwich-like composite with 5 mg ml^{-1} SWCNT dispersion is optical transparent and shows the optical transmittance of 62% in the visible range (Figure 2.3g).

In terms of three-dimensional composite, nanofillers are often distributed in a foam composite that is usually obtained by freezing drying method [87, 88]. For example, Huang et al. [87] fabricated a novel aligned porous CNT/thermoplastic polyurethane (TPU) foam composite by using a directional-freezing method. During the freezing–drying process, the solvent of the mixture would form directional crystal due to the low temperature and then the ice crystal of the solvent would be sublimated, leaving aligned interconnected pores.

2.3.2 Selective Distribution of Nanofillers on the Interface

To reduce the content of conductive fillers in polymer matrix and at the same time maintain a relatively high conductivity of the CPCs, researchers try to locate conductive fillers on the interfaces of the polymer granule (i.e. segregated structure) or on the skeleton (surface coating) of the porous materials. Also, when used as sensors, the specially distributed conductive paths are easier to destruct upon external stress compared with conventional CPCs with relatively strong and dense conductive paths.

2.3.2.1 Segregated Structure

The study about construction of segregated structure was first reported in 1971 [89], and to date much work have been done on this topic [15, 30, 90]. In fact, segregated structure is a unique dispersion state of the conductive fillers in the polymer matrix, at which conductive fillers are dispersed at the interfaces between polymer particles. Mechanical blending and hot compression molding technique is usually applied to fabricate CPCs with a segregated structure [91, 92]. Generally, conductive fillers such as CNTs and graphene were adsorbed to the surface of polymer microspheres by chemical or physical methods, and then the temperature and pressure of hot pressing were controlled, guaranteeing that the conductive fillers were only distributed at the interface between polymer microspheres instead of evenly dispersed in the whole polymer matrix [15, 93, 94].

For example, Wu et al. [95] added amino-functionalized PS microspheres suspension into the GO solution. Graphene was tightly coated on the surface of PS microspheres after a series process of flocculation, filtration, washing, and hydroiodic acid reduction. The composite with a low percolation value of 0.15 vol% was obtained after hot press of the graphene coated PS microsphere. Also, the conductivity of the composite could reach as high as 1024.8 S m^{-1} when the volume content of graphene is 4.8%, which is much higher than that of PS/graphene and PS/CNT composite made by solvent blending.

An ultralow percolation threshold of 0.047 vol% was achieved by Cui and Zhou [92]. In their work, the conductive PS/graphene and PS/MWCNTs composites with segregated structures were obtained by hot press surface sulfonated PS microspheres and protonated triethylenetetramine functionalized perylene

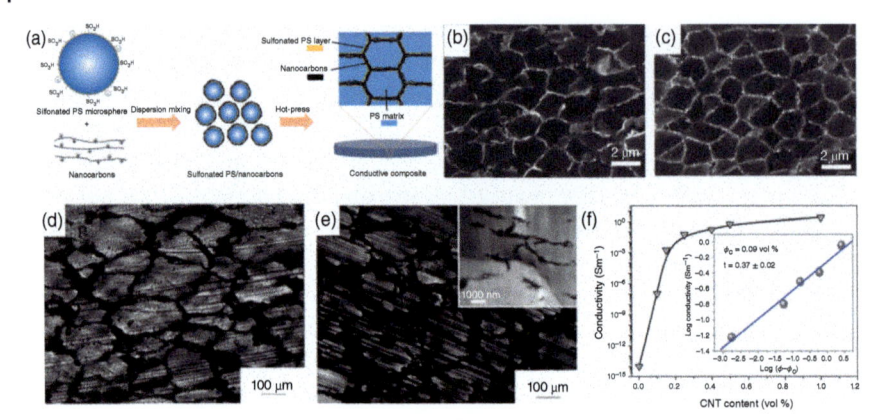

Figure 2.4 (a) The fabrication process of PS−nanocarbon composite with interconnected networks. Cross-sectional SEM images of PS composites with (b) 0.94 vol% graphene sheets and (c) 0.94 vol% MWCNTs. Source: (a)–(c) Reproduced with permission. [92] Copyright 2017, The Royal Society of Chemistry. Morphologies of double-segregated (d) CNT/PMMA/UHMWPE (0.2/7.8/92.0 by volume) and (e) CNT/PMMA/UHMWPE (0.5/16.2/83.3 by volume) composites. The inset transmission electron micrograph (TEM) image in (e) shows the state of the segregated CNT conductive network in the CNT/PMMA layers. (f) The variation of electrical conductivity for the double segregated CNT/PMMA/UHMWPE composites with different CNT content. Source: (d)–(f) Reproduced with permission. [96] Copyright 2013, The Royal Society of Chemistry.

bisimide (HTAPBI)-stabilized nanocarbon. The formation of conductive PS composites with a segregated network is schematically demonstrated in Figure 2.4a. The SEM micrographs of the fracture surface of PS composites containing 0.94 vol% graphene sheets and 0.94 vol% MWCNTs are displayed in Figure 2.4b,c, respectively. Due to the strong electrostatic attraction between the negatively charged PS and positively charged nanocarbon, the interconnected conductive pathways could be preserved regardless of a high hot press temperature, leading to an ultralow percolation threshold.

Pang et al. reported CPCs with double-segregated structure [96, 97]. CNTs wrapped around the small-sized PMMA particles to form the first layer of segregated structure, and CNTs/PMMA were distributed at the interface of the large-sized ultrahigh molecular weight polyethylene (UHMWPE) particles to construct a second layer of segregated conductive network throughout the whole system. The optical micrographs of CNT/PMMA/UHMWPE composite with 0.2 and 0.5 vol% CNTs are shown in Figure 2.4d,e, respectively, showing perfect double-segregated structure. Figure 2.4f shows the electrical conductivity of CNT/PMMA/UHMWPE composites as a function of CNT content. The percolation value of the CPC was calculated to be only 0.09 vol% [96].

Another kind of segregated structure can be constructed by infiltration of flexible elastomers into conductive foams. Carbon based foams (e.g. graphene and CNTs) and carbonized polymer foams are usually selected as the conductive skeleton and polydimethylsiloxane (PDMS) ink was impregnated in the foam to prepare the CPCs [98–100].

2.3.2.2 Surface Coating

To prepare CPCs with a low percolation, a high conductivity at a low filler content, and meanwhile no evident fillers aggregation in polymer matrix, decoration of conductive nanofillers on the skeleton (out surface) of the polymer material tends to be an effective solution, which includes dip coating [101, 102], spray coating [103, 104], and ultrasonication [105, 106]. Note that the polymer materials usually possess a porous structure (e.g. fabric and foam), which facilitates the nanofillers penetration into the interior of the material during the surface coating.

In dip coating, the polymer scaffold was immersed into the ink of conductive fillers (graphene, CNTs, AgNWs) [102, 107, 108] or the precursor solution [109, 110] followed by drying or in situ reduction and drying. Cui and coworkers [108] produced highly conductive textiles with conductivity reaching 125 S cm^{-1} by immersing textile into single-walled carbon nanotube (SWNT) ink (Figure 2.5a). The color of the textile becomes black after immersed in SWNT ink (Figure 2.5b), suggesting that the SWNT has been successfully decorated on the surface of textile fibers. Also, SWNT can be observed on the surface of textile from SEM image in Figure 2.5c. The interaction of van der Waals forces and hydrogen bonding guarantees the tightly binding of single walled CNTs to the cellulose. A conductive multifilament fiber with conductive Ag shell and elastomer polymer core was reported by Lee et al. [111]. The multifilament elastomer fiber was first dipped in the AgCF$_3$COO solution to adsorb Ag precursor through ion–dipole interaction. Then the Ag$^+$ ions absorbed in the fibers was in situ reduced by the reduction solution. The microstructure of the core–shell structure can be observed in Figure 2.5d,e. The interfacial interactions like electrostatic interaction [111], hydrophobic interaction, and hydrogen bonding [108, 114] are considered to be the leading driving force for the successful decoration of conductive nanofillers on the polymer substrate surface.

In addition to the aforementioned interfacial interaction, ultrasonication induced nanofiller anchored onto the nanofiber surface or the skeleton of the foam is wildly used to prepare conductive polymer sponge [106, 115] or nanofiber [116] composite. Its working mechanism is as follows: under ultrasonication, a large number of cavitation bubbles will be generated in the solution, and these cavitation bubbles will be blasted generating micro-jets [117], which will break up the nanofiller aggregates in the solution, and at the same time, the dispersed nanofillers can rush to polymer scaffold at high speed. The viscoelastic elastomer at the interface between fiber and CNTs would be soften or even partially melt under the instantaneous high temperature and high pressure. At this point, the well dispersed nanofillers under ultrasonication would be anchored in situ on the surface of elastomer surface. Gao et al. [118] first reported this technique to prepare the CNTs decorated TPU nanofiber composite. Recently, they [112] prepared multifunctional nanofiber composite by dip coating one PDMS layer onto the CNTs/TPU surface. The fluffy CNTs forms conductive shell on the fiber surface as shown in Figure 2.5f,g. The PDMS layer acts not only as adhesion promoter enhancing the interfacial interaction but also as low surface energy material endowing the CPCs surface with low surface energy. CPCs with dual conductive networks are also reported by Gao and his coworkers by combination of ultrasonication and dip-coating [113]. The acidified carbon nanotubes

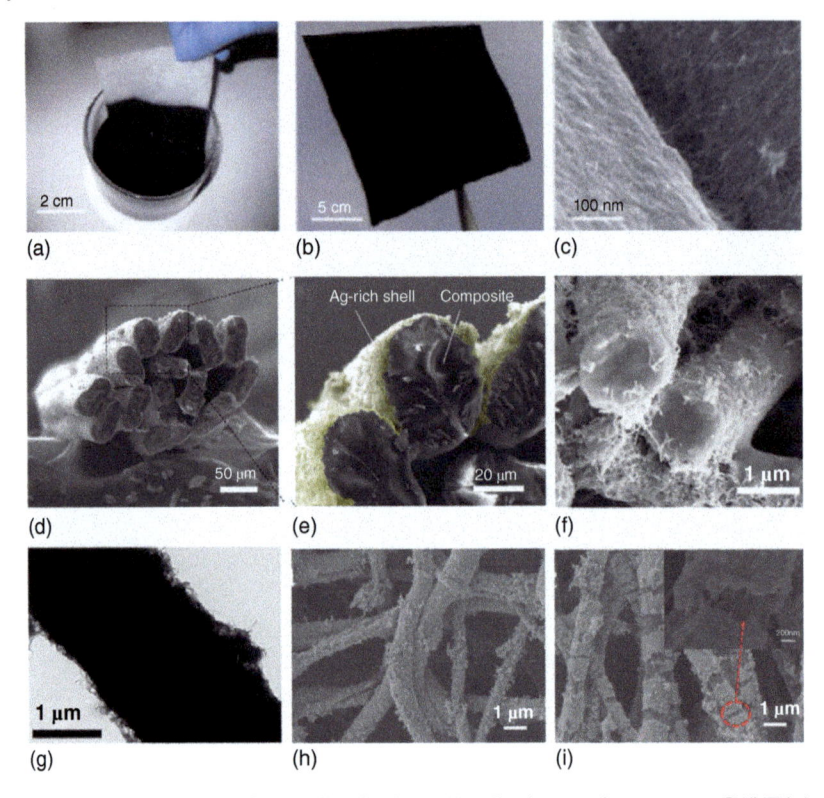

Figure 2.5 (a) Conductive textiles fabricated by dipping coating aqueous SWNT ink. (b) Picture of obtained conductive textile. (c) SEM image showing SWNT coating on the surface of fabric fibers. Source: (a)–(c) Reproduced with permission. [108] Copyright 2010, American Chemical Society. (d) Cross-sectional SEM images of the electrically conductive elastomer fiber composite. (e) Magnified SEM image of (d) showing the conductive Ag-rich shell on the outer surface of elastomer fiber. Source: (d)–(e) Reproduced with permission. [111] Copyright 2018, American Chemical Society. (f) Cross sectional SEM image and (g) TEM image of the TPU/CNTs/PDMS nanofiber composite. SEM images of the nanofiber composite at different strain during stretching, Source: (f)–(g) Reproduced with permission. [112] Copyright 2019, Elsevier B.V. (h) $\varepsilon = 10\%$, (i) $\varepsilon = 70\%$. Source: (h)–(i) Reproduced with permission. [113] Copyright 2019, Elsevier B.V.

(ACNTs) decorated nanofiber composite was immersed in Ag precursor solution, and then AgNPs were generated after the precursor was reduced. PDMS was also used to improve the interfacial adhesion between the AgNPs. The dual conductive structures on the fiber surface can be observed in Figure 2.5h. The outer surface of the CPC is surrounded by silver nanoparticles, and ACNTs in the interlayer could be observed after stretching the fiber composite as shown in Figure 2.5i.

2.4 Application in Sensors

CPCs have been widely used for different sensors, due to its light weight, portability, flexibility, and low cost [90, 119]. The sensing mechanism of CPC sensors is based on the resistance variation of CPCs when experiencing external stimulus like mechanical deformation [120, 121], temperature variation [122, 123], and adsorption of vapors or organic solvent [124, 125]. The resistance of CPCs is affected by the evolution of conductive network under external stimulus [5]. These transient resistance change can be detected as a signal for sensing purpose.

2.4.1 Strain Sensor

As a device for detecting object deformation, CPC based strain sensors have been widely used for health monitoring, electronic skin, wearable electronics, etc. For the purpose of monitoring, the resistance (conductive networks) should response upon external strain/stress [119]. Thus, CPCs should possess excellent resilience such that the material could be elongated under external strain (destruction of conductive networks) and recovered immediately after the removal of the stress (reconstruction of conductive networks).

Conductive natural rubber/elastomer composites decorated with different conductive nanofillers are explored as promising candidates of wearable strain sensors [110, 126, 127]. The sensors show broad workable sensing range, high sensitivity, and excellent recoverability. Among these materials, conductive fabric composite strain sensors have shown promising applications in wearable electronics, thanks to their breathability, skin affinity, and so on [128, 129]. For example, they can be incorporated into clothing or gloves to detect human body motions [111, 114] and detect the motions of human body as shown in Figure 2.6. Figure 2.6a–d demonstrate that the fiber strain sensor could be used for human–machine interfaces by simply incorporating the fiber composites onto a glove, demonstrating the potential application in remote control of hand robot. Figure 2.6e–j show sensing performance of the smart integrated glove in monitoring motions like head-forward, shoulder imbalance, kyphosis, etc. In order to improve the environmental suitability of the sensors, self-protective CPCs with superhydrophobicity have attracted great interests [128, 130] from the academia. This superhydrophobicity endows the sensor with self-cleaning behavior and can prevent the water or even corrosive solution diffusion into the materials, which can extend their applications in some harsh conditions. To reveal the strain sensing mechanism, the evolution of conductive network of the CPC under different strains is observed by using the SEM [126].

Hydrogel as a new type of conductive polymer composite demonstrating preferable bio-compatibility, self-healing, and self-adhesiveness is an ideal material for strain sensing usage [131, 132]. Zhang et al. [133] developed a MXene ($Ti_3C_2T_x$)/PVA hydrogel sensor with outstanding sensing performance by mixing MXene nanosheet with PVA hydrogel. The hydrogel composite shows outstanding stretchability, self-healing property, and strong adhesiveness to skin, which can adhere to human skin without the assistance of bonding materials to detect subtle motions including facial expression, vocal signals, handwriting, and finger bending, demonstrating high accuracy and sensitivity.

2.4.2 Piezoresistive Sensor

In recent years, CPC based pressure sensors as a branch of smart material, which could respond to compressive deformation and transform mechanical forces into

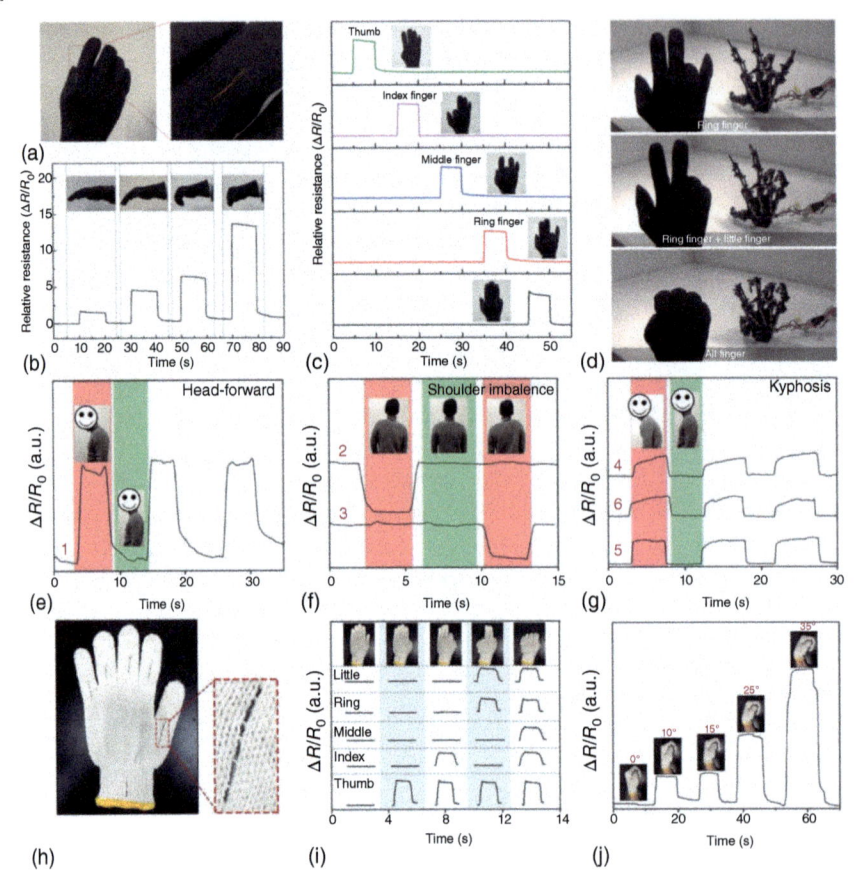

Figure 2.6 (a) Photograph showing the smart glove integrated with the fiber composites. (b, c) Strain sensing behavior of the smart glove for (b) fingers bending and (c) a full bending stimulation of each finger. (d) Photographs showing the hand robot operated by the smart glove. Source: (a)–(d) Reproduced with permission. [111] Copyright 2018, American Chemical Society. Relative resistance change of the sensor responding to various postures: (e) head-forward, (f) shoulder imbalance, and (g) kyphosis, respectively. (h) Photos showing a smart glove incorporated with fiber composite. Strain sensing behavior of the smart glove for (i) different gestures of the five fingers and (j) different bending angels of the wrist. Source: (e)–(j) Reproduced with permission. [114] Copyright 2019, The Royal Society of Chemistry.

electrical signals, have been extensively developed [134]. To meet the demands of different applications, innovated material structures and fabrication strategies are developed to prepare flexible and wearable pressure sensors with excellent sensing performance. CPCs containing an elastic matrix and conductive nanofillers are the most popular candidates of the piezoresistive sensor [135]. Textiles with fiber networks are often used as flexible scaffolds of piezoresistive sensors by carbonization or coating conductive fillers onto the fiber surface [136, 137]. Li et al. [137] fabricated flexible and electrically conductive carbon cotton (CC)/PDMS composites by infiltrating PDMS glue into the CC scaffold. The CC/PDMS demonstrates a sensitivity of 6.04 kPa^{-1}, a working pressure of 700 kPa and durability over

1000 cycles. CPCs with a porous structure (foam, aerogels) are another ideal active materials of piezoresistive sensors because of their highly reversibility and hence reusability after large-deformation cycles [138, 139]. Zhong and coworkers [140] successfully fabricated lightweight MXene (Ti_3C_2) aerogel with ultra-stable lamellar structure by freeze drying the mixture of bacterial cellulose fiber and MXene nanosheets. The carbon aerogel used for pressure sensor demonstrates ultrahigh compressibility and superelasticity, a wide linear range and low detection limits.

To enhance the response intensity and low detection limit, CPCs with delicate structures like fingerprint pattern [141, 142], microdome or micropillar array [143, 144], hair of human skin [145], plant leaves [146, 147], hollow spheres [148], etc. were studied. Park et al. [149] have reported flexible piezoresistive sensor by building interlocked microdome arrays on the elastomer composite film. The flexible film with regular microdome arrays can be observed by SEM as shown in Figure 2.7a. The inset picture demonstrates its excellent flexibility. The schematic diagram (top) and SEM image (bottom) of the fracture surface for the composite films are shown in Figure 2.7b, and an interlocked structure can be observed. The operating principle of the sensing device is also schematically illustrated in Figure 2.7c. Under the stress of external pressure, the microdomes would deform and the contact area between interlocked microdomes was increased, which in turn influenced the tunneling resistance of the sensor. Figure 2.7d shows the pressure sensing performance (relative electrical resistance (R/R_0)) of composite film with different structures. The film with interlocked microdomes demonstrates a sharp decrease in R/R_0 when the applied pressure increases from 0 to 10 kPa, while the sensing signals for films with single microdome arrays and planar structure are much weaker. Also, the sensing curve (log–log plot) of R/R_0 vs. pressure demonstrates a high linearity (Figure 2.7e), indicating an exponential dependence of resistance on the applied pressure.

Bae et al. [144] prepared highly linear and sensitive graphene/PDMS composite films with microdome structures. They found that pressure sensor made of hierarchical graphene/PDMS array with wrinkles on the surface of the microdome was more sensitive to external stimuli than sensor made of smooth-dome graphene/PDMS array. This novel type of pressure sensor exhibits a superb endurance of 10 000 cycles and a low detection limit of 1 Pa.

2.4.3 Gas Sensor

CPCs based gas sensors have also aroused tremendous attention for application in detecting or quantifying organic vapors or liquids chemicals [150, 151]. The electrical resistance of the CPC usually undergoes change when the sensor is exposed to organic vapor or liquid. The sensing mechanism of the gas sensor can be discussed on the basis of percolation theory. Upon exposure to vapor or liquid, the interaction between polymer matrix and gas would cause polymer expansion, resulting in the redistribution of the conductive nanofillers hence change of resistivity. Once the sensor is removed from the vapor stimuli, the resistance of the CPC recovers to its original value due to the desorption of vapor [152, 153]. Li et al. [154] proved a schematic illustration of the conductive networks formed by different fillers

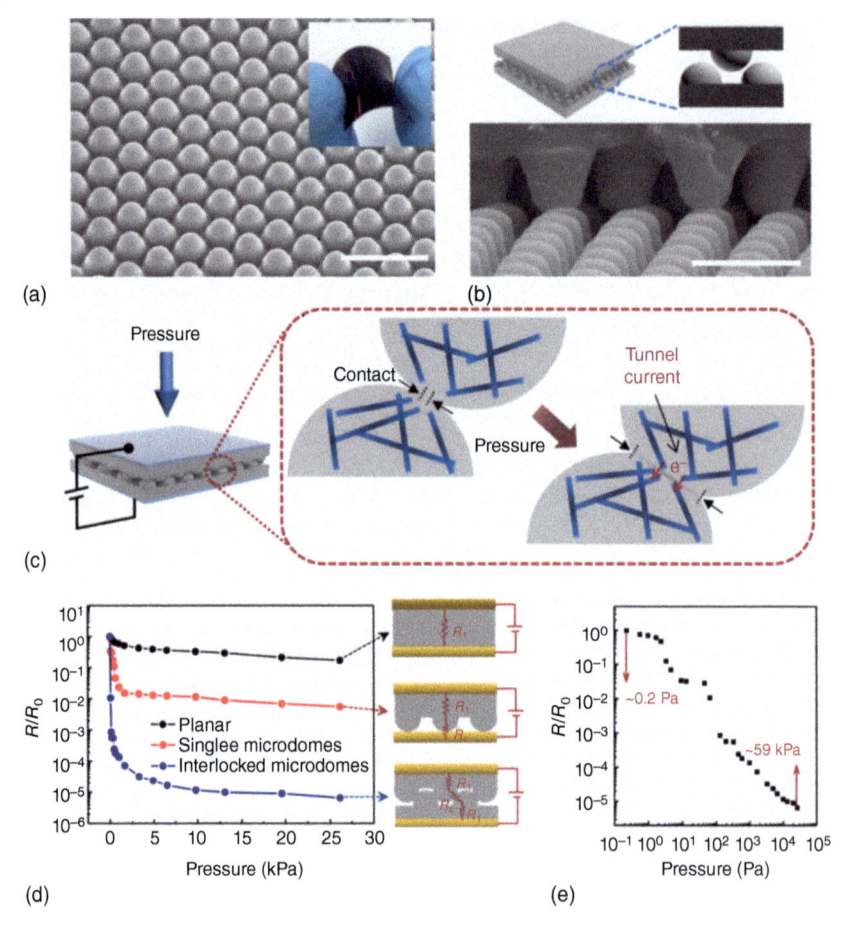

Figure 2.7 (a) SEM image showing the surface morphology of the composite elastomers (scale bar: 10 µm); Inset photograph shows the flexibility of the film composite. (b) Schematic diagram (top) and SEM image (bottom) of the fracture surface for the composite films (scale bar: 5 µm). (c) Schematic illustration demonstrating the operation principle of the film composite. (d) The pressure sensing performance of composite film with different structures. (e) Relative resistance (R/R_0) as a function of applied pressure (log–log plot) for the sensor with interlocked microdome arrays (8 wt% CNTs). Source: (a)–(e) Reproduced with permission. [149] Copyright 2014, American Chemical Society.

in sensing process shown in Figure 2.8a–d. It can be observed that the dispersion state of conductive nanoparticles (CB, CNTs) in poly(lactic acid) (PLA) matrix (Figure 2.8a,c) are quite different from that of ramie fiber (RF)/PLA (Figure 2.8b,d), resulting in different sensing mechanism. They found the addition of RF greatly tailored the vapor sensing behaviors of CB/PLA and CNTs/PLA composite.

Many efforts have been paid to modify the sensitivity, selectivity, and recoverability of vapor sensors. Gao et al. [116] reported a superhydrophobic PU@SEBS/carbon nanofiber (CNF) composite vapor sensor to improve water proof and corrosion resistance. The superhydrophobic membrane could be easily wetted by different solvents, corresponding to a rapid response time. The sensor is also sensitive to

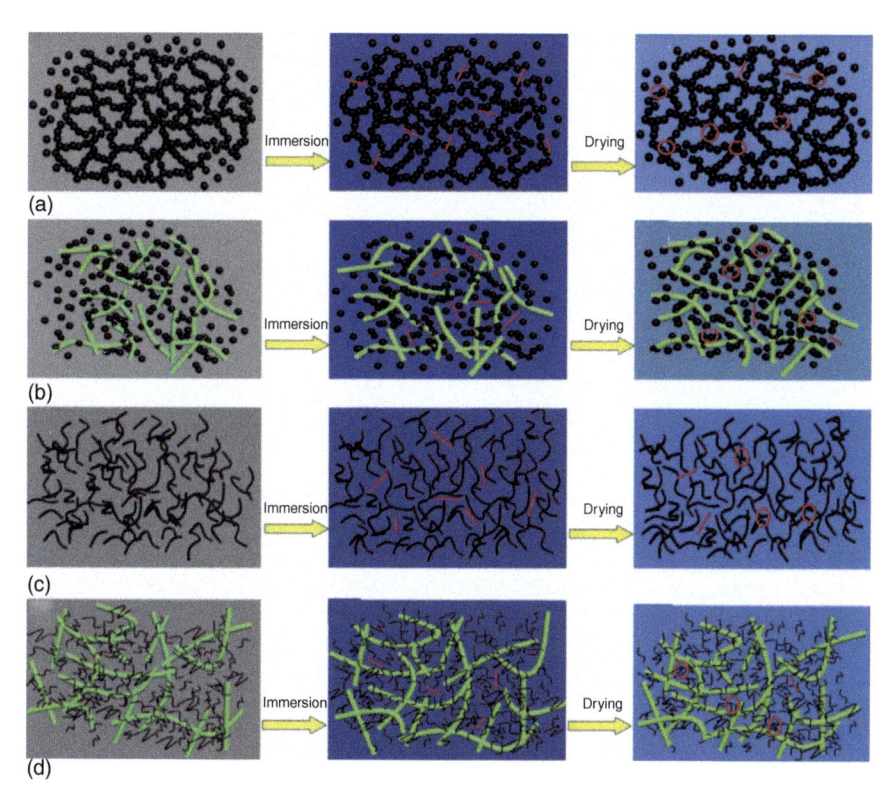

Figure 2.8 Schematic illustration of conductive networks formed by carbonaceous filler in PLA during immersion-drying runs. (a–d) Illustrating the evolution of conductive networks in CB/PLA, CB/RF/PLA, CNTs/PLA, and CNTs/RF/PLA, respectively (the red lines and circles suggest the changes of conductive paths). Source: (a)–(d) Reproduced with permission. [154] Copyright 2015, Elsevier B.V.

both polar and non-polar vapors. It is obvious that the PU@SEBS/CNF composites show different sensing behavior to tetrahydrofuran (THF), toluene, and heptane as shown in Figure 2.9a–c.

Usually, CPCs with disordered dispersion of conductive nanofillers show poor sensing repeatability, because the swelled conductive network can't perfectly recover when the materials is taken out from the vapor in most cases. To solve this problem, CPCs with segregated structures have become alternatives [156, 157], because organic vapors or liquids could more easily be permeated in the CPC by the capillary effect generated from the interfaces of polymer matrix.

Sensor that could detect both humidity and chemical vapors with high sensitivity and low detection limit is reported by Huang et al. [155]. The obtained conductive ACNT/PU nanofiber membrane shows excellent superhydrophilicity and underwater superoleophobicity, and it is sensitive to both humidity and chemical vapors. The membrane composite can be integrated into masks to detect human respiration showing potential usage in monitoring biomarker gases from human breath (Figure 2.9d–f).

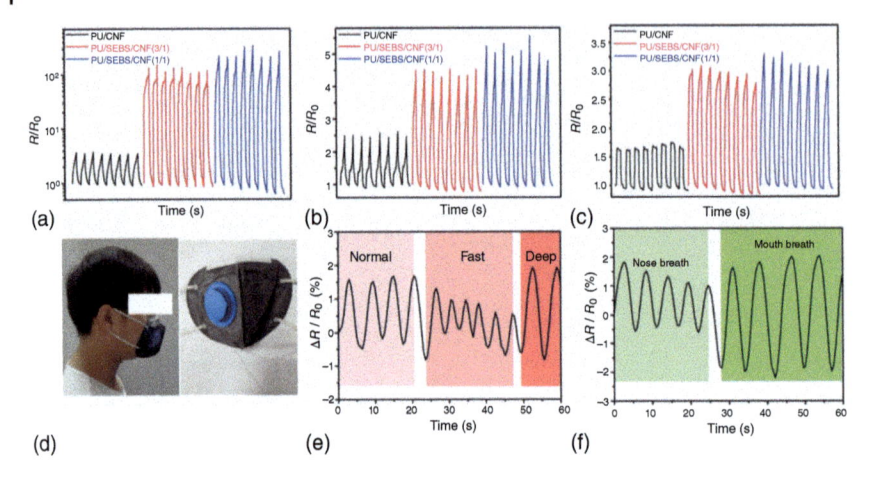

Figure 2.9 Cyclic vapor sensing behavior of the nanofiber composite with different SEBS contents to different vapors including: (a) THF, (b) toluene, and (c) heptane. Source: (a)–(c) Reproduced with permission. [116] Copyright 2018, The Royal Society of Chemistry.
(d) Photos of a volunteer wearing the integrated mask, and the mask incorporated with CNC-based humidity sensor. (e) Vapor sensing behavior of the sensor to fast, normal, and deep breathing and (f) mouth and nose breathing. Source: (d)–(f) Reproduced with permission. [155] Copyright 2019, American Chemical Society.

2.4.4 Temperature Sensor

CPCs served as temperature sensors show many popular applications for thermal detectors, over-temperature protection devices and self-regulating heaters, etc. [158]. The temperature sensing behavior of CPCs is based on positive temperature coefficient (PTC) effect and negative temperature coefficient (NTC) effect. Conceptually, PTC and NTC of thermistors display a conspicuous increase and a succeeding gradual decrease in electrical resistance with increasing temperature, respectively [122]. It is believed that the PTC effect arises from the expansion or the melting of the matrix, which would increase the distance between conductive elements. While a flocculated structure formed at elevated temperatures, resulting in NTC effect.

Chlorinated poly(propylene carbonate) (CPPC)/CB foam composite with a nearly-linear NTC effect of resistance is reported by Cui et al. [159]. As shown in Figure 2.10a, the relative resistance of solid CB/CPPC composites demonstrates both PTC and NTC effect with increasing the temperature from 25 to 70 °C. However, the resistivity-temperature performance of CPPC/CB foam composite exhibits a significant NTC effect with temperature variation (Figure 2.10b). The CB content also shows influence on the sensing behavior of CPPC/CB foam and the NTC effect declines with the increase of CB content. The schematic diagram of temperature sensing mechanism for the CPPC/CB foam composite is demonstrated in Figure 2.10c. The cell walls of the foam were squeezed by expanded gas and the thickness of the cell walls was reduced during the heating. Therefore, the distance between adjacent CB particles was decreased and more conductive pathways formed in polymer matrices, leading to the decrease of the resistance.

Lee and coworkers [160] fabricated graphite nanofibers (GNF) and CB filled high density polyethylene (HDPE) hybrid composites with PTC effect and explored

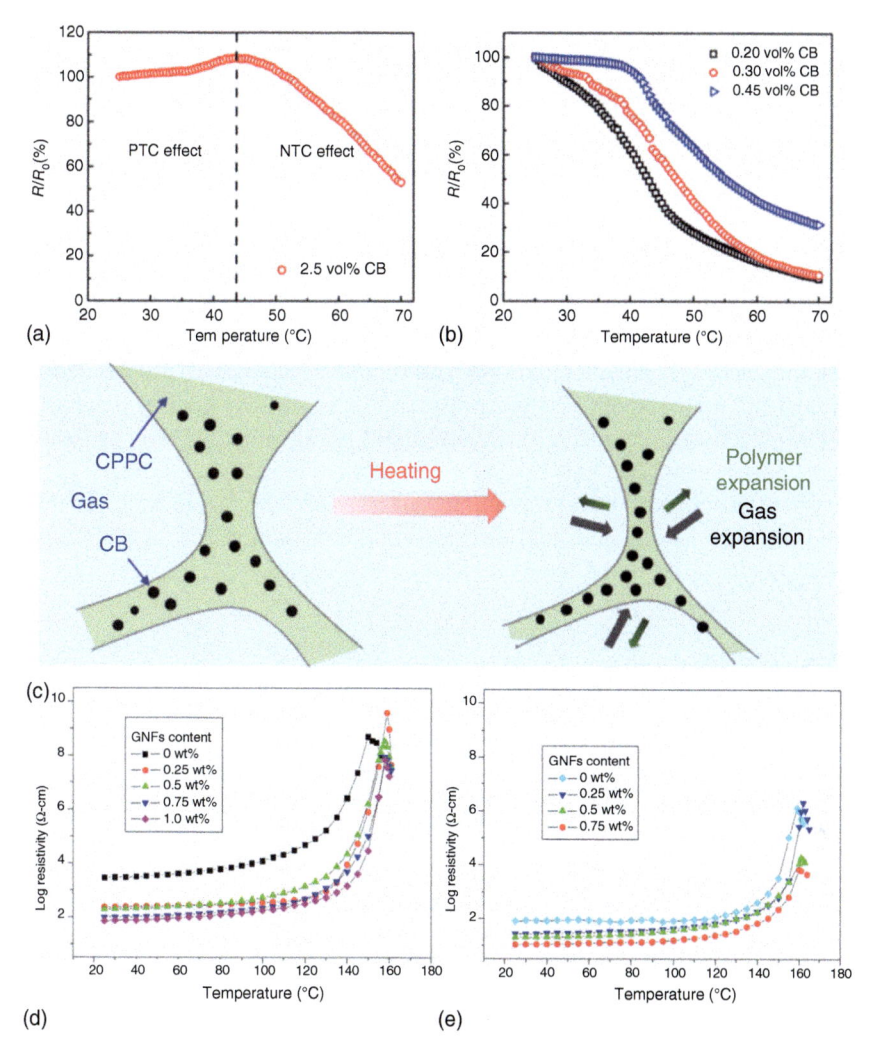

Figure 2.10 (a) The normalized resistance (R/R_0) variation with temperature from 25 to 70 °C for the solid CB/CPPC composites with 2.5 vol% CB. (b) The normalized resistance (R/R_0) variation with temperature from 25 to 70 °C for CPPC/CB foams with 0.20, 0.30, and 0.45 vol% CB. (c) Schematic illustration of microstructural evolution in CPPC/CB foams during heating process. Source: (a)–(c) Reproduced with permission. [159] Copyright 2018, The Royal Society of Chemistry. (d) The curve of resistivity vs. temperature for the HDPE/CB (80/20) with different content of GNFs. (e) The curve of resistivity vs. temperature for the HDPE/CB (75/25) with different content of GNFs. Source: (d)–(e) Reproduced with permission. [160] Copyright 2009, Elsevier Ltd.

the influence of conductive filler content on the PTC behavior of the hybrid composite. The relationship between resistivity and temperature of the HDPE/CB (80/20) and the HDPE/CB (75/25) with various GNFs contents is demonstrated in Figure 2.10d,e, respectively. It is found that the incorporation of a small quantity of GNFs into HDPE/CB composite shows significant enhancement in reproducibility and PTC intensity. The incorporation of GNFs would hinder the migration of conductive nanoparticles and the shape deformation of the polymer resin, thus resulting in this phenomenon.

The temperature-resistivity intensity (intensity of positive temperature coefficient [I_{PTC}] and intensity of negative temperature coefficient [I_{NTC}]) of CPCs is an essential index for temperature sensors. A low I_{NTC} value is required for a preferable temperature sensor to output large response toward temperature stimuli [161]. To increase the I_{PTC} intensity, many efforts have been made to remove the NTC effect. It is accepted that the NTC effect is due to the reaggregation of the conductive nanofillers in the polymer matrix and recovery of disconnected conductive pathways [162]. The usual method is using crosslinking agent or radiations, which could increase the viscosity of polymer matrix and prevent the reaggregation of conductive fillers, thus eliminating the NTC effect [163, 164]. Lu et al. [162] fabricated nylon6 (PA6)/PS/(poly(styrene-co-maleic anhydride) (SMA)–CB) composite with especial interface morphology. The PA6/PS/(SMA–CB) composites showed stronger I_{PTC} than PA6/PS/CB and NTC effect was eliminated. The special interfacial morphologies and low percolation threshold are responsible for elimination of NTC and stronger I_{PTC}.

2.5 Conclusion

Due to its inherent properties of lightweight, low cost, easy fabrication, and controllable resistance, CPCs have been the research hotspot in the past decades. The incorporation of different conductive nanofillers like CNTs, graphene, silver nanoparticles, and silver nanowires greatly enhances the electrical property of CPCs. However, fabricating high performance CPCs still remains a great challenge, because conductive fillers, especially nano-sized conductive particles, are easy to aggregate in the polymer due to their high-aspect ratio, resulting in uneven distribution of fillers.

Nanofiller aggregations would affect or even worsen the performance of CPC. Thus, even dispersion of conductive nanofillers in the polymer matrix is a vital issue. Here, surface modification of nanofillers and special processing technique are raised: (i) the physical blending, (ii) in situ polymerization, (iii) chemical modification of conductive filler, and (iv) introduced surfactant. All of the aforementioned methods can improve the dispersion of conductive particles in polymer, but there are still many disadvantages. Method (i) the polymer may be partially degraded under high shear strength. In addition, when the external force stops, the conductive particles will reunite. In method (ii), the solvent should be carefully chosen, that is, the selected solvent can not only dissolve the polymer monomer and its initiator, but also disperse the conductive filler well. In method (iii), the chemical modification of conductive particles is complicated and the yield is low. The introduction of surfactants in method (iv) may have adverse effects on the mechanical and other properties of polymer materials. Therefore, improving the dispersion of conductive particles in polymer is still a key problem in CPC preparation.

In addition to dispersion, the concentration of conductive filler in polymer matrix is also an important factor affecting the preparation and performance of CPC. Although some methods have been used to decrease the percolation value, there are many limitations for each structure design of the CPCs. For instance, pressure and temperature should be delicately controlled for construction of the segregated structure. Furthermore, CPCs with a segregated structure usually exhibit inferior mechanical properties. Coating conductive nanofillers on the surface of polymer nanofibers or the skeleton of the polymer foams could avoid filler aggregation and increase the electrical conductivity, but how to achieve the good interfacial interaction between nanofillers and polymer scaffold and ensure the surface stability and durability is still challenging.

The flexibility, light weight, and controllable network structure endow the CPCs with potential applications in sensing including strain sensors, piezoresistive sensors, gas sensors, and temperature sensors. With the development of artificial intelligence (e.g. electronic skin and human–machine interface), skin adhesive and mechanically flexible CPCs with multi-functionality will become a new hotspot.

References

1 Zhang, Y., Pan, T., and Yang, Z. (2020). Flexible polyethylene terephthalate/polyaniline composite paper with bending durability and effective electromagnetic shielding performance. *Chem. Eng. J.* 389: 124433.
2 Lim, Y.W., Jin, J., and Bae, B.S. (2020). Optically transparent multiscale composite films for flexible and wearable electronics. *Adv. Mater.* 32: 1907143.
3 Chen, J., Yu, Q., Cui, X. et al. (2019). An overview of stretchable strain sensors from conductive polymer nanocomposites. *J. Mater. Chem. C* 7: 11710–11730.
4 Shirakawa, H. (2001). The discovery of polyacetylene film: the dawning of an era of conducting polymers (Nobel lecture). *Angew. Chem. Int. Ed.* 40: 2574–2580.
5 Tang, C., Chen, N., and Hu, X. (2017). Conducting polymer nanocomposites: recent developments and future prospects. In: *Conducting Polymer Hybrids*, 1–44. Springer International Publishing.
6 Erdem, E., Karakışla, M., and Sacak, M. (2004). The chemical synthesis of conductive polyaniline doped with dicarboxylic acids. *Eur. Polym. J.* 40: 785–791.
7 Han, M.G. and Im, S.S. (2001). Dielectric spectroscopy of conductive polyaniline salt films. *J. Appl. Polym. Sci.* 82: 2760–2769.
8 Kim, B.R., Lee, H.-K., Park, S., and Kim, H.-K. (2011). Electromagnetic interference shielding characteristics and shielding effectiveness of polyaniline-coated films. *Thin Solid Films* 519: 3492–3496.
9 An, K.H., Jeong, S.Y., Hwang, H.R., and Lee, Y.H. (2004). Enhanced sensitivity of a gas sensor incorporating single-walled carbon nanotube–polypyrrole nanocomposites. *Adv. Mater.* 16: 1005–1009.
10 Molapo, K.M., Ndangili, P.M., Ajayi, R.F. et al. (2012). Electronics of conjugated polymers (I): polyaniline. *Int. J. Electrochem. Sci.* 7: 11859–11875.

11 Chiu, H.-T., Chiang, T.-Y., Chang, C.-Y., and Kuo, M.-T. (2011). Carbon black/polypyrrole/nitrile rubber conducting composites: synergistic properties and compounding conductivity effect. *E-Polymers* 11: 037.

12 Choi, S., Han, S.I., Kim, D. et al. (2019). High-performance stretchable conductive nanocomposites: materials, processes, and device applications. *Chem. Soc. Rev.* 48: 1566–1595.

13 Chen, J., Cui, X., Sui, K. et al. (2017). Balance the electrical properties and mechanical properties of carbon black filled immiscible polymer blends with a double percolation structure. *Compos. Sci. Technol.* 140: 99–105.

14 Tung, T.T., Karunagaran, R., Tran, D.N. et al. (2016). Engineering of graphene/epoxy nanocomposites with improved distribution of graphene nanosheets for advanced piezo-resistive mechanical sensing. *J. Mater. Chem. C* 4: 3422–3430.

15 Pang, H., Xu, L., Yan, D.-X., and Li, Z.-M. (2014). Conductive polymer composites with segregated structures. *Prog. Polym. Sci.* 39: 1908–1933.

16 Jeon, K.S., Nirmala, R., Navamathavan, R., and Kim, H.Y. (2013). Mechanical behavior of electrospun nylon66 fibers reinforced with pristine and treated multi-walled carbon nanotube fillers. *Ceram. Int.* 39: 8199–8206.

17 Chen, G.-X., Li, Y., and Shimizu, H. (2007). Ultrahigh-shear processing for the preparation of polymer/carbon nanotube composites. *Carbon* 45: 2334–2340.

18 Ren, F., Zhu, G., Ren, P. et al. (2014). In situ polymerization of graphene oxide and cyanate ester–epoxy with enhanced mechanical and thermal properties. *Appl. Surf. Sci.* 316: 549–557.

19 Wang, L.T., Chen, Q., Hong, R.Y., and Kumar, M.R. (2015). Preparation of oleic acid modified multi-walled carbon nanotubes for polystyrene matrix and enhanced properties by solution blending. *J. Mater. Sci. - Mater. Electron.* 26: 8667–8675.

20 Mazinani, S., Ajji, A., and Dubois, C. (2009). Morphology, structure and properties of conductive PS/CNT nanocomposite electrospun mat. *Polymer* 50: 3329–3342.

21 Balogun, Y.A. and Buchanan, R.C. (2010). Enhanced percolative properties from partial solubility dispersion of filler phase in conducting polymer composites (CPCs). *Compos. Sci. Technol.* 70: 892–900.

22 Dubson, M.A. and Garland, J.C. (1985). Measurement of the conductivity exponent in two-dimensional percolating networks: square lattice versus random-void continuum. *Phys. Rev. B* 32: 7621–7623.

23 Du, J., Zhao, L., Zeng, Y. et al. (2011). Comparison of electrical properties between multi-walled carbon nanotube and graphene nanosheet/high density polyethylene composites with a segregated network structure. *Carbon* 49: 1094–1100.

24 Sumfleth, J., Buschhorn, S.T., and Schulte, K. (2011). Comparison of rheological and electrical percolation phenomena in carbon black and carbon nanotube filled epoxy polymers. *J. Mater. Sci.* 46: 659–669.

25 Kuilla, T., Bhadra, S., Yao, D. et al. (2010). Recent advances in graphene based polymer composites. *Prog. Polym. Sci.* 35: 1350–1375.

References

26 Ameli, A., Kazemi, Y., Wang, S. et al. (2017). Process-microstructure-electrical conductivity relationships in injection-molded polypropylene/carbon nanotube nanocomposite foams. *Compos. Part A: Appl. Sci. Manuf.* 96: 28–36.

27 Deng, H., Zhang, R., Bilotti, E. et al. (2009). Conductive polymer tape containing highly oriented carbon nanofillers. *J. Appl. Polym. Sci.* 113: 742–751.

28 Zhao, P., Luo, Y., Yang, J. et al. (2014). Electrically conductive graphene-filled polymer composites with well organized three-dimensional microstructure. *Mater. Lett.* 121: 74–77.

29 Pang, H., Yan, D.-X., Bao, Y. et al. (2012). Super-tough conducting carbon nanotube/ultrahigh-molecular-weight polyethylene composites with segregated and double-percolated structure. *J. Mater. Chem.* 22: 23568–23575.

30 Deng, H., Lin, L., Ji, M. et al. (2014). Progress on the morphological control of conductive network in conductive polymer composites and the use as electroactive multifunctional materials. *Prog. Polym. Sci.* 39: 627–655.

31 Nakayama, Y., Takeda, E., Shigeishi, T. et al. (2011). Melt-mixing by novel pitched-tip kneading disks in a co-rotating twin-screw extruder. *Chem. Eng. Sci.* 66: 103–110.

32 Deng, L., Xu, C., Ding, S. et al. (2019). Processing a supertoughened polylactide ternary blend with high heat deflection temperature by melt blending with a high screw rotation speed. *Ind. Eng. Chem. Res.* 58: 10618–10628.

33 Wu, H.-Y., Zhang, Y.-P., Jia, L.-C. et al. (2018). Injection molded segregated carbon nanotube/polypropylene composite for efficient electromagnetic interference shielding. *Ind. Eng. Chem. Res.* 57: 12378–12385.

34 Qu, Y., Dai, K., Zhao, J. et al. (2014). The strain-sensing behaviors of carbon black/polypropylene and carbon nanotubes/polypropylene conductive composites prepared by the vacuum-assisted hot compression. *Colloid. Polym. Sci.* 292: 945–951.

35 Strååt, M., Rigdahl, M., and Hagström, B. (2012). Conducting bicomponent fibers obtained by melt spinning of PA6 and polyolefins containing high amounts of carbonaceous fillers. *J. Appl. Polym. Sci.* 123: 936–943.

36 Devaux, E., Koncar, V., Kim, B. et al. (2016). Processing and characterization of conductive yarns by coating or bulk treatment for smart textile applications. *Trans. Inst. Meas. Control* 29: 355–376.

37 Kim, J.Y. (2009). The effect of carbon nanotube on the physical properties of poly(butylene terephthalate) nanocomposite by simple melt blending. *J. Appl. Polym. Sci.* 112: 2589–2600.

38 Shen, B., Zhai, W., Chen, C. et al. (2011). Melt blending in situ enhances the interaction between polystyrene and graphene through π–π stacking. *ACS Appl. Mater. Interfaces* 3: 3103–3109.

39 Pan, Y., Li, L., Chan, S.H., and Zhao, J. (2010). Correlation between dispersion state and electrical conductivity of MWCNTs/PP composites prepared by melt blending. *Composites Part A* 41: 419–426.

40 Jiang, S., Gui, Z., Bao, C. et al. (2013). Preparation of functionalized graphene by simultaneous reduction and surface modification and its polymethyl

methacrylate composites through latex technology and melt blending. *Chem. Eng. J.* 226: 326–335.

41 You, F., Wang, D., Cao, J. et al. (2014). In situ thermal reduction of graphene oxide in a styrene-ethylene/butylene-styrene triblock copolymer via melt blending. *Polym. Int.* 63: 93–99.

42 Maiti, S., Suin, S., Shrivastava, N.K., and Khatua, B.B. (2013). Low percolation threshold in polycarbonate/multiwalled carbon nanotubes nanocomposites through melt blending with poly(butylene terephthalate). *J. Appl. Polym. Sci.* 130: 543–553.

43 Sharma, M., Sharma, S., Abraham, J. et al. (2014). Flexible EMI shielding materials derived by melt blending PVDF and ionic liquid modified MWNTs. *Mater. Res. Express* 1: 035003.

44 Soroudi, A. and Skrifvars, M. (2010). Melt blending of carbon nanotubes/polyaniline/polypropylene compounds and their melt spinning to conductive fibres. *Synth. Met.* 160: 1143–1147.

45 Yu, F., Deng, H., Zhang, Q. et al. (2013). Anisotropic multilayer conductive networks in carbon nanotubes filled polyethylene/polypropylene blends obtained through high speed thin wall injection molding. *Polymer* 54: 6425–6436.

46 Fan, Z. and Advani, S.G. (2007). Rheology of multiwall carbon nanotube suspensions. *J. Rheol.* 51: 585–604.

47 Pan, H., Zhang, Y., Hang, Y. et al. (2012). Significantly reinforced composite fibers electrospun from silk fibroin/carbon nanotube aqueous solutions. *Biomacromolecules* 13: 2859–2867.

48 Li, T., Zhao, G., and Wang, G. (2018). Effect of preparation methods on electrical and electromagnetic interference shielding properties of PMMA/MWCNT nanocomposites. *Polym. Compos.* 40: E1786–E1800.

49 Ramanujam, B.T.S. and Radhakrishnan, S. (2014). Solution-blended polyethersulfone–graphite hybrid composites. *J. Thermoplast. Compos. Mater.* 28: 835–848.

50 Gu, J., Gu, H., Zhang, Q. et al. (2018). Sandwich-structured composite fibrous membranes with tunable porous structure for waterproof, breathable, and oil–water separation applications. *J. Colloid Interface Sci.* 514: 386–395.

51 Kim, Y., Le, T.-H., Kim, S. et al. (2018). Single-walled carbon nanotube-in-binary-polymer nanofiber structures and their use as carbon precursors for electrochemical applications. *J. Phys. Chem. C* 122: 4189–4198.

52 Zhang, S., Li, D., Kang, J. et al. (2018). Electrospinning preparation of a graphene oxide nanohybrid proton-exchange membrane for fuel cells. *J. Appl. Polym. Sci.* 135: 46443.

53 Jin, L., Hu, B., Kuddannaya, S. et al. (2018). A three-dimensional carbon nanotube–nanofiber composite foam for selective adsorption of oils and organic liquids. *Polym. Compos.* 39: E271–E277.

54 Wang, K., Gu, M., Wang, J.-J. et al. (2012). Functionalized carbon nanotube/polyacrylonitrile composite nanofibers: fabrication and properties. *Polym. Adv. Technol.* 23: 262–271.

55 Dhakshnamoorthy, M., Ramakrishnan, S., Vikram, S. et al. (2014). In-situ preparation and characterization of acid functionalized single walled carbon nanotubes with polyimide nanofibers. *J. Nanosci. Nanotechnol.* 14: 5011–5018.

56 Bekyarova, E., Itkis, M.E., Cabrera, N. et al. (2005). Electronic properties of single-walled carbon nanotube networks. *J. Am. Chem. Soc.* 127: 5990–5995.

57 Kim, J.H., Kataoka, M., Jung, Y.C. et al. (2013). Mechanically tough, electrically conductive polyethylene oxide nanofiber web incorporating DNA-wrapped double-walled carbon nanotubes. *ACS Appl. Mater. Interfaces* 5: 4150–4154.

58 Hirsch, A. (2002). Functionalization of single-walled carbon nanotubes. *Angew. Chem. Int. Ed.* 41: 1853–1859.

59 Li, Y., Zhou, B., Zheng, G. et al. (2018). Continuously prepared highly conductive and stretchable SWNT/MWNT synergistically composited electrospun thermoplastic polyurethane yarns for wearable sensing. *J. Mater. Chem. C* 6: 2258–2269.

60 Ntim, S.A., Sae-Khow, O., Witzmann, F.A., and Mitra, S. (2011). Effects of polymer wrapping and covalent functionalization on the stability of MWCNT in aqueous dispersions. *J. Colloid Interface Sci.* 355: 383–388.

61 Khazaee, M., Ye, D., Majumder, A. et al. (2016). Non-covalent modified multi-walled carbon nanotubes: dispersion capabilities and interactions with bacteria. *Biomed. Phys. Eng. Express* 2: 055008.

62 Amirilargani, M., Tofighy, M.A., Mohammadi, T., and Sadatnia, B. (2014). Novel poly(vinyl alcohol)/multiwalled carbon nanotube nanocomposite membranes for pervaporation dehydration of isopropanol: poly(sodium-4-styrenesulfonate) as a functionalization agent. *Ind. Eng. Chem. Res.* 53: 12819–12829.

63 Lee, J.Y., Kang, T.H., Choi, J.H. et al. (2018). Improved electrical conductivity of poly(ethylene oxide) nanofibers using multi-walled carbon nanotubes. *AIP Adv.* 8: 035024.

64 Tu, X., Hight Walker, A.R., Khripin, C.Y., and Zheng, M. (2011). Evolution of DNA sequences toward recognition of metallic armchair carbon nanotubes. *J. Am. Chem. Soc.* 133: 12998–13001.

65 Kim, J.H., Kataoka, M., Fujisawa, K. et al. (2011). Unusually high dispersion of nitrogen-doped carbon nanotubes in DNA solution. *J. Phys. Chem. B* 115: 14295–14300.

66 Imai, Y., Fueki, T., Inoue, T., and Kakimoto, M.A. (1998). A new direct preparation of electroconductive polyimide/carbon black composite via polycondensation of nylon–salt-type monomer/carbon black mixture. *J. Polym. Sci., Part A: Polym. Chem.* 36: 1031–1034.

67 Li, Y., Pan, D., Chen, S. et al. (2013). In situ polymerization and mechanical, thermal properties of polyurethane/graphene oxide/epoxy nanocomposites. *Mater. Des.* 47: 850–856.

68 Li, J., Zhang, G., Deng, L. et al. (2014). In situ polymerization of mechanically reinforced, thermally healable graphene oxide/polyurethane composites based on Diels–Alder chemistry. *J. Mater. Chem. A* 2: 20642–20649.

69 Xu, Z. and Gao, C. (2010). In situ polymerization approach to graphene-reinforced nylon-6 composites. *Macromolecules* 43: 6716–6723.

70 Zeng, H., Gao, C., Wang, Y. et al. (2006). In situ polymerization approach to multiwalled carbon nanotubes-reinforced nylon 1010 composites: mechanical properties and crystallization behavior. *Polymer* 47: 113–122.

71 Wang, X., Hu, Y., Song, L. et al. (2011). In situ polymerization of graphene nanosheets and polyurethane with enhanced mechanical and thermal properties. *J. Mater. Chem.* 21: 4222–4227.

72 Fim, F.d.C., Basso, N.R.S., Graebin, A.P. et al. (2013). Thermal, electrical, and mechanical properties of polyethylene–graphene nanocomposites obtained by in situ polymerization. *J. Appl. Polym. Sci.* 128: 2630–2637.

73 Zhu, J., Lim, J., Lee, C.-H. et al. (2014). Multifunctional polyimide/graphene oxide composites via in situ polymerization. *J. Appl. Polym. Sci.* 131: 40177.

74 Potts, J.R., Lee, S.H., Alam, T.M. et al. (2011). Thermomechanical properties of chemically modified graphene/poly(methyl methacrylate) composites made by in situ polymerization. *Carbon* 49: 2615–2623.

75 Lee, J.K.Y., Chen, N., Peng, S. et al. (2018). Polymer-based composites by electrospinning: preparation & functionalization with nanocarbons. *Prog. Polym. Sci.* 86: 40–84.

76 Mamunya, E., Davidenko, V., and Lebedev, E. (1995). Percolation conductivity of polymer composites filled with dispersed conductive filler. *Polym. Compos.* 16: 319–324.

77 Zhou, J., Xu, X., Xin, Y., and Lubineau, G. (2018). Coaxial thermoplastic elastomer-wrapped carbon nanotube fibers for deformable and wearable strain sensors. *Adv. Funct. Mater.* 28: 1705591.

78 Wang, X., Sun, H., Yue, X. et al. (2018). A highly stretchable carbon nanotubes/thermoplastic polyurethane fiber-shaped strain sensor with porous structure for human motion monitoring. *Compos. Sci. Technol.* 168: 126–132.

79 Li, J., Zhang, D., Yang, T. et al. (2018). Nanofibrous membrane of graphene oxide-in-polyacrylonitrile composite with low filtration resistance for the effective capture of PM2.5. *J. Membr. Sci.* 551: 85–92.

80 Yu, S., Wang, X., Xiang, H. et al. (2018). Superior piezoresistive strain sensing behaviors of carbon nanotubes in one-dimensional polymer fiber structure. *Carbon* 140: 1–9.

81 Roh, E., Hwang, B.-U., Kim, D. et al. (2015). Stretchable, transparent, ultrasensitive, and patchable strain sensor for human–machine interfaces comprising a nanohybrid of carbon nanotubes and conductive elastomers. *ACS Nano* 9: 6252–6261.

82 Zheng, Y., Li, Y., Dai, K. et al. (2018). A highly stretchable and stable strain sensor based on hybrid carbon nanofillers/polydimethylsiloxane conductive composites for large human motions monitoring. *Compos. Sci. Technol.* 156: 276–286.

83 Xu, H., Qu, M., and Schubert, D.W. (2019). Conductivity of poly(methyl methacrylate) composite films filled with ultra-high aspect ratio carbon fibers. *Compos. Sci. Technol.* 181: 107690.

84 Duan, S., Wang, Z., Zhang, L. et al. (2018). A highly stretchable, sensitive, and transparent strain sensor based on binary hybrid network consisting of hierarchical multiscale metal nanowires. *Adv. Mater. Technol.* 3: 1800020.

85 Fan, X., Wang, N., Yan, F. et al. (2018). A transfer-printed, stretchable, and reliable strain sensor using PEDOT:PSS/Ag NW hybrid films embedded into elastomers. *Adv. Mater. Technol.* 3: 1800030.

86 Joo, Y., Byun, J., Seong, N. et al. (2015). Silver nanowire-embedded PDMS with a multiscale structure for a highly sensitive and robust flexible pressure sensor. *Nanoscale* 7: 6208–6215.

87 Huang, W., Dai, K., Zhai, Y. et al. (2017). Flexible and lightweight pressure sensor based on carbon nanotube/thermoplastic polyurethane-aligned conductive foam with superior compressibility and stability. *ACS Appl. Mater. Interfaces* 9: 42266–42277.

88 Liu, H., Dong, M., Huang, W. et al. (2017). Lightweight conductive graphene/thermoplastic polyurethane foams with ultrahigh compressibility for piezoresistive sensing. *J. Mater. Chem. C* 5: 73–83.

89 Malliaris, A. and Turner, D.T. (1971). Influence of particle size on the electrical resistivity of compacted mixtures of polymeric and metallic powders. *J. Appl. Phys.* 42: 614–618.

90 Liu, H., Li, Q., Zhang, S. et al. (2018). Electrically conductive polymer composites for smart flexible strain sensors: a critical review. *J. Mater. Chem. C* 6: 12121–12141.

91 Ma, M., Zhu, Z., Wu, B. et al. (2017). Preparation of highly conductive composites with segregated structure based on polyamide-6 and reduced graphene oxide. *Mater. Lett.* 190: 71–74.

92 Cui, J. and Zhou, S. (2018). Facile fabrication of highly conductive polystyrene/nanocarbon composites with robust interconnected network via electrostatic attraction strategy. *J. Mater. Chem. C* 6: 550–557.

93 Xie, L. and Zhu, Y. (2018). Tune the phase morphology to design conductive polymer composites: a review. *Polym. Compos.* 39: 2985–2996.

94 Tang, C., Long, G., Hu, X. et al. (2014). Conductive polymer nanocomposites with hierarchical multi-scale structures via self-assembly of carbon-nanotubes on graphene on polymer-microspheres. *Nanoscale* 6: 7877–7888.

95 Wu, C., Huang, X., Wang, G. et al. (2013). Highly conductive nanocomposites with three-dimensional, compactly interconnected graphene networks via a self-assembly process. *Adv. Funct. Mater.* 23: 506–513.

96 Pang, H., Bao, Y., Xu, L. et al. (2013). Double-segregated carbon nanotube–polymer conductive composites as candidates for liquid sensing materials. *J. Mater. Chem. A* 1: 4177–4181.

97 Pang, H., Bao, Y., Yang, S.-G. et al. (2014). Preparation and properties of carbon nanotube/binary-polymer composites with a double-segregated structure. *J. Appl. Polym. Sci.* 131: 39789.

98 Luo, W., Charara, M., Saha, M.C., and Liu, Y. (2019). Fabrication and characterization of porous CNF/PDMS nanocomposites for sensing applications. *Appl. Nanosci.* 9: 1309–1317.

99 Cho, E.-C., Chang-Jian, C.-W., Hsiao, Y.-S. et al. (2016). Three-dimensional carbon nanotube based polymer composites for thermal management. *Composites Part A* 90: 678–686.

100 Zhao, S., Yan, Y., Gao, A. et al. (2018). Flexible polydimethylsilane nanocomposites enhanced with a three-dimensional graphene/carbon nanotube bicontinuous framework for high-performance electromagnetic interference shielding. *ACS Appl. Mater. Interfaces* 10: 26723–26732.

101 Hu, X., Tian, M., Xu, T. et al. (2020). Multiscale disordered porous fibers for self-sensing and self-cooling integrated smart sportswear. *ACS Nano* 14: 559–567.

102 Zhang, S., Liu, H., Yang, S. et al. (2019). Ultrasensitive and highly compressible piezoresistive sensor based on polyurethane sponge coated with a cracked cellulose nanofibril/silver nanowire layer. *ACS Appl. Mater. Interfaces* 11: 10922–10932.

103 Mates, J.E., Bayer, I.S., Palumbo, J.M. et al. (2015). Extremely stretchable and conductive water-repellent coatings for low-cost ultra-flexible electronics. *Nat. Commun.* 6: 8874.

104 Gao, J., Wu, L., Guo, Z. et al. (2019). A hierarchical carbon nanotube/SiO_2 nanoparticle network induced superhydrophobic and conductive coating for wearable strain sensors with superior sensitivity and ultra-low detection limit. *J. Mater. Chem. C* 7: 4199–4209.

105 Ren, M., Zhou, Y., Wang, Y. et al. (2019). Highly stretchable and durable strain sensor based on carbon nanotubes decorated thermoplastic polyurethane fibrous network with aligned wave-like structure. *Chem. Eng. J.* 360: 762–777.

106 Shi, H., Shi, D., Yin, L. et al. (2014). Ultrasonication assisted preparation of carbonaceous nanoparticles modified polyurethane foam with good conductivity and high oil absorption properties. *Nanoscale* 6: 13748–13753.

107 Park, J.J., Hyun, W.J., Mun, S.C. et al. (2015). Highly stretchable and wearable graphene strain sensors with controllable sensitivity for human motion monitoring. *ACS Appl. Mater. Interfaces* 7: 6317–6324.

108 Hu, L., Pasta, M., Mantia, F.L. et al. (2010). Stretchable, porous, and conductive energy textiles. *Nano Lett.* 10: 708–714.

109 Gao, J., Luo, J., Wang, L. et al. (2019). Flexible, superhydrophobic and highly conductive composite based on non-woven polypropylene fabric for electromagnetic interference shielding. *Chem. Eng. J.* 364: 493–502.

110 Wang, L., Wang, H., Huang, X.-W. et al. (2018). Superhydrophobic and superelastic conductive rubber composite for wearable strain sensors with ultrahigh sensitivity and excellent anti-corrosion property. *J. Mater. Chem. A* 6: 24523–24533.

111 Lee, J., Shin, S., Lee, S. et al. (2018). Highly sensitive multifilament fiber strain sensors with ultrabroad sensing range for textile electronics. *ACS Nano* 12: 4259–4268.

112 Wang, L., Chen, Y., Lin, L. et al. (2019). Highly stretchable, anti-corrosive and wearable strain sensors based on the PDMS/CNTs decorated elastomer nanofiber composite. *Chem. Eng. J.* 362: 89–98.

113 Lin, L., Wang, L., Li, B. et al. (2020). Dual conductive network enabled superhydrophobic and high performance strain sensors with outstanding electro-thermal performance and extremely high gauge factors. *Chem. Eng. J.* 385: 123391.

114 Pu, J.-H., Zhao, X., Zha, X.-J. et al. (2019). Multilayer structured AgNW/WPU-MXene fiber strain sensors with ultrahigh sensitivity and a wide operating range for wearable monitoring and healthcare. *J. Mater. Chem. A* 7: 15913–15923.

115 Zhai, W., Xia, Q., Zhou, K. et al. (2019). Multifunctional flexible carbon black/polydimethylsiloxane piezoresistive sensor with ultrahigh linear range, excellent durability and oil/water separation capability. *Chem. Eng. J.* 372: 373–382.

116 Gao, J., Wang, H., Huang, X. et al. (2018). A super-hydrophobic and electrically conductive nanofibrous membrane for a chemical vapor sensor. *J. Mater. Chem. A* 6: 10036–10047.

117 Flint, E.B. and Suslick, K.S. (1991). The temperature of cavitation. *Science* 253: 1397–1399.

118 Gao, J., Hu, M., and Li, R.K.Y. (2012). Ultrasonication induced adsorption of carbon nanotubes onto electrospun nanofibers with improved thermal and electrical performances. *J. Mater. Chem.* 22: 10867–10872.

119 Trung, T.Q. and Lee, N.E. (2016). Flexible and stretchable physical sensor integrated platforms for wearable human-activity monitoring and personal healthcare. *Adv. Mater.* 28: 4338–4372.

120 Rinaldi, A., Tamburrano, A., Fortunato, M. et al. (2016). Highly sensitive pressure sensor based on a PDMS foam coated with graphene nanoplatelets. *Sensors* 16: 2148.

121 Yang, H., Yao, X., Yuan, L. et al. (2019). Strain-sensitive electrical conductivity of carbon nanotube-graphene-filled rubber composites under cyclic loading. *Nanoscale* 11: 578–586.

122 Cao, X., Lan, Y., Wei, Y. et al. (2015). Tunable resistivity–temperature characteristics of an electrically conductive multi-walled carbon nanotubes/epoxy composite. *Mater. Lett.* 159: 276–279.

123 Wang, W., Wang, C., Yue, X. et al. (2019). Raman spectroscopy and resistance-temperature studies of functionalized multiwalled carbon nanotubes/epoxy resin composite film. *Microelectron. Eng.* 214: 50–54.

124 Li, K., Dai, K., Xu, X. et al. (2013). Organic vapor sensing behaviors of carbon black/poly(lactic acid) conductive biopolymer composite. *Colloid. Polym. Sci.* 291: 2871–2878.

125 Li, J.R., Xu, J.R., Zhang, M.Q., and Rong, M.Z. (2003). Carbon black/polystyrene composites as candidates for gas sensing materials. *Carbon* 41: 2353–2360.

126 Wang, L., Luo, J., Chen, Y. et al. (2019). Fluorine-free superhydrophobic and conductive rubber composite with outstanding deicing performance for highly sensitive and stretchable strain sensors. *ACS Appl. Mater. Interfaces* 11: 17774–17783.

127 Boland, C.S., Khan, U., Backes, C. et al. (2014). Sensitive, high-strain, high-rate bodily motion sensors based on graphene–rubber composites. *ACS Nano* 8: 8819–8830.

128 Zhang, L., He, J., Liao, Y. et al. (2019). A self-protective, reproducible textile sensor with high performance towards human–machine interactions. *J. Mater. Chem. A* 7: 26631–26640.

129 Gao, J., Wang, L., Guo, Z. et al. (2020). Flexible, superhydrophobic, and electrically conductive polymer nanofiber composite for multifunctional sensing applications. *Chem. Eng. J.* 381: 122778.

130 Li, L., Bai, Y., Li, L. et al. (2017). A superhydrophobic smart coating for flexible and wearable sensing electronics. *Adv. Mater.* 29: 1702517.

131 Liu, S. and Li, L. (2017). Ultrastretchable and self-healing double-network hydrogel for 3D printing and strain sensor. *ACS Appl. Mater. Interfaces* 9: 26429–26437.

132 Kim, S.H., Jung, S., Yoon, I.S. et al. (2018). Ultrastretchable conductor fabricated on skin-like hydrogel-elastomer hybrid substrates for skin electronics. *Adv. Mater.* 30: e1800109.

133 Zhang, Y.-Z., Lee, K.H., Anjum, D.H. et al. (2018). MXenes stretch hydrogel sensor performance to new limits. *Sci. Adv.* 4: eaat0098.

134 Zhu, D., Handschuh-Wang, S., and Zhou, X. (2017). Recent progress in fabrication and application of polydimethylsiloxane sponges. *J. Mater. Chem. A* 5: 16467–16497.

135 Huang, Y., Fan, X., Chen, S.C., and Zhao, N. (2019). Emerging technologies of flexible pressure sensors: materials, modeling, devices, and manufacturing. *Adv. Funct. Mater.* 29: 1808509.

136 Nie, B., Huang, R., Yao, T. et al. (2019). Textile-based wireless pressure sensor array for human-interactive sensing. *Adv. Funct. Mater.* 29: 1808786.

137 Li, Y., Samad, Y.A., and Liao, K. (2015). From cotton to wearable pressure sensor. *J. Mater. Chem. A* 3: 2181–2187.

138 Xue, F., Lu, Y., Qi, X.-d. et al. (2019). Melamine foam-templated graphene nanoplatelet framework toward phase change materials with multiple energy conversion abilities. *Chem. Eng. J.* 365: 20–29.

139 Dong, X., Wei, Y., Chen, S. et al. (2018). A linear and large-range pressure sensor based on a graphene/silver nanowires nanobiocomposites network and a hierarchical structural sponge. *Compos. Sci. Technol.* 155: 108–116.

140 Chen, Z., Hu, Y., Zhuo, H. et al. (2019). Compressible, elastic, and pressure-sensitive carbon aerogels derived from 2D titanium carbide nanosheets and bacterial cellulose for wearable sensors. *Chem. Mater.* 31: 3301–3312.

141 Sun, Q.J., Zhao, X.H., Zhou, Y. et al. (2019). Fingertip-skin-inspired highly sensitive and multifunctional sensor with hierarchically structured conductive graphite/polydimethylsiloxane foams. *Adv. Funct. Mater.* 29: 1808829.

142 Xia, K., Wang, C., Jian, M. et al. (2017). CVD growth of fingerprint-like patterned 3D graphene film for an ultrasensitive pressure sensor. *Nano Res.* 11: 1124–1134.

143 Wu, N., Chen, S., Lin, S. et al. (2018). Theoretical study and structural optimization of a flexible piezoelectret-based pressure sensor. *J. Mater. Chem. A* 6: 5065–5070.

144 Bae, G.Y., Pak, S.W., Kim, D. et al. (2016). Linearly and highly pressure-sensitive electronic skin based on a bioinspired hierarchical structural array. *Adv. Mater.* 28: 5300–5306.

145 Liu, Y.-F., Huang, P., Li, Y.-Q. et al. (2019). A biomimetic multifunctional electronic hair sensor. *J. Mater. Chem. A* 7: 1889–1896.

146 Shi, J., Wang, L., Dai, Z. et al. (2018). Multiscale hierarchical design of a flexible piezoresistive pressure sensor with high sensitivity and wide linearity range. *Small* 14: 1800819.

147 Jian, M., Xia, K., Wang, Q. et al. (2017). Flexible and highly sensitive pressure sensors based on bionic hierarchical structures. *Adv. Funct. Mater.* 27: 1606066.

148 Pan, L., Chortos, A., Yu, G. et al. (2014). An ultra-sensitive resistive pressure sensor based on hollow-sphere microstructure induced elasticity in conducting polymer film. *Nat. Commun.* 5: 3002.

149 Park, J., Lee, Y., Hong, J. et al. (2014). Giant tunneling piezoresistance of composite elastomers with interlocked microdome arrays for ultrasensitive and multimodal electronic skins. *ACS Nano* 8: 4689–4697.

150 Li, Y., Zheng, Y., Zhan, P. et al. (2018). Vapor sensing performance as a diagnosis probe to estimate the distribution of multi-walled carbon nanotubes in poly(lactic acid)/polypropylene conductive composites. *Sens. Actuators, B* 255: 2809–2819.

151 Dai, K., Zhao, S., Zhai, W. et al. (2013). Tuning of liquid sensing performance of conductive carbon black (CB)/polypropylene (PP) composite utilizing a segregated structure. *Composites Part A* 55: 11–18.

152 Li, Y., Pionteck, J., Pötschke, P., and Voit, B. (2020). Thermal annealing to influence the vapor sensing behavior of co-continuous poly(lactic acid)/polystyrene/multiwalled carbon nanotube composites. *Mater. Des.* 187: 108383.

153 Gao, J., Wang, H., Huang, X. et al. (2018). Electrically conductive polymer nanofiber composite with an ultralow percolation threshold for chemical vapour sensing. *Compos. Sci. Technol.* 161: 135–142.

154 Li, Y., Liu, H., Dai, K. et al. (2015). Tuning of vapor sensing behaviors of eco-friendly conductive polymer composites utilizing ramie fiber. *Sens. Actuators, B* 221: 1279–1289.

155 Huang, X., Li, B., Wang, L. et al. (2019). Superhydrophilic, underwater superoleophobic, and highly stretchable humidity and chemical vapor sensors for human breath detection. *ACS Appl. Mater. Interfaces* 11: 24533–24543.

156 Feller, J.F., Lu, J., Zhang, K. et al. (2011). Novel architecture of carbon nanotube decorated poly(methyl methacrylate) microbead vapour sensors assembled by spray layer by layer. *J. Mater. Chem.* 21: 4142–4149.

157 Zhao, S., Zhai, W., Li, N. et al. (2014). Liquid sensing properties of carbon black/polypropylene composite with a segregated conductive network. *Sensor. Actuat. A: Phys* 217: 13–20.

158 Liu, X., Guo, Y., Ma, Y. et al. (2014). Flexible, low-voltage and high-performance polymer thin-film transistors and their application in photo/thermal detectors. *Adv. Mater.* 26: 3631–3636.

159 Cui, X., Chen, J., Zhu, Y., and Jiang, W. (2018). Lightweight and conductive carbon black/chlorinated poly(propylene carbonate) foams with a remarkable negative temperature coefficient effect of resistance for temperature sensor applications. *J. Mater. Chem. C* 6: 9354–9362.

160 Li, Q., Siddaramaiah, N.H., Kim, G.-H., and Yoo, J.H.L. (2009). Positive temperature coefficient characteristic and structure of graphite nanofibers reinforced high density polyethylene/carbon black nanocomposites. *Composites Part B* 40: 218–224.

161 Zhang, X., Zheng, X., Ren, D. et al. (2016). Unusual positive temperature coefficient effect of polyolefin/carbon fiber conductive composites. *Mater. Lett.* 164: 587–590.

162 Lu, C., Hu, X.-n., He, Y.-x. et al. (2012). Triple percolation behavior and positive temperature coefficient effect of conductive polymer composites with especial interface morphology. *Polym. Bull.* 68: 2071–2087.

163 Xi, Y., Yamanaka, A., Bin, Y., and Matsuo, M. (2007). Electrical properties of segregated ultrahigh molecular weight polyethylene/multiwalled carbon nanotube composites. *J. Appl. Polym. Sci.* 105: 2868–2876.

164 Asare, E., Basir, A., Tu, W. et al. (2016). Effect of mixed fillers on positive temperature coefficient of conductive polymer composites. *Nanocomposites* 2: 58–64.

3

Biodegradable Polymer Nanocomposites for Electronics

Wei Wu

Key Laboratory of Polymer Processing Engineering of Ministry of Education, Guangdong Provincial Key Laboratory of Technique and Equipment for Macromolecular Advanced Manufacturing, School of Mechanical and Automotive Engineering, South China University of Technology, Road Wushan, Guangzhou, Guangdong 510640, China

3.1 Introduction

With the fast development of modern technology, a variety of electric and electronic equipment has entered into our daily life, such as telecommunication, entertainment, and healthcare. However, the discard and disposal of electronic devices in the case of traditional petroleum-derived polymers have caused serious environmental problems. The traditional polymers in electronic devices, such as polypropylene (PP), polystyrene (PS), poly(ethylene terephthalate) (PET), polyimide (PI), and epoxy, take hundreds of years to degrade in the natural environment. The accumulation of these non-biodegradable polymers in electronic waste (E-waste) will not only require landfill to disposal and/or discard but also trend to leak some hazardous chemicals, which are toxic to humans, animals, and plants [1]. Considering the increasing consumption of polymers in electronic industry, the application of biodegradable polymers in electronic devices is one possible effective way to ease the issue of growing environmental problem [2, 3]. The biodegradability of electronic devices will bring positive effects on environmental protection as well as the reduction of health risks resulting from recycling operations [4]. On the other hand, most of the biodegradable polymers possess the advantages of biocompatibility and metabolization, indicating that they are suitable for the implantable chips, electronic skin, biomedical diagnosis, and wearable devices. Thus, the exploration of "green" electronic devices based on biodegradable polymer nanocomposites will benefit to the development of next generation of electronics.

Biodegradable polymers are first introduced in the 1980s. They are a special type of polymers that living organisms can convert them rapidly into carbon dioxide, water, biomass, and humus under appropriate conditions. Based on the definition by the standard of EN 13432 in Europe, biodegradable materials need to be degraded more than 90% of the weight within six months by biological actions. Biodegradable polymers are generally classified into two types based on their sources: natural

Polymer Nanocomposite Materials: Applications in Integrated Electronic Devices, First Edition.
Edited by Ye Zhou and Guanglong Ding.
© 2021 WILEY-VCH GmbH. Published 2021 by WILEY-VCH GmbH.

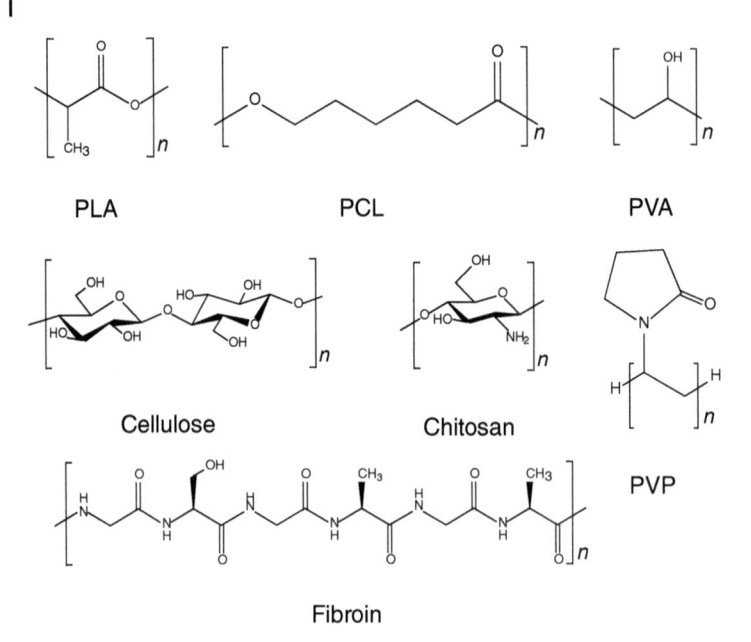

Figure 3.1 Chemical structures of biodegradable polymers utilized for electronic devices: polylactide (PLA), polycaprolactone (PCL), poly(vinyl alcohol) (PVA), polyvinylpyrrolidone (PVP), cellulose, chitosan, and silk (fibroin).

polymers and synthetic polymers. Natural polymers are usually derived from renewable resources, including plants, animals, and microorganisms. The most common types of natural biodegradable polymers are cellulose, chitosan, lignin, starch, protein, and so on. Meanwhile, the synthetic biodegradable polymers are manufactured from fossil oil, such as poly(ε-caprolactone) (PCL) and poly(butylene succinate). With the fast development of technology, more and more synthetic biodegradable polymer monomers can be obtained from microorganisms, like polylactide (PLA) and polyhydroxyalkanoates. Figure 3.1 shows the chemical structure of common types of biodegradable polymers. The shortcomings of these biodegradable materials such as low thermal stability, poor electrical, and/or mechanical properties restrict their broad applications in electronic devices. A number of essential methods, including copolymerization, blending, and cross-linking have been utilized to improve the performance of biodegradable polymers. Among these different methods, the addition of nano-fillers to prepare polymer nanocomposites seems to be a simple and effective strategy. It has been widely accepted that a small amount of nano-fillers can significantly enhance the thermal, electrical, and mechanical properties of biodegradable polymers. The tunable properties of biodegradable polymer nanocomposites will be beneficial for the bloom development of advanced electronics because they can be designed for different components in electronic devices (Figure 3.2) with desired properties [5]. This chapter will present a comprehensive overview of biodegradable polymer nanocomposites for the applications of electronics devices and their potential to enable easy waste disposal (Table 3.1).

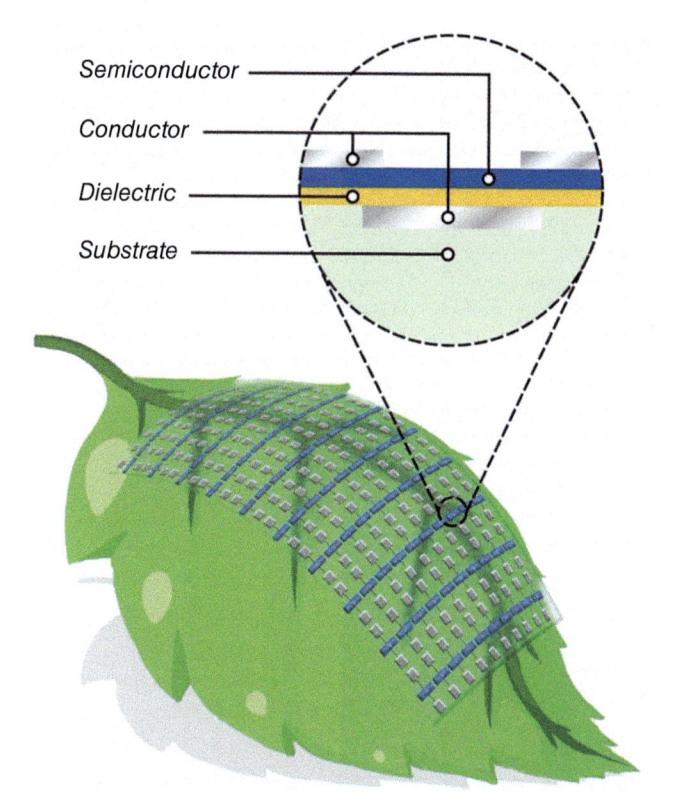

Figure 3.2 A typical biodegradable thin film transistor (TFT). Source: Feig et al. [5].

3.2 Biodegradable Polymer Nanocomposites in Electronics

3.2.1 Polylactide

PLA is a type of linear aliphatic polyester, which can be synthesized by ring opening polymerization (ROP) of lactide in the presence of tin(II) octoate as catalyst. It is the most promising biodegradable polymer to replace non-degradable material in electronics owning to its advantages of biodegradability, renewability from plants (such as corn and wheat), and excellent transparency. It is a tough transparent material with a tensile strength around 50 MPa, a Young's modulus of around 3 GPa, and an elongation at break of 4–5% [20, 21]. PLA has been utilized as substrates and dielectric layers for fabrication of memristors and transistors [22–25]. In order to integrate PLA into different device configurations, various nanofillers have been explored to endow PLA with improved thermal resistance, conductivity, and dielectric performances.

Carbon materials, such as carbon nanotubes (CNTs), graphene, and carbon quantum dots (CQDs), have received great attention in electronic devices due to their

Table 3.1 The PLA nanocomposite for electronic devices.

Biodegradable polymers	Filler type	Preparation method	Device type and application	References
PLA	CNTs	Melt spinning	Textile sensor for humidity	[6]
PLA	CNTs	Solution mixing and micromolding	Biosensor for ascorbic acid	[7]
PLA	CNTs	Solution blow spinning	Biosensor for *Pseudomonas putida*	[8, 9]
PLA	CQDs	Electrospinning	Nanogenerator	[10]
PLA	Graphene	3D printing	Electrochemical sensor for 2,4,6-trinitrotoluene (TNT)	[11]
PLA	Carbon black	3D printing	Sensors for dichloromethane, chloroform, tetrahydrofuran, acetone, ethyl acetate, and ethanol	[12]
PLA	Graphite	Solution mixing	Electrochemical sensor for Pb^{2+} and Cd^{2+}	[13]
PLA	Gold nanoparticles	Solution mixing	Electrochemical biosensor for leukemia cancer cells	[14]
PLA	Silver nanowire	Transferring	Organic light-emitting diodes (OLEDs) as electrode	[15]
PLA	TiO_2	Spin coating	Sensor for humidity	[16]
PLA	$BaTiO_3$ nanowire	3D printing	Energy harvesters	[17]
PLA	NiTi nanowire	3D printing	Sensor for temperature and strain	[18]
PLA	Boron dye	Electrospinning	Sensor for oxygen	[19]

large aspect ratio, outstanding conductivity, and high thermal stability. Devaux et al. mixed 4 wt% CNTs with PLA by melt spinning method to obtain electric conductive PLA yarns [6]. Due to the adjustable electrical conductivity, the PLA nanocomposites based flexible textile sensor could detect the presence of moisture. A microneedle array (MNA) dermal biosensor based on PLA/CNTs nanocomposites has also been reported [7]. In order to reduce the aggregation of CNTs, carboxylated CNTs in combination with sonication was utilized to achieve homogenization dispersion of CNTs in PLA matrix. The MNA biosensor exhibited good mechanical properties for skin penetration and could detect rapidly the ascorbic acid in ex vivo porcine skin. In addition, the PLA/CNTs nanocomposite fiber mats were prepared by solution blow spinning method. By detecting the presence and concentration of *Pseudomonas putida* in vitro [8], this impedance change biosensor could be used for diagnostic the wound application. This impedance change biosensor could be used for diagnostic the wound application. CQDs were also employed to prepare novel biodegradable PLA hybrid nanofibers through electrospinning [10]. The presence of CQDs endowed the PLA nanofibers with desirable excitation-light dependent multicolor luminescence as well as enhanced thermal stability. The nanogenerator based on PLA/CQDs nanofibers exhibited a maximum open-circuit voltage of 74.2 V cm^{-3} and a short-circuit current output density of 4.9 μA cm^{-3}. Researchers proposed to prepare electrochemical sensors by 3D printing technologies to detect different types of molecules. A commercially-available PLA/graphene hybrid filament was used for the preparation of sensors, which was demonstrated to trace and identify the 2,4,6-trinitrotoluene (TNT) in the presence of other nitroaromatic species as well as other nitro-explosives [11]. Owing to the similarity in solubility parameter, Sathies et al. exhibited 3D printed PLA/carbon black composite based sensors were sensitive to different solvent, including dichloromethane, chloroform, tetrahydrofuran, acetone, ethyl acetate, and ethanol [12]. Another interesting work with PLA/graphite nanocomposites as electrodes demonstrated the electrochemical sensor can fast simultaneous determinate Pb^{2+} and Cd^{2+} in jewelry with the merits of inexpensive and simple construction [13].

Another example of conductive nanofillers employed to improve the conductivity of PLA was gold nanoparticle. The PLA/gold nanoparticles hybrid nanofibers endowed the indium tin oxide (ITO) electrode with a relatively hydrophilic interface. This novel electrochemical biosensor could selectively identify different leukemia cancer cells rapidly [14]. To improve the thermal stability of PLA, the stereocomplex PLA were prepared by blending poly(L-lactide) (PLLA) with its enantiomer poly(D-lactide) (PDLA). Then the silver nanowires were transformed into the stereocomplex PLA matrix, serving as electrodes for the organic light-emitting diodes (OLEDs) [15]. As expected, the PLA composites containing silver nanowires exhibited highly transparency and electrical conductivity, even after 10 000 cycles bending or 100 tape test cycles.

Apart from conductive materials, different types of dielectric materials were also mixed with PLA in the fabrication of various electronic devices. A humidity sensor based on PLA/TiO$_2$ nanocomposite could detect the humidity ranging from 20% to 90% [16]. Malakooti et al. dispersed high aspect ratio BaTiO$_3$ nanowires in the PLA solution to produce printable energy harvesters by 3D printing method [17]. The well-aligned BaTiO$_3$ nanowire harvesters exhibited 273% higher power generation capacity as compared with those of conventional cast nanocomposites with randomly oriented nanowires. Besides, NiTi nanowires connected network were built in the PLA matrix by 3D printing method [18]. The cascaded sensor can measure temperature and strain simultaneously. Moreover, spatiotemporal oxygen sensors were developed by combine a single component boron dye with PLA, which could be a powerful tool to detect the real-time oxygen gradients in tissue scaffolds applications [19]. The presence of boron-dye could monitor the dissolved low concentration of oxygen (less than 15 ppm) in scaffold.

Based on the aforementioned discussion, PLA exhibits a good processability that it can be mixed with different types of nanofillers by conventional processing equipment, as well as some new technologies, such as 3D printing, electrospinning.

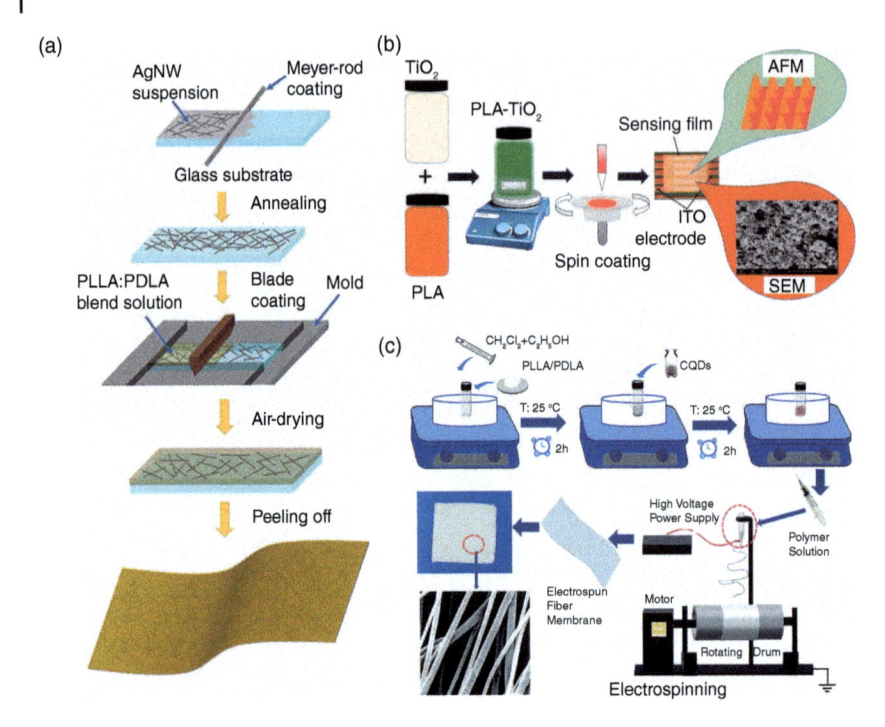

Figure 3.3 Various preparation methods of PLA based nanocomposites for electronic devices. (a) Schematic illustration of the fabrication process of the silver nanowires/PLLA:PDLA composite film. Source: Wang et al. [15]. Licensed under CC-BY-4.0. (b) Procedures of the spin coating for the humidity sensors. Source: Mallick et al. [16]. (c) Schematic diagram of electrospinning of PDLA/PLLA/CQD nanofiber membranes. Source: Xu et al. [10].

Due to the intrinsic insulating properties of PLA, it was mainly utilized as substrate or dielectric material in electronic devices. The big challenge for application of PLA in electronic device is low thermal resistance that it is not suitable for deposit other functional layer by thermal evaporation or sputtering. The appearance of stereocomplex PLA by mixing PDLA and PLLA together (molar ratio 1 : 1), as shown in Figure 3.3a, can overcome this big problem. The heat distortion temperature (HDT) of stereocomplex PLA can increase from 60 to 190 °C as compared with that of common PLA. It opens a new window to renewable and environmentally benign PLA-based materials for biodegradable electronic devices.

3.2.2 PCL

PCL is a type of biodegradable linear aliphatic polyester, which can be obtained by ROP of ε-caprolactone [26]. It is hydrophobic with a high degree of crystallinity. PCL has been proved by United States Food and Drug Administration (USFDA) to utilize for the tissue engineering applications due to its excellent biodegradability and biocompatibility. In addition, PCL exhibits good chemical resistance with a high ductility. It can be easily processed by conventional processing method due to its low melting point (~65 °C) and low viscosity.

An obstacle for PCL in the application of electronic device is its low Young's modulus and strength. For this reason, efforts have been made to blend PCL with stiff polymers, such as PLA and PP. Besides, further reinforcement of PCL matrix can be achieved by the introduction of nanofillers. The nanofillers will selective dispersed in the soft PCL phase due to the low viscosity of PCL. A new type of electrochemical sensor consisting of PCL/CNTs nanocomposites was developed by grafting PCL on the surface of CNTs through in situ polymerization [27]. This novel conductive polymer nanocomposite sensor could quantitative identify different organic vapor (methanol, toluene, tetrahydrofuran, and chloroform) in a wide range of concentration. Pötschke et al. prepared co-continuous PCL/PP blends containing CNTs by melt compounding method. The conductive PCL phase containing 3 wt% CNTs exhibited a significant change in electrical resistance so that the nanocomposites could detect various solvents, such as n-hexane, ethanol, methanol, water, toluene, chloroform, and tetrahydrofuran [28]. With the weight ratio of the conductive PCL phase increased over PP phase in the nanocomposite, more obvious electrical resistance changes were observed. Similarly, PCL nanocomposites containing 4 wt% CNTs/PLA blends were melt-spun to prepare conductive multifilament fibers [29]. The fibers exhibited a profound electrical resistance response to ethyl acetate and acetone, respectively, which had a promising potential application as a textile sensor. In addition, a dispersion of CNTs in PCL was also utilized as sensing phase to monitor the temperature in a range of 20–80 °C [30]. The sensing phase was blended with PP to fabricate nonwoven fabric by melt-blow method. Rana et al. employed spray layer-by-layer technique to prepare chitosan-co-polycaprolactone grafted CNTs transducer films [31]. The ternary composites were sensitive to volatile organic compounds and toluene vapors. They observed that the variety of relative resistance for the conductive transducers was increased as follow order: toluene < chloroform < ethanol < methanol. Aside from these electrochemical sensors, PCL together with graphene and silver nanowires were used to prepare self-healing strain sensor (Figure 3.4a) [32]. The good synergetic effect of PCL, graphene, and silver nanowires formed a stable conductive network in the polydimethylsiloxane (PDMS), exhibiting a high sensitivity (0.26 rad^{-1}) and good durability (>2400 bending cycles). In addition, the strain sensor could self-heal at 80 °C within 3 minutes. In Figure 3.4b, PCL nanocomposite fibrous with various content of graphene oxide (GO) were prepared by electrospinning [33]. The presence of polar groups of GO and the nanocomposite fibrous structure contributed to enhance the negative charges on PCL. The maximum open circuit voltage, short circuit current, and load power of PCL/GO based triboelectric nanogenerator (TENG) achieved 120 V, 4 µA, and 116 µW, respectively, which could directly light 21 pieces of LEDs.

3.2.3 PVA

Poly(vinyl alcohol) (PVA) is a water-soluble synthetic biodegradable polymer. PVA can be synthesized by the hydrolysis of poly(vinyl acetate) rather than directly

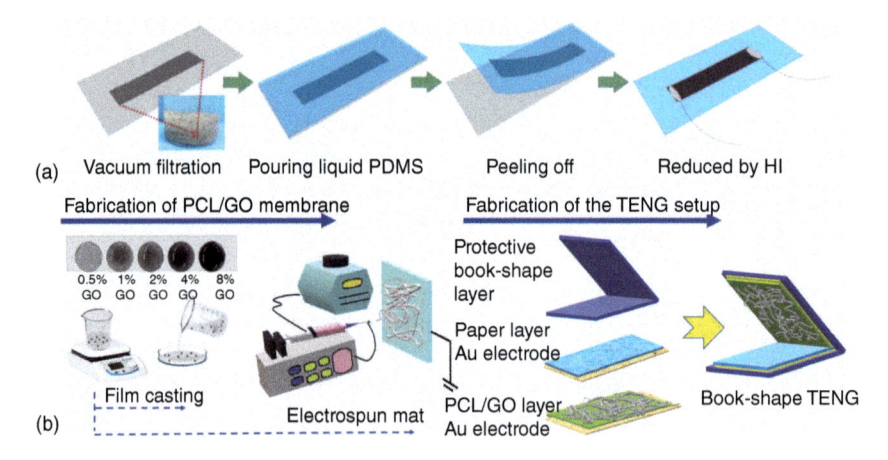

Figure 3.4 Preparation of (a) Schematic diagram of the preparation procedures of the strain sensor. Source: Liu et al. [32], (b) Schematic diagram of the fabrication of PCL/GO-cellulose paper based TENG. Source: Parandeh et al. [33].

polymerization of the monomer vinyl alcohol [34]. It has high transparent, high strength, flexibility, and good biocompatibility. The dielectric constant of PVA is around 5–8 in the range of 10–10^6 Hz, which is relative higher than those other biodegradable polymers [35, 36]. However, the high density of polar groups (−OH) in PVA makes it difficult to process by conventional melting compounding methods.

Researchers usually mix PVA aqueous solution with inorganic nano-fillers to serve as dielectric layers [37–39]. This method can bring some benefits that reduce the process complexity of electronic device, because the PVA layer has a good chemical resistance that can avoid being damage by most of the organic solvents. Canimkurbey et al. incorporated 70 wt% Al_2O_3 into PVA layer for field-effect transistors (FETs) application [35]. The dielectric constant and dielectric strength of PVA nanocomposite films increased with the increasing content of Al_2O_3. Zhang et al. [36] integrated PVA/SiO_2 hybrids to prepare flexible thin film transistors (TFTs). The flexible TFTs exhibited excellent mechanical robust that could work normally after 2000 cycles bending at 3 mm bending radius. Afsharimani and his colleague demonstrated that the presence of SiO_2 in the PVA dielectric layer could result in the larger on–off ratio and output conduction of TFTs as well as considerable leakage current [40]. Hashim investigated the effects of In_2O_3 and Cr_2O_3 nanoparticles on the optical, electronic, and electrical properties of PVA/poly(vinyl pyrrolidone) (PVP) blend [41]. The energy gap value of PVA/PVP blend could be reduced from 2.8 to 1.2 eV (In_2O_3) and 1.77 eV (Cr_2O_3), respectively, exhibiting promising potential applications in flexible optoelectronics fields. Besides the traditional solution casting and spinning coating methods, Wang et al. built a nozzle jet printing system to print PVA/GO nanocomposite dielectric layer [42]. The printing technique exhibited high deposition yield, low cost, and high efficiency with good compatibility with viscous PVA/GO ink.

PVA nanocomposites have also been reported to serve as active layers for resistance random access memory (RRAM). Different types of materials including metal nanoparticles, metal oxides, and graphene have incorporated in to PVA matrix, and exhibited resistive switching effect. Pham et al. embedded the TiO$_2$ nanotube in the PVA matrix, and the PVA-based RRAM exhibited an excellent retention of 10^4 seconds and large ON/OFF ratio >10^4 [43]. Hmar prepared resistive switching memristive devices by mixing ZnO nanoparticles with PVA and PEDOT:PSS [44]. The fabricated hybrid devices with good repeatability had a switching current ratio larger than 5 orders of magnitude. Nguyen et al. fabricated bipolar resistive switching memory based on PVA/Fe$_2$O$_3$ nanocomposites [45]. With the addition of SrTiO$_3$ nanoparticles, the active layer of PVA/SrTiO$_3$ hybrid exhibited bipolar, rewritable, and nonvolatile memristive characteristics [46]. Conductive fillers were also utilized as charge trappers to tune the properties of PVA. Kim et al. built multi-stacked PVA/GO insulating layer-based RRAM [47]. The memristor devices exhibited good retention (>2×10^3 seconds) and low setting and resetting voltages (<4 V). Rehman et al. prepared all printed memory with PVA/MoS$_2$ active layer [48]. PVA served as dielectric material and a supportive layer to transfer the MoS$_2$ onto the bottom electrodes. This PVA/MoS$_2$ based memory devices exhibited good mechanical robustness for 1500 cycles at various bending radium (1–25 mm).

3.2.4 PVP

PVP is a type of water-soluble synthetic polymer that has many advantages, such as excellent chemical resistance, good processability, and low cost. Besides, it also exhibits good transparency and biocompatibility, which is attractive for the fabrication of wearable electronic devices. PVP can be obtained by radical polymerization of the monomer N-vinylpyrrolidone [49].

PVP has been widely used as substrates for electronic devices [50, 51]. However, the low dielectric constant of PVP is insufficient to meet demands of the current high performance of electronic devices. In order to further improve the dielectric properties, various nanocomposite dielectrics were prepared [52–55]. Lee et al. successfully prepared TFTs based on PVP/TiO$_2$ nanocomposites by spinning coating method [56]. Similarly, Kim et al. crosslinked PVP/TiO$_2$ polymer nanocomposite dielectrics with poly(melamine-co-formaldehyde) to obtain low operating voltage TFTs [57]. They found the addition of TiO$_2$ in the dielectric layer could bring in large on/off ratio, high mobility, and reduced sub-threshold swing for TFTs. Soltani et al. synthesized Al$_2$O$_3$ by sol–gel method, and then mixed with PVP in acetylacetone [58]. The PVP/Al$_2$O$_3$ nanocomposites were utilized as gate dielectrics. Higher dielectric constant and lower leakage current of PVP nanocomposites can be achieved by the introduction of inorganic ceramic nanofillers, thus simultaneously improving organic thin-film transistor (OTFT) performance. The solution processing method can endow the nanocomposite dielectrics with tuning and tailoring of the dielectric constant for different applications [53]. Moon et al. embedded silicon nanowire into PVP layer by transform implantation method [59]. The silicon nanowire with random directions enhanced the gate coupling and bending stability of TFTs. Li et al. developed a novel strategy to fabricate proton-conducting synaptic transistors by using PVP/graphitic carbon nitride (g-C$_3$N$_4$) aqueous-solution [60]. This synapse device achieved a high field effect mobility of 75.4 cm^2 V^{-1} s^{-1} at a low operating voltage of 0.33 V.

Besides the applications as dielectric layers or substrates for TFTs, researchers also employed PVP as active materials for memristor devices. Kaur and Tripathi in situ synthesized cadmium selenide (CdSe) nanoparticles in PVP aqueous solution, and then casted on Al electrode [61]. The resultant PVP/CdSe based memristor exhibited bistability behavior. Zhang et al. demonstrated different phase of MoS$_2$ would result in different nonvolatile memory effects in the PVP/MoS$_2$ memory devices [62]. The active PVP composites layer containing 1T@2H-MoS$_2$ and 2H-MoS$_2$ exhibited write-once read-many times and rewritable memory effects, respectively. To improve the fabrication efficiency, Ali et al. used electro-hydrodynamic (EHD) method to prepare all printed memory device based on PVP/graphene quantum dots [63]. The printed memory had a long stable retention time over 30 days as well as mechanical robust without destroy after 1000 cycles bend at the diameter of 8 mm.

3.2.5 Cellulose

Cellulose is the most abundant of biopolymer from nature, such as trees, herbs, cottons, and marine algae. It is a linear polysaccharide consisting of β-1,4-linked D-glucose units and has a large amount of hydroxyl groups that can form strong hydrogen-bond networks (Figure 3.1) [64]. Over the past few decades, cellulose has attracted considerable attention as a nanomaterial [65]. Nanocellulose can be generally classified as cellulose nanofibers (CNFs, crystallinity ~60–80%) and cellulose nanocrystals (CNCs, crystallinity >85%) by crystallinity.

Cellulose has been explored as substrates for different types of electronic devices, such as sensors, memristors, and transistors [66–69]. The paper based electronic devices exhibited excellent flexibility and mechanical robust. Fortunato et al. showed that the cellulose fiber utilized as a dielectric layer had no significant difference in the electronic performance of transistor [70]. Gaspar et al. applied CNC as substrate and dielectric simultaneously to prepare flexible TFTs [71]. However, the cellulose based devices require large operating voltage. To further improve the performance of cellulose, various types of nanofillers were introduced into the cellulosic composites, as shown in Table 3.2. Faraji et al. mixed cyanoethyl cellulose with barium strontium titanate (BST) nanoparticles in N,N-dimethylformamide solution [72]. The results revealed that the high dielectric constant cyanoethyl cellulose based nanocomposite were suitable for the preparation of ultralow power electronics with high on/off ratio and low operating voltage. Wang and Yu combined ion gel with cellulose fibers as dielectrics [73]. Then the cellulose/ion gel nanocomposites were coated on the cellulose substrate to achieve a low-threshold voltage transistor.

Cellulose-based electro-active paper (EAPap) has also been utilized as a smart material for actuators. Cai et al. reported an unprecedented bilayer-structured actuator based on MXene/cellulose composites [74]. This actuator with a sophisticated leaf structure could harvest and converse energy, which has a potential application in soft robot system. The actuator consisting of cellulose, chitosan, and

Table 3.2 The cellulose based nanocomposite for electronic devices.

Biodegradable polymers	Filler type	Preparation method	Device type and application	References
Cyanoethyl cellulose	Barium strontium titanate	Spin coating	Transistor	[72]
Cellulose fiber	Ion gel	Stencil printing	Transistor	[73]
Cellulose	MXene	Filtering method	Actuator	[74]
Cellulose	Chitosan/rGO	—	Actuator	[75]
Cellulose	ZnO	Leaching method	Actuator	[76]
Cellulose nanowhiskers	PVDF	Electrospinning	Actuator	[77]
Carboxymethyl cellulose	CNT	Hydrogel method	Sensor for humidity and temperature	[78]
Cellulose nanocrystal	Fe_2O_3	—	Sensor for NO_2	[79]
Cellulose acetate	Amino acid ionic liquids	Phase inversion	Sensor for NH_3	[80]
Carboxymethyl cellulose	PVA/PAA/silver nanowire	Freeze drying	Sensor for strain	[81]
Cellulose	Polyaniline	—	Sensor for acids	[82]
Carboxy-methyl cellulose	Carbon black and gelatin	Freeze drying	Pressure sensor	[83]
Carboxymethyl cellulose	Graphene oxide	Spin coating	Write once and read many (WORM) memory	[84]

reduced graphene oxide (rGO) showed significant enhancement in peak to peak displacement, which was 3.64 times increased than the traditional EAPap actuator at 5 V 0.1 Hz [75]. The actuator made by the leaching method of ZnO nanoparticles on a cellulose paper showed 5.8 times larger bending deformation as compared to the pure cellulose based EAPap [76]. Moreover, electrospinning method can contribute to the alignment of the cellulose film, thus, enhancing mechanical performance of the actuator, as well as improving electromechanical efficiency of the EAPap actuator [77].

Cellulose and its derivatives can also be used for constructing intelligent sensors. Functional nanofillers such as metals, metal oxides, and carbon materials can be mixed with cellulosic materials to fabricate flexible sensors. Li et al. mixed CNT with carboxymethyl cellulose by hydrogel method to prepare humidity and temperature sensors [78]. The results revealed that cellulose was an excellent humidity sensitive material for optical fiber sensors. In addition, the presence of CNT can effectively increase the humidity sensitivity of the hydrogel film. Sadasivuni et al. prepared a flexible NO_2 sensor from CNC/iron oxide nanocomposite by in situ growth of iron oxide on a paper sheet [79]. The as-prepared sensor could detect the NO_2 gas with highly sensitive (ppm level) and fully recoverable at room temperature. A cellulose based sensor for measuring NH_3 has also been reported [80]. The gas sensor with ionic liquid/cellulose measured traces of NH_3 effectively with reliability in the range of 1–100 ppm at room temperature. Dong's group incorporated silver nanowires into carboxymethyl cellulose matrix together with PVA and poly(acrylic acid) (PAA) to get binary-networked hydrogels [81]. These wearable epidermal sensors could detect the dynamic strains in the range of 4–3000%, which was ascribed to the formation of covalent bond in the binary network. Souza et al. deposited polyaniline nanoparticles into cellulosic paper so as to detect acidic concentration [82]. The results revealed that the color of the polyaniline modified paper sheets exhibited RGB color changes. In addition, this change is more obvious when the acid concentration was in the range of 0–500 ppm. Meng et al. proposed a flexible pressure sensor based on carboxymethyl cellulose/gelatin/carbon black ternary composite [83]. The porous sponge was obtained by freeze drying, and could monitor the strain in the range from 0% to 140%. This pressure sensor exhibited repeatability after 3000 cycles repeated presses and kept its high sensitivity. Liu et al. investigated a flexible nonvolatile resistive switching memory devices based on carboxymethyl cellulose and GO [84]. The memory configuring with Al/carboxymethyl cellulose-GO/Al/SiO_2 structure exhibited brilliant write-once-read-many-times resistance change characteristic with high ON/OFF current ratio (~10^5), and low switching voltage (2.22 V).

3.2.6 Chitosan

Chitosan is a linear polysaccharide composed of randomly distributed β-(1-4)-linked D-glucosamine and N-acetyl-d-glucosamine. It can be obtained from chitin by removing its acetyl groups in alkaline environment [85]. It is a unique cationic polysaccharide with amino groups. Due to the presence of amine and hydroxyl groups in the molecular chains, chitosan has an abundant of strong inter- and intra-molecular hydrogen bonding, resulting in a rigid crystalline structure [86].

Chitosan can form high quality film with good mechanical strength and biodegradability. These properties make it suitable for the applications of flexible electronics [87]. Feng et al. used chitosan as a self-supporting layer to fabricate an aluminum-doped zinc oxide-based TFT [88]. Wan's group demonstrated chitosan composite film containing GO had an excellent proton conductivity, and the capacitance of the dielectric composite layer could reach ~3.2 µF cm^{-2} at 1.0 Hz [89]. To achieve better electric and dielectric performance of chitosan, Du et al. introduced high-K material Y_2O_3 into the chitosan as the gate dielectrics. The thin film consisting of chitosan and Y_2O_3 nanoparticles decreased the leakage current and the depth of the pinholes as compared with that of pure chitosan. Similarly, Hosseini and Lee mixed silver nanoparticles with chitosan as dielectric gate to fabricate transparent resistive switching memory, which showed a high biocompatibility and low-power operation capabilities [90]. The chitosan based nanocomposites also exhibited resistive switching effects under the applied voltage.

Li and his colleague fabricated an environment-friendly multi-bit biomemory based on the chitosan/GO nanocomposites [91]. The solution processed nonvolatile biomemory device with ITO/chitosan-GO/Ni structure had a high current ratio and long data retention characteristics (>10^4 seconds). The presence of GO in the active layer acted as trapping sites, which could be responsible for the biomemory mechanism.

In the field of biosensors, many devices based on chitosan nanocomposites have been reported. Chen et al. developed a novel ultrahigh sensitive strain sensor by layer-by-layer assembly method [92]. The flexible sensor had an ultrahigh sensitivity with a gauge factor ~359 and detection limit of $\varepsilon = 0.5\%$. Hosseini et al. embedded glycine microspheres into amorphous chitosan matrix as a free-standing layer, and then the pressure sensor could produce a output voltage of 190 mV under 60 kPa pressure with a piezoelectric sensitivity of 2.82 ± 0.2 mV kPa^{-1} [93]. The sensor has a stable repeatability after 9000 cycles. Molla-Abbasi and his colleague found the application of chitosan decorated CNT surfaces for polar chemical vapors [94]. They demonstrated that the chitosan@CNT heterostructures could monitor the humidity in the ambient environment. Velmurugan et al. deposited chitosan hydrogel on the surface of CNTs by a facile ultrasonic-assisted method, and then mixed with hydroxyapatite nanoparticles [95]. The chitosan/CNTs/hydroxyapatite nanocomposites exhibited a highly selective and sensitive detection to nitrofurantoin. The electrochemical sensor had a detection limit of 1.3 nM in the linear range of 0.005–982.1 µM for the nitrofurantoin detection. An electrochemical sensor based on MXene/multiwalled carbon nanotube (MWCNT)/chitosan ternary nanocomposite thin film was also developed to detect ifosfamide, acetaminophen, domperidone, and sumatriptan [96]. The sensor was successfully utilized for analysis in urine and blood serum samples with recoveries >95.21%. A sensor with the structure of chitosan/gold nanoparticles/graphene modified glassy carbon electrode fabricated by Wu et al. was developed via surface imprinting and then utilized for the selective extraction and determination of trace sensing cadmium(II) ions in drinking water and milk samples [97]. The linear range of the sensor for tracing cadmium(II) ions ranged from 0.1 to 0.9 µM and the limit of detection was 1.62×10^{-4} µM. Kushwaha et al. have reported a ZnO-encapsulated polyaniline grafted chitosan based electrochemical sensor for urea detection [98]. The self-activating potentiometric sensor could detect urea in the range of 20–500 ppm with a sensitivity of 187.5 µV ppm^{-1} cm^{-2}, response time of 3 minutes, and recovery time of 30 seconds. Diouf et al. synthesized chitosan capped with gold nanoparticles for potential aspirin sensing through self-assembling method [99]. This chitosan based sensor was suitable for efficient sensing of aspirin in the concentration range between 1 pg ml^{-1} and 1 µg ml^{-1} with good reproducibility.

3.2.7 Silk

Silk fibril (SF) is a type of ancient protein biomaterial. It has attracted considerable attentions due to its biodegradability, biocompatibility, large-scale production, and unique mechanical strength [100]. SF is easy to be processed into versatile material formats, such as hydrogels, powders, nanofibers, or membranes [101]. Recently, SF-based electronics including biosensors, memristor, transistor, and actuator have been developed [102–105]. However, it is still at a preliminary stage to fabricate SF-based electronics due to the intrinsic brittleness of SF as well as the solubility in water.

To overcome the instability of electrochemical biosensors, Zhao et al. developed a micro-needle based biosensor with silk/D-sorbitol composites [106]. The glucose selective enzyme was immobilized in silk/D-polyols matrix and displayed high stability. This biosensor could monitor the glucose concentration within 1.7–10.4 mM l^{-1}. Huang and coworkers reported an SF/PVA blend as the gate dielectric to prepare TFTs [107]. The presence of PVA in the dielectric layer exhibited a higher bias stability than that with pure SF dielectric. Tsukruk's group reported flexible, strong, and tough GO/SF nanocomposites prepared by seriography-based approach [108]. The strong interfacial interactions between the constituent GO and SF endowed the nanocomposites based sensor with a strong chemical stability and do not undergo water plasticization. Thanks to the strong interfacial interactions between SF and polyurethane, Huang et al. built stable SF/polyurethane nanocomposite membranes with high transparency (>90%), good stretchability (>200%), and excellent heat resistance (up to 160 °C) [109]. The SF/polyurethane composite based sensors exhibited high temperature sensitivity (0.205% °C^{-1}), good reliability, and fast response (<2 seconds), which had potential application in personal thermal management.

The remarkable mechanical properties of SF also render it favorable for pressure or strain biosensor applications. Inspired by the unique structure of the sunflowers, Lu et al. achieved a hierarchical structured hybrid material by growing vertically aligned MoS$_2$ nanosheets on carbonized SF, which was further utilized as a sensing material in pressure sensors [110]. Reizabal et al. integrated CNTs into SF matrix by solvent-casting method to explore the application in piezoresistive sensor [111].

3.3 Challenges and Prospects

We have reviewed the representative processing methods of biodegradable polymer nanocomposites that may be suitable for the preparation of next generation of green electronic devices. These biodegradable polymer nanocomposites can serve as important parts in the devices, such as substrates, dielectric layers, and active layers. These biodegradable polymers, especially PLA, PCL, cellulose, chitosan, and silk, exhibit excellent biocompatibility that have great potential application to replace traditional petroleum based polymers in the components of wearable devices. However, there are still many challenges to overcome to integrate these biodegradable polymer nanocomposites into electronic devices. The big issue is the duration of these biodegradable electronic devices. It is well known that the biodegradable polymer will degradable in the ambient environment, and some of them are sensitive to moisture in the atmosphere, resulting in the change of electric properties of these nanocomposites. Thus, it is important to control the degradation rate to keep the duration of the electronic devices. The second issue is thermal stability of biodegradable polymer. The natural biopolymers, such as cellulose, chitosan, and silk, contain a large amount of hydroxyl groups. Besides, the synthetic polymers, such as PLA and PCL, have low heat distortion temperature. They are easy to be destroyed or affected during the preparation process of electronic devices. The third issue of biodegradable polymer nanocomposites is difficult to achieve uniformly dispersion of these functional nano-fillers in the polymer matrix due to high surface energy of nano-fillers will result in aggregation.

Nevertheless, despite these aforementioned issues, biodegradable polymer nanocomposites are continuous receiving more and more attention from both academia and industry due to the world are urgent to require sustainable electric and electronic industry. It is worthy note that the recent researches have integrated various biodegradable polymer nanocomposites into electronic devices to replace petroleum based polymer. These breakthroughs contribute to the development of next generation of biodegradable electronics.

List of Abbreviations

CNCs	cellulose nanocrystals
CNFs	cellulose nanofibers
CNTs	carbon nanotubes
CQDs	carbon quantum dots
EDH	electro-hydrodynamic
GO	graphene oxide
HDT	heat distortion temperature
PAA	poly(acrylic acid)
PCL	polycaprolactone
PET	poly(ethylene terephthalate)
PI	polyimide
PLA	polylactide
PP	polypropylene
PS	polystyrene
PVA	poly(vinyl alcohol)
PVP	poly(vinyl pyrrolidone)
RRAM	resistance random access memory
SF	silk fibril
TENG	triboelectric nanogenerator
TFT	thin film transistor

References

1 Wang, Z.H., Zhang, B., and Guan, D.B. (2016). Take responsibility for electronic-waste disposal. *Nature* 536: 23–25.

2 Irimia-Vladu, M. (2014). "Green" electronics: biodegradable and biocompatible materials and devices for sustainable future. *Chem. Soc. Rev.* 43: 588–610.
3 Irimia-Vladu, M., Glowacki, E.D., Voss, G. et al. (2012). Green and biodegradable electronics. *Mater. Today* 15: 340–346.
4 Haider, T.P., Völker, C., Kramm, J. et al. (2019). Plastics of the future? The impact of biodegradable polymers on the environment and on society. *Angew. Chem. Int. Ed.* 58: 50–62.
5 Feig, V.R., Tran, H., and Bao, Z. (2018). Biodegradable polymeric materials in degradable electronic devices. *ACS Cent. Sci.* 4: 337–348.
6 Devaux, E., Aubry, C., Campagne, C., and Rochery, M. (2011). PLA/carbon nanotubes multifilament yarns for relative humidity textile sensor. *J. Eng. Fibers Fabr.* 6: 13–24.
7 Skaria, E., Patel, B.A., Flint, M.S., and Ng, K.W. (2019). Poly(lactic acid)/carbon nanotube composite microneedle arrays for dermal biosensing. *Anal. Chem.* 91: 4436–4443.
8 Miller, C., Stiglich, M., Livingstone, M., and Gilmore, J. (2019). Impedance-based biosensing of *Pseudomonas putida* via solution blow spun PLA: MWCNT composite nanofibers. *Micromachines* 10: 876.
9 Miller, C.L., Stafford, G., Sigmon, N., and Gilmore, J.A. (2019). Conductive nonwoven carbon nanotube–PLA composite nanofibers towards wound sensors via solution blow spinning. *IEEE Trans. Nanobiosci.* 18: 244–247.
10 Xu, Y.L., Jin, L., He, X.B. et al. (2019). Glowing stereocomplex biopolymers are generating power: polylactide/carbon quantum dot hybrid nanofibers with high piezoresponse and multicolor luminescence. *J. Mater. Chem. A* 7: 1810–1823.
11 Cardoso, R.M., Castro, S.V.F., Silva, M.N.T. et al. (2019). 3D-printed flexible device combining sampling and detection of explosives. *Sens. Actuators B, Chem.* 292: 308–313.
12 Sathies, T., Senthil, P., and Prakash, C. (2019). Application of 3D printed PLA-carbon black conductive polymer composite in solvent sensing. *Mater. Res. Express* 6: 115349.
13 Silva, A.L., Correa, M.M., de Oliveira, G.C. et al. (2018). Development and application of a routine robust graphite/poly(lactic acid) composite electrode for the fast simultaneous determination of Pb^{2+} and Cd^{2+} in jewelry by square wave anodic stripping voltammetry. *New J. Chem.* 42: 19537–19547.
14 Wu, X.J., Jiang, H., Zheng, J.S. et al. (2011). Highly sensitive recognition of cancer cells by electrochemical biosensor based on the interface of gold nanoparticles/polylactide nanocomposites. *J. Electroanal. Chem.* 656: 174–178.
15 Wang, J.J., Yu, J.S., Bai, D.Y. et al. (2020). Biodegradable, flexible, and transparent conducting silver nanowires/polylactide film with high performance for optoelectronic devices. *Polymers* 12: 604.
16 Mallick, S., Ahmad, Z., Touati, F. et al. (2018). PLA–TiO_2 nanocomposites: thermal, morphological, structural, and humidity sensing properties. *Ceram. Int.* 44: 16507–16513.

17 Malakooti, M.H., Jule, F., and Sodano, H.A. (2018). Printed nanocomposite energy harvesters with controlled alignment of barium titanate nanowires. *ACS Appl. Mater. Interfaces* 10: 38359–38367.

18 Nascimento, M., Inacio, P., Paixao, T. et al. (2020). Embedded fiber sensors to monitor temperature and strain of polymeric parts fabricated by additive manufacturing and reinforced with NiTi wires. *Sensors* 20: 1122.

19 Bowers, D.T., Tanes, M.L., Das, A. et al. (2014). Spatiotemporal oxygen sensing using dual emissive boron dye polylactide nanofibers. *ACS Nano* 8: 12080–12091.

20 Raquez, J.M., Habibi, Y., Murariu, M., and Dubois, P. (2013). Polylactide (PLA)-based nanocomposites. *Prog. Polym. Sci.* 38: 1504–1542.

21 Hamad, K., Kaseem, M., Ayyoob, M. et al. (2018). Polylactic acid blends: the future of green, light and tough. *Prog. Polym. Sci.* 85: 83–127.

22 Bettinger, C.J. and Bao, Z.N. (2010). Organic thin-film transistors fabricated on resorbable biomaterial substrates. *Adv. Mater.* 22: 651–655.

23 Wu, X.H., Ma, Y., Zhang, G.Q. et al. (2015). Thermally stable, biocompatible, and flexible organic field-effect transistors and their application in temperature sensing arrays for artificial skin. *Adv. Funct. Mater.* 25: 2138–2146.

24 Wu, W., Han, S.T., Venkatesh, S. et al. (2018). Biodegradable skin-inspired nonvolatile resistive switching memory based on gold nanoparticles embedded alkali lignin. *Org. Electron.* 59: 382–388.

25 Fang, S.L., Liu, W.H., Li, X. et al. (2019). Biodegradable transient resistive random-access memory based on $MoO_3/MgO/MoO_3$ stack. *Appl. Phys. Lett.* 115: 244102.

26 Labet, M. and Thielemans, W. (2009). Synthesis of polycaprolactone: a review. *Chem. Soc. Rev.* 38: 3484–3504.

27 Castro, M., Lu, J.B., Bruzaud, S. et al. (2009). Carbon nanotubes/poly(ε-caprolactone) composite vapour sensors. *Carbon* 47: 1930–1942.

28 Potschke, P., Kobashi, K., Villmow, T. et al. (2011). Liquid sensing properties of melt processed polypropylene/poly(ε-caprolactone) blends containing multi-walled carbon nanotubes. *Compos. Sci. Technol.* 71: 1451–1460.

29 Rentenberger, R., Cayla, A., Villmow, T. et al. (2011). Multifilament fibres of poly(ε-caprolactone)/poly(lactic acid) blends with multiwalled carbon nanotubes as sensor materials for ethyl acetate and acetone. *Sens. Actuators B, Chem.* 160: 22–31.

30 Krucinska, I., Surma, B., Chrzanowski, M. et al. (2013). Application of melt-blown technology for the manufacture of temperature-sensitive nonwoven fabrics composed of polymer blends PP/PCL loaded with multiwall carbon nanotubes. *J. Appl. Polym. Sci.* 127: 869–878.

31 Rana, V.K., Akhtar, S., Chatterjee, S. et al. (2014). Chitosan and chitosan-*co*-poly(ε-caprolactone) grafted multiwalled carbon nanotube transducers for vapor sensing. *J. Nanosci. Nanotechnol.* 14: 2425–2435.

32 Liu, S.Q., Lin, Y., Wei, Y. et al. (2017). A high performance self-healing strain sensor with synergetic networks of poly(ε-caprolactone) microspheres, graphene and silver nanowires. *Compos. Sci. Technol.* 146: 110–118.

33 Parandeh, S., Kharaziha, M., and Karimzadeh, F. (2019). An eco-friendly triboelectric hybrid nanogenerators based on graphene oxide incorporated polycaprolactone fibers and cellulose paper. *Nano Energy* 59: 412–421.

34 Peixoto, L.S., Silva, F.M., Niemeyer, M.A. et al. (2006). Synthesis of poly(vinyl alcohol) and/or poly(vinyl acetate) particles with spherical morphology and core–shell structure and its use in vascular embolization. *Macromol. Symp.* 243: 190–199.

35 Canimkurbey, B., Cakirlar, C., Mucur, S.P. et al. (2019). Influence of Al_2O_3 nanoparticles incorporation on the dielectric properties of solution processed PVA films for organic field effect transistor applications. *J. Mater. Sci. - Mater. Electron.* 30: 18384–18390.

36 Zhang, Z.W., Du, C.H., Jiao, H.X., and Zhang, M. (2020). Polyvinyl alcohol/SiO_2 hybrid dielectric for transparent flexible/stretchable all-carbon-nanotube thin-film-transistor integration. *Adv. Electron. Mater.* 6: 1901133.

37 de la Rosa, C.J.L., Nourbakhsh, A., Heyne, M. et al. (2017). Highly efficient and stable MoS_2 FETs with reversible *n*-doping using a dehydrated poly(vinyl-alcohol) coating. *Nanoscale* 9: 258–265.

38 Xiong, W., Zhu, L.Q., Ye, C. et al. (2020). Flexible poly(vinyl alcohol)–graphene oxide hybrid nanocomposite based cognitive memristor with pavlovian-conditioned reflex activities. *Adv. Electron. Mater.* 6: 1901402.

39 Midya, A., Gogurla, N., and Ray, S.K. (2015). Flexible and transparent resistive switching devices using Au nanoparticles decorated reduced graphene oxide in polyvinyl alcohol matrix. *Curr. Appl. Phys.* 15: 706–710.

40 Afsharimani, N. and Nysten, B. (2019). Hybrid gate dielectrics: a comparative study between polyvinyl alcohol/SiO_2 nanocomposite and pure polyvinyl alcohol thin-film transistors. *Bull. Mater. Sci.* 42: 26.

41 Hashim, A. (2020). Enhanced structural, optical, and electronic properties of In_2O_3 and Cr_2O_3 nanoparticles doped polymer blend for flexible electronics and potential applications. *J. Inorg. Organomet. Polym. Mater.* 30: 3894–3906.

42 Wang, X.C., Wei, L., Mou, P.L. et al. (2020). A printable GO–PVA composite dielectric for EDL gating of metal-oxide TFTs. *Flex. Print. Electron.* 5: 015002.

43 Pham, N.K., Vu, N.H., Pham, V.V. et al. (2018). Comprehensive resistive switching behavior of hybrid polyvinyl alcohol and TiO_2 nanotube nanocomposites identified by combining experimental and density functional theory studies. *J. Mater. Chem. C* 6: 1971–1979.

44 Hmar, J.J.L. (2018). Flexible resistive switching bistable memory devices using ZnO nanoparticles embedded in polyvinyl alcohol (PVA) matrix and poly(3,4-ethylenedioxythiophene) polystyrene sulfonate (PEDOT:PSS). *RSC Adv.* 8: 20423–20433.

45 Nguyen, H.H., Ta, H.K.T., Park, S. et al. (2020). Resistive switching effect and magnetic properties of iron oxide nanoparticles embedded-polyvinyl alcohol film. *RSC Adv.* 10: 12900–12907.

46 Khalid, M.A.U., Kim, S.W., Lee, J. et al. (2020). Resistive switching device based on $SrTiO_3$/PVA hybrid composite thin film as active layer. *Polymer* 189: 122183.

47 Kim, T., Kim, D.K., Kim, J., and Pak, J.J. (2019). Resistive switching behaviour of multi-stacked PVA/graphene oxide plus PVA composite/PVA insulating layer-based RRAM devices. *Semicond. Sci. Technol.* 34: 065006.

48 Rehman, M.M., Siddiqui, G.U., Gul, J.Z. et al. (2016). Resistive switching in all-printed, flexible and hybrid MoS$_2$-PVA nanocomposite based memristive device fabricated by reverse offset. *Sci. Rep.* 6: 36195.

49 Koczkur, K.M., Mourdikoudis, S., Polavarapu, L., and Skrabalak, S.E. (2015). Polyvinylpyrrolidone (PVP) in nanoparticle synthesis. *Dalton Trans.* 44: 17883–17905.

50 Kumar, R., Rahman, H., Ranwa, S. et al. (2020). Development of cost effective metal oxide semiconductor based gas sensor over flexible chitosan/PVP blended polymeric substrate. *Carbohydr. Polym.* 239: 116213.

51 Lan, J.L., Wan, C.C., and Wang, Y.Y. (2008). Mechanistic study of Ag/Pd-PVP nanoparticles and their functions as catalyst for electroless copper deposition. *J. Electrochem. Soc.* 155: 77–83.

52 Bubel, S., Mechau, N., and Schmechel, R. (2011). Electronic properties of polyvinylpyrrolidone at the zinc oxide nanoparticle surface PVP in ZnO dispersions and nanoparticulate ZnO thin films for thin film transistors. *J. Mater. Sci.* 46: 7776–7783.

53 Zhou, Y., Han, S.T., Xu, Z.X., and Roy, V.A.L. (2012). Polymer–nanoparticle hybrid dielectrics for flexible transistors and inverters. *J. Mater. Chem.* 22: 4060–4065.

54 Saini, P., Sharma, B., Singh, M. et al. (2019). Electrical properties of self sustained layer of graphene oxide and polyvinylpyriodine composite. *Integr. Ferroelectr.* 202: 197–203.

55 Chen, X., Zhang, H., Zhang, Y. et al. (2020). Low-power flexible organic field-effect transistors with solution-processable polymer–ceramic nanoparticle composite dielectrics. *Nanomaterials* 10: 518.

56 Lee, W.H., Wang, C.C., and Ho, J.C. (2009). Influence of nano-composite gate dielectrics on OTFT characteristics. *Thin Solid Films* 517: 5305–5310.

57 Kim, J., Lim, S.H., and Kim, Y.S. (2010). Solution-based TiO$_2$-polymer composite dielectric for low operating voltage OTFTs. *J. Am. Chem. Soc.* 132: 14721–14723.

58 Soltani, B., Babaeipour, M., and Bahari, A. (2017). Studying electrical characteristics of Al$_2$O$_3$/PVP nano-hybrid composites as OFET gate dielectric. *J. Mater. Sci. - Mater. Electron.* 28: 4378–4387.

59 Moon, K.J., Lee, T.I., Choi, J.H. et al. (2011). One-dimensional semiconductor nanostructure based thin-film partial composite formed by transfer implantation for high-performance flexible and printable electronics at low temperature. *ACS Nano* 5: 159–164.

60 Li, J., Yang, Y.H., Chen, Q. et al. (2020). Aqueous-solution-processed proton-conducting carbon nitride/polyvinylpyrrolidone composite electrolytes for low-power synaptic transistors with learning and memory functions. *J. Mater. Chem. C* 8: 4065–4072.

61 Kaur, R. and Tripathi, S.K. (2015). Study of conductivity switching mechanism of CdSe/PVP nanocomposite for memory device application. *Microelectron. Eng.* 133: 59–65.

62 Zhang, P., Gao, C.X., Xu, B.H. et al. (2016). Structural phase transition effect on resistive switching behavior of MoS$_2$-polyvinylpyrrolidone nanocomposites films for flexible memory devices. *Small* 12: 2077–2084.

63 Ali, S., Bae, J., Lee, C.H. et al. (2015). All-printed and highly stable organic resistive switching device based on graphene quantum dots and polyvinylpyrrolidone composite. *Org. Electron.* 25: 225–231.

64 Kargarzadeh, H., Huang, J., Lin, N. et al. (2018). Recent developments in nanocellulose-based biodegradable polymers, thermoplastic polymers, and porous nanocomposites. *Prog. Polym. Sci.* 87: 197–227.

65 Calvino, C., Macke, N., Kato, R., and Rowan, S.J. (2020). Development, processing and applications of bio-sourced cellulose nanocrystal composites. *Prog. Polym. Sci.* 103: 101221.

66 Jung, Y.H., Chang, T.H., Zhang, H.L. et al. (2015). High-performance green flexible electronics based on biodegradable cellulose nanofibril paper. *Nat. Commun.* 6: 7170.

67 Li, X.K., Li, M.J., Xu, J. et al. (2019). Evaporation-induced sintering of liquid metal droplets with biological nanofibrils for flexible conductivity and responsive actuation. *Nat. Commun.* 10: 3514.

68 Conti, S., Martinez-Domingo, C., Lay, M. et al. (2020). Nanopaper-based organic inkjet-printed diodes. *Adv. Mater. Technol.* 5: 1900773.

69 Yoo, C., Kaium, G., Hurtado, L. et al. (2020). Wafer-scale two-dimensional MoS$_2$ layers integrated on cellulose substrates toward environmentally friendly transient electronic devices. *ACS Appl. Mater. Interfaces* 12: 25200–25210.

70 Fortunato, E., Barquinha, P., and Martins, R. (2012). Oxide semiconductor thin-film transistors: a review of recent advances. *Adv. Mater.* 24: 2945–2986.

71 Gaspar, D., Fernandes, S.N., de Oliveira, A.G. et al. (2014). Nanocrystalline cellulose applied simultaneously as the gate dielectric and the substrate in flexible field effect transistors. *Nanotechnology* 25: 094008.

72 Faraji, S., Danesh, E., Tate, D.J. et al. (2016). Cyanoethyl cellulose-based nanocomposite dielectric for low-voltage, solution-processed organic field-effect transistors (OFETs). *J. Phys. D: Appl. Phys.* 49: 185102.

73 Wang, X. and Yu, C.J. (2020). Flexible low-voltage paper transistors harnessing ion gel/cellulose fiber composites. *J. Mater. Res.* 35: 940–948.

74 Cai, G.F., Ciou, J.H., Liu, Y.Z. et al. (2019). Leaf-inspired multiresponsive MXene-based actuator for programmable smart devices. *Sci. Adv.* 5: 7956.

75 Sun, Z.Z., Du, S.Q., Li, F. et al. (2018). High-performance cellulose based nanocomposite soft actuators with porous high-conductivity electrode doped by graphene-coated carbon nanosheet. *Cellulose* 25: 5807–5819.

76 Wang, F., Jin, Z., Zheng, S.H. et al. (2017). High-fidelity bioelectronic muscular actuator based on porous carboxylate bacterial cellulose membrane. *Sens. Actuators B, Chem.* 250: 402–411.

77 Kim, S.S. and Kee, C.D. (2014). Electro-active polymer actuator based on PVDF with bacterial cellulose nano-whiskers (BCNW) via electrospinning method. *Int. J. Precis. Eng. Manuf.* 15: 315–321.

78 Li, J.Z., Zhang, J.Q., Sun, H. et al. (2020). An optical fiber sensor based on carboxymethyl cellulose/carbon nanotubes composite film for simultaneous measurement of relative humidity and temperature. *Opt. Commun.* 467: 125740.

79 Sadasivuni, K.K., Ponnamma, D., Ko, H.U. et al. (2016). Flexible NO_2 sensors from renewable cellulose nanocrystals/iron oxide composites. *Sens. Actuators B, Chem.* 233: 633–638.

80 Mehta, P., Vedachalam, S., Sathyaraj, G. et al. (2020). Fast sensing ammonia at room temperature with proline ionic liquid incorporated cellulose acetate membranes. *J. Mol. Liq.* 305: 112820.

81 Zhao, W., Qu, X.Y., Xu, Q. et al. (2020). Ultrastretchable, self-healable, and wearable epidermal sensors based on ultralong Ag nanowires composited binary-networked hydrogels. *Adv. Electron. Mater.* 6: 2000267.

82 Souza, F.G., Oliveira, G.E., Anzai, T. et al. (2009). A sensor for acid concentration based on cellulose paper sheets modified with polyaniline nanoparticles. *Macromol. Mater. Eng.* 294: 739–748.

83 Meng, J.J., Pan, P., Yang, Z.C. et al. (2020). Degradable and highly sensitive CB-based pressure sensor with applications for speech recognition and human motion monitoring. *J. Mater. Sci.* 55: 10084–10094.

84 Liu, T., Wu, W., Liao, K.N. et al. (2019). Fabrication of carboxymethyl cellulose and graphene oxide bio-nanocomposites for flexible nonvolatile resistive switching memory devices. *Carbohydr. Polym.* 214: 213–220.

85 Islam, S., Bhuiyan, M.A.R., and Islam, M.N. (2017). Chitin and chitosan: structure, properties and applications in biomedical engineering. *J. Polym. Environ.* 25: 854–866.

86 Wei, D.W., Sun, W.Y., Qian, W.P. et al. (2009). The synthesis of chitosan-based silver nanoparticles and their antibacterial activity. *Carbohydr. Res.* 344: 2375–2382.

87 Li, B.Q., Cheng, Y.F., Xu, F. et al. (2015). Biosensor based on chitosan nanocomposite. *Adv. Bioelectron. Mater.*: 277–307.

88 Feng, G.D., Zhao, Y.H., and Jiang, J. (2019). Lightweight flexible indium-free oxide TFTs with and logic function employing chitosan biopolymer as self-supporting layer. *Solid-State Electron.* 153: 16–22.

89 Feng, P., Du, P.F., Wan, C.J. et al. (2016). Proton conducting graphene oxide/chitosan composite electrolytes as gate dielectrics for new-concept devices. *Sci. Rep.* 6: 34065.

90 Hosseini, N.R. and Lee, J.S. (2015). Biocompatible and flexible chitosan-based resistive switching memory with magnesium electrodes. *Adv. Funct. Mater.* 25: 5586–5592.

91 Li, L. and Li, G. (2020). Multi-bit biomemory based on chitosan: graphene oxide nanocomposite with wrinkled surface. *Micromachines* 11: 580.

92 Chen, Z.M., Liu, X.H., Wang, S.M. et al. (2018). A bioinspired multilayer assembled microcrack architecture nanocomposite for highly sensitive strain sensing. *Compos. Sci. Technol.* 164: 51–58.

93 Hosseini, E.S., Manjakkal, L., Shakthivel, D., and Dahiya, R. (2020). Glycine-chitosan-based flexible biodegradable piezoelectric pressure sensor. *ACS Appl. Mater. Interfaces* 12: 9008–9016.

94 Molla-Abbasi, P. and Ghaffarian, S.R. (2014). Decoration of carbon nanotubes by chitosan in a nanohybrid conductive polymer composite for detection of polar vapours. *RSC Adv.* 4: 30906–30913.

95 Velmurugan, S., Palanisamy, S., Yang, T.C.K. et al. (2020). Ultrasonic assisted functionalization of MWCNT and synergistic electrocatalytic effect of nano-hydroxyapatite incorporated MWCNT-chitosan scaffolds for sensing of nitrofurantoin. *Ultrason. Sonochem.* 62: 104863.

96 Kalambate, P.K., Dhanjai, A., Sinha, Y.K. et al. (2020). An electrochemical sensor for ifosfamide, acetaminophen, domperidone, and sumatriptan based on self-assembled MXene/MWCNT/chitosan nanocomposite thin film. *Microchim. Acta* 187: 402.

97 Wu, S.P., Li, K.H., Dai, X.Z. et al. (2020). An ultrasensitive electrochemical platform based on imprinted chitosan/gold nanoparticles/graphene nanocomposite for sensing cadmium(II) ions. *Microchem. J.* 155: 104710.

98 Kushwaha, C.S., Singh, P., Abbas, N.S., and Shukla, S.K. (2020). Self-activating zinc oxide encapsulated polyaniline-grafted chitosan composite for potentiometric urea sensor. *J. Mater. Sci. - Mater. Electron.* 31: 11887–11896.

99 Diouf, A., Moufid, M., Bouyahya, D. et al. (2020). An electrochemical sensor based on chitosan capped with gold nanoparticles combined with a voltammetric electronic tongue for quantitative aspirin detection in human physiological fluids and tablets. *Mater. Sci. Eng., C* 110: 110665.

100 Liu, B., Song, Y.W., Jin, L. et al. (2015). Silk structure and degradation. *Colloids Surf., B* 131: 122–128.

101 Kim, S.H., Nam, Y.S., Lee, T.S., and Park, W.H. (2003). Silk fibroin nanofiber. Electrospinning, properties, and structure. *Polym. J.* 35: 185–190.

102 Kook, G., Jeong, S., Kim, M.K. et al. (2020). Fabrication of highly dense silk fibroin biomemristor array and its resistive switching characteristics. *Adv. Mater. Technol.* 5: 1900991.

103 Jian, M.Q., Zhang, Y.Y., and Liu, Z.F. (2020). Natural biopolymers for flexible sensing and energy devices. *Chin. J. Polym. Sci.* 38: 459–490.

104 Liu, L.X., Chen, W., Zhang, H.B. et al. (2019). Flexible and multifunctional silk textiles with biomimetic leaf-like MXene/silver nanowire nanostructures for electromagnetic interference shielding, humidity monitoring, and self-derived hydrophobicity. *Adv. Funct. Mater.* 29: 1905197.

105 Zhu, B.W., Wang, H., Leow, W.R. et al. (2016). Silk fibroin for flexible electronic devices. *Adv. Mater.* 28: 4250–4265.

106 Zhao, L., Wen, Z.Z., Jiang, F.J. et al. (2020). Silk/polyols/GOD microneedle based electrochemical biosensor for continuous glucose monitoring. *RSC Adv.* 10: 6163–6171.

107 Zhuang, X.M., Huang, W., Yang, X. et al. (2016). Biocompatible/degradable silk fibroin: poly(vinyl alcohol)-blended dielectric layer towards high-performance organic field-effect transistor. *Nanoscale Res. Lett.* 11: 439.

108 Ma, R.L. and Tsukruk, V.V. (2017). Seriography-guided reduction of graphene oxide biopapers for wearable sensory electronics. *Adv. Funct. Mater.* 27: 1604802.

109 Huang, J.N., Xu, Z.J., Qiu, W. et al. (2020). Stretchable and heat-resistant protein-based electronic skin for human thermoregulation. *Adv. Funct. Mater.* 30: 1910547.

110 Lu, W.D., Yu, P., Jian, M.Q. et al. (2020). Molybdenum disulfide nanosheets aligned vertically on carbonized silk fabric as smart textile for wearable pressure-sensing and energy devices. *ACS Appl. Mater. Interfaces* 12: 11825–11832.

111 Reizabal, A., Goncalves, S., Brito-Pereira, R. et al. (2019). Optimized silk fibroin piezoresistive nanocomposites for pressure sensing applications based on natural polymers. *Nanoscale Adv.* 1: 2284–2292.

4

Polymer Nanocomposites for Photodetectors

Raj Wali Khan, Zheng Wen, and Zhenhua Sun

Shenzhen University, College of Physics and Optoelectronic Engineering, South China Sea Road 3688, Shenzhen, Nanshan District 518060, China

4.1 Introduction

Photodetection has been considered as a key technology in modern information society. The basic function of a photodetector (PD) is to convert a light signal to a manipulable signal, which is mostly an electrical signal. To do so, the core part of a PD, which is normally called the active layer, must comprise of materials, which can fulfill the function efficiently and swiftly. Currently, the engaged commercial PDs mainly utilize crystalline inorganic elemental semiconductors, such as silicon and germanium, or compounds, such as III–V semiconductors, due to their small exciton binding energy, high stability, charge-carrier mobility, etc. The success of the inorganic crystalline based PDs has highly promoted the progress of human society and has made drastic changes in human life due to the wide application of PDs regarding a wide range of spectrum from X-rays to far-infrared (FIR). The sensing of light signal in the different spectrum draws forth various applications, such as medical imaging, light controlled switcher, different sensors for physical health and safety monitoring, information and communication, and display technology [1]. Likewise, to harness light with different wavelengths for a specific application, different materials should be utilized in the PD. Figure 4.1 shows the electromagnetic spectrum range from X-rays to mid-infrared (MIR) and the semiconductors used in the commercial PDs for different spectral regions. Notice that the spectrum shown in this figure covers the regions tightly related to the human daily life, instead of the all spectrum in the scope of human ability. Figure 4.1 also shows the main application area in each spectral region, as well as the inorganic semiconductors used in commercial PDs for each spectrum region. Among the shown spectrum in Figure 4.1, the visible and near-infrared (NIR) regions are particularly important for their application in cameras and communications. The complementary metal-oxide-semiconductor (CMOS) PD, which is the most common product to sense visible and NIR light, took up more than 80% market share in 2013 [2]. Despite of these successes, with the rapid progress of digital technologies, PDs with better performance regarding

Polymer Nanocomposite Materials: Applications in Integrated Electronic Devices, First Edition.
Edited by Ye Zhou and Guanglong Ding.
© 2021 WILEY-VCH GmbH. Published 2021 by WILEY-VCH GmbH.

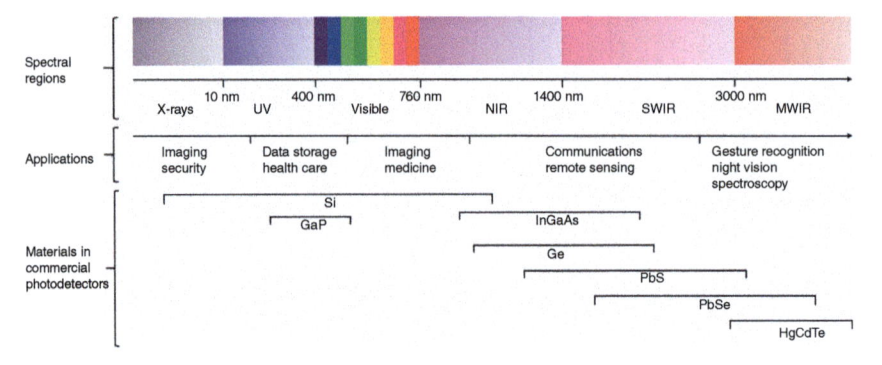

Figure 4.1 Mostly studied spectral regions for photodetection, their corresponding application examples, and the materials used in commercial PDs for each region.

responsivity, sensitivity, speed, etc. are getting more eager. What's more, with new application scenarios for optoelectronic devices, like electronic skins, virtual reality, and wearable devices, are brought forth, device features including low cost, flexibility, integrated circuit (IC) integrability, and transparency are being considered more and more in the development of PDs. All these new demands are exhausting the capability of conventional semiconductor materials, and novel materials with appropriate properties should be introduced into the PDs.

In the quest of next-generation PDs, novel semiconductors including nanomaterials [3, 4], organics [2, 5], two dimensional (2D) materials [6, 7], and metal-halide perovskites (MHP) [1] have been introduced and well investigated. The reported PDs based on these materials have demonstrated, promising but diverse performance, which can be mostly attributed to the different properties of materials [8]. Toward photodetection application, these materials are prepared and processed using methods different from conventional semiconductor. Solution method is applicable for them except for 2D materials. Therefore, it is easy to mix different materials forming a composite material system [1, 9]. Moreover, it is found that by a rational design, the composite of two or more different materials can be endowed with the virtues of all the included components. As a result, the PD based on a composite material can perform better than the device based on each single component. Among all different composite forms, polymer nanocomposite (PoNa) is of particular interest for its unique properties. A PoNa is defined here as a composite material system in which the components in nanoscale hybridized and comprise of at least one polymer. The PoNa is normally prepared by blending all components into one disposable solution. In a PoNa the polymer acts like a matrix in both the pre-deposited solution and the deposited film. Therefore, the flexibility and solution-processability of the polymer is well preserved in the composite, which would enable the corresponding features for the final device. The photoelectric properties of all the components in a PoNa can be utilized for the optoelectronic application, i.e. photodetection in this chapter. What's more, the hybridization of different materials introduces massive interface within PoNa polymer nanocrystal (NC), which would facilitate the dissociation of photo-induced

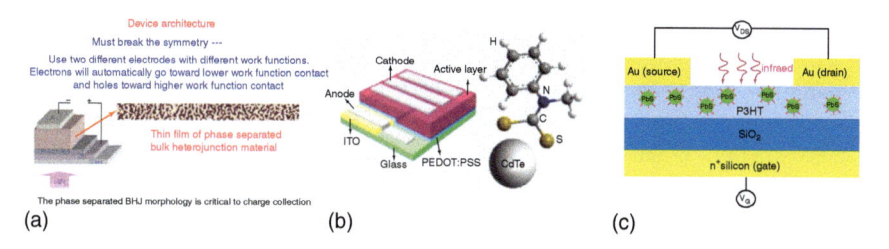

Figure 4.2 (a) A typical structure of bulk heterojunction organic solar cells based on the P3HT/PCBM composite. Source: Heeger [10]. (b) A reported PDi using P3HT/PCBM composite as the active layer. Source: Chen et al. [11]. (c) A reported photoconductor using P3HT/PbS nanocrystals composite as the active layer. Source: Sun et al. [12].

exciton and the extraction of minority carriers [1, 9]. Therefore, the PoNa find huge application potential in optoelectronic devices like solar cells and PDs [9, 10]. For instance, the PoNa of poly(3-hexylthiophene-2,5-diyl) (P3HT) and [6,6]-phenyl C_{61} butyric acid methyl ester (PCBM) is a widely investigated composite system used in organic solar cells (Figure 4.2a) [10], and PDs (Figure 4.2b) [11]. The PoNa of P3HT and lead sulfide (PbS) inorganic NCs is proved valid to produce high-performing photoconductors, as shown in Figure 4.2c [12]. There are already a lot of reports on PoNa-based photodetectors, which are still under intense study nowadays. A careful discussion on this topic will benefit the development in this field. This chapter will present the main research progress of PDs based on PoNa since 1996. Relevant reports are categorized systematically and analyzed specifically. An outlook on this field is finally provided, hoping to shed light on the forward path of the researchers.

4.2 Photodetector Brief

There are two different physical mechanisms underlying the various photodetection technologies, i.e. photo-electric effect and photo-thermal-electric effect. In the former, light signal is converted to electric signal directly. In the latter one, light signal is converted to heat energy, and then an electric signal is produced as a result. The photo-electric detector possesses a particularly important position in the current market as well as the research in scientific community, due to its advantages in performance, including sensitivity, responsivity, and response speed. And among all kinds of photo-electric photodetection devices, semiconductor-based PDs have the merits of compact structure, facile to be integrated, and low driving voltage, so they are winning out over the other fragile and high energy consumption devices like photomultipliers. There are many types of semiconductor-based PDs, like avalanche photodiode, PIN photodiode, photoresistor, charge couple device, and so on. These technologies can be divided into two general categories: photodiode (PDi) and photoconductor (PC), on the basis of their different working mechanisms. Their basic device structures are shown in Figure 4.2.

4.2.1 Photodiode

A PDi comprises a heterojunction formed between two different semiconductors, which normally are p-type and n-type, respectively. A built-in electric field exists in the heterojunction. When a light shines on the semiconductor layer, excitons would be generated under the excitation of incident photons. These excitons would diffuse to the heterojunction region and then dissociate to be electrons and holes there, which are known as the photo-induced carriers. The electrons and holes would drift to the n-type and p-type semiconductor, respectively. The photo-induced carriers are extracted to the external circuit forming a photocurrent. Regardless of the multiple excitation effect, one photon can generate one pair of electron and hole, which will recombine in the external circuit, giving an electric signal. Therefore, no gain (G) can be expected in this process, inducing a limited responsivity (R) and an external quantum efficiency (EQE) smaller than one. However, a PDi normally has a fast response speed due to the short carrier lifetime, and low dark current due to the existence of the heterojunction, enabling a high detectivity (D^*) and large bandwidth (B). The basic structure of a PDi is shown in Figure 4.3a. Besides the basic PDi, there are several derivatives making use of the heterojunction structure as well, including PIN PDi, avalanche PDi, and bipolar phototransistors. The latter two kinds of devices have internal gain at the cost of dark current and response speed.

4.2.2 Photoconductor

A photoconductor takes advantages of a whole semiconductor layer. Its simplest form is a photoresistor, as shown in Figure 4.3b. Notice there is no junction in the semiconductor layer but a uniform layer. The incident photons excite excitons, which would dissociate into majority carriers and minority carriers in the semiconductor layer. Only one type of carrier can contribute to the electric current, with the other one trapped. The lifetime of a contributing carrier (τ_0) can be much larger than the time used for its travel through the conductive channel. In appearance, the carrier circulates in the circuit for several times before it recombines with its counterpart trapped in the semiconductor layer. This process can be narrated as that

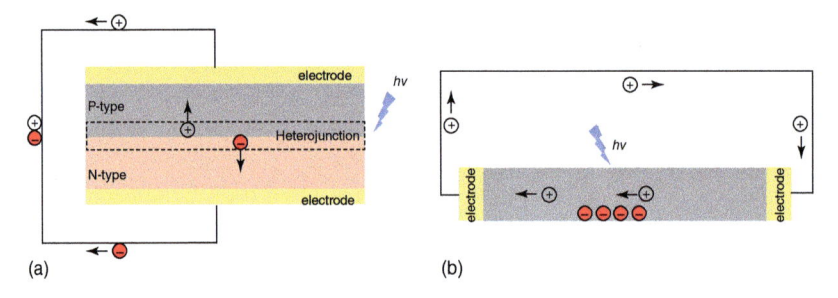

Figure 4.3 (a) Simplified device structure schematic of PDi. The flow of the photo-induced electron and holes is illustrated in figure; (b) simplified device structure schematic of photoconductor. The flow of the photo-induced carriers (hole in this example) is illustrated in figure.

one carrier contributes to electric current for several times, or several carriers contribute to the electric current once. The times number is exact in the internal gain (G). From this viewpoint, one incident photon can generate more than one carrier, meaning an EQE larger than one. The photo-responsivity of the device can be very large. On the other hand, in a photoconductor, the response time would be long and the dark current can be large. An important derivative of the photoconductor is the field effect phototransistor, which have a third electrode-the gate electrode-to tune the photo-response through the field effect. The field effect phototransistor has been proven a feasible and facile platform to exploit the optoelectronic properties of novel materials and to seek for high-performing PDs.

There is a battery of parameters used to evaluate the performance of a PD. Table 4.1 shows some most important parameters, their definitions, and calculation equations. Among them, the R and specific detectivity (D^*) are two parameters explicitly describing the sensing ability of the device to light. R represents the capability of a device to convert incident photons to charge carriers. Apparently, R will vary along with wavelength (λ) and light modulation frequency (f), which give $R(\lambda)$ and $R(f)$, respectively. The R is also tightly related to the EQE and the G.

Table 4.1 Main parameters used to evaluate the performance of a photodetector.

Performance parameters	Definition	Calculation formula	Unit
Responsivity (R)	Ratio of the photocurrent (I_{ph})/photovoltage (V_{ph}) to the incident optical power (P_{in})	$R = \frac{I_{ph} \text{ or } V_{ph}}{P_{in}}$	A W^{-1} (V W^{-1})
External quantum efficiency (EQE)	Number of carriers contributing to the photocurrent induced by an incident photon	$EQE = 100 \frac{I_{ph}/e}{P_{in}/h\nu}$	%
Gain (G)	Number of charge carrier generated by a utilized photon	$G = \frac{\tau_0}{\tau_T}$ (for PC)	Null
−3 dB bandwidth (B)	Modulation frequency at which the responsivity of the device is half of that at steady state conditions	Measured	Hz
Noise current (I_N)	Root mean square of current fluctuation divided by the electrical bandwidth (B)	Measured	A Hz$^{-1/2}$
Noise-equivalent power (NEP)	Optical power that would be required to generate the given noise current density divided by the electrical bandwidth (B)	$NEP = \frac{I_N}{RB^{1/2}}$	W Hz$^{-1/2}$
Specific detectivity (D^*)	NEP normalized to the area of the device (A)	$D^* = \frac{A^{1/2}}{NEP}$	cm Hz$^{1/2}$ W^{-1} (Jones)
Linear dynamic range (LDR)	Range of incident optical powers for which the detector responds linearly	Measured	dB

Note that in the working process of a photodetector, the incident photons, absorbed photons, and utilized photons are different, which are diminishing along the same sequence. The amount ratios of the generated carriers to those three give EQE, internal quantum efficiency (IQE), and G, respectively. In a device with $G = 1$, normally the EQE is smaller than one. If there is any gain process existing, like multiple excitations, avalanche effect, or photoconductive gain, the G is larger than one. Consequently, there is a possibility that EQE is larger than one. Definitionally, the R is linearly dependent on EQE. Therefore, a very large R can be expected in a PD with gain inside. In a photoconductor, one utilized photon would generate only one carrier, but the carrier may circulate in the channel many times during its lifetime (τ_0), inducing a photoconductive gain, as discussed in the previous "Photoconductor" paragraph. Under this condition, the large R is in the cost of the response speed, as the photocurrent would keep increasing during the τ_0. As a result, the modulation frequency would affect R seriously, inducing a concept of bandwidth (B). B is normally determined directly in experiment. In principle, B can be described as $B = \sqrt{\left(\frac{3.5}{2\pi\tau_T}\right)^2 + \left(\frac{1}{2\pi RC}\right)^2}$, where R and C are the resistance and capacitance of the circuit, respectively. Note that in some reports, the rise time (τ_{rise}) and decay time (τ_{decay}) are also used to describe the response speed. They are the times used for the photocurrent to rise or decay to a certain percentage (80% or 90%) of the overall value during a light pulse shines on a PD. Besides, as the carrier recombination rate in a semiconductor would vary along with the incident optical power (P_{in}), R is strongly dependent on P_{in}. Therefore, despite of its importance, R is not a perfect parameter to compare different PDs. Instead, according to the definition, D^* is determined with the light sensitive area, bandwidth, dark current, and R normalized. Comparison on D^* between different PDs would be fairer to conclude a better device. The research on novel PDs is paying more and more attention to the D^*.

4.3 Photodetectors Based on Novel Semiconductors

The quest to improve the performance of PDs has never ceased since it was invented. Besides the performance, features of low cost, flexibility, and transparency are concerned more and more in the development of PDs for some emerging application like flexible electronics. The realization of these new features is possible because novel semiconductors come forth in last several decades. The most investigated novel materials include 2D materials, MHP, inorganic NCs, and organic semiconductors. 2D materials are materials with ultrathin thickness of one or several atomic layers. They have been under intensive investigation since the discovery of graphene in 2004, mostly due to their unique material properties in such a thin thickness. PDs based on 2D materials have shown outstanding performance in EQE, speed, wavelength selection, device flexibility, etc. Nevertheless, the ultrathin property limits the light absorption ability of 2D materials. The delicate growth and processing skills would severely hinder the application of 2D materials under the

4.3 Photodetectors Based on Novel Semiconductors

current technology condition. For example, as early as 2010, Thomas Mueller et al. reported a NIR PD fabricated using a typical 2D material, graphene (Figure 4.4a). The device has a very wide bandwidth of 16 GHz, corresponding to a data link speed of 10 Gbit s^{-1} when applied in optical communication. The shortcoming of this device is its limited photo-responsivity, which is 6.1 mA W^{-1}. The fast response speed of this device is due to the high carrier mobility in graphene, while the low R can be attributed to its very thin thickness [13]. MHP is another research hotspot in optoelectronic devices, especially photovoltaic devices. Perovskites have been discovered as a kind of ceramic for a long history since 1839. This material has found no significant application in optoelectronic device until 2009, when Akihiro Kojima et al. applied an MHP material in a photovoltaic device. With the promising potential of MPH in photovoltaic device revealed in that work, a race of development of MHP for solar cell set out, promoting the rapid increase of the power conversion efficiency to a certified value of 25%. MHP possess many excellent optoelectronic properties, including strong absorbance, low exciton binding energy, long carrier diffusion length, and high carrier mobility [15]. These properties also make MHP a good material for PD. A variety of PDs have been reported using MHP as the light sensitive material [14, 16–18]. For example, in 2015, Feng Li et al. reported a phototransistor using $CH_3NH_3PbI_3$ as the active layer [14]. The device demonstrated a wide response spectrum across the whole visible region, and a high R of 320 A W^{-1}. The response speed is fast with a rise time of 6.5 μs and a decay time of 5 μs. The device structure and its spectral response behavior are shown in Figure 4.4b. The shortcomings of MHP include: (i) MHPs is unstable in atmosphere, especially to the humidity; (ii) The bandgap of MHPs lies in visible light range and can be slightly tuned according to composition [15]. These severely limit the application of MHPs in a PD.

The NCs in this chapter refer to inorganic semiconductor particles with size under 100 nm. Due to the small size, the NCs demonstrate unique properties different from their bulk materials. The NCs have large specific surface area, which means massive surface states facilitate the dissociation of photon-induced excitons. The NCs have a big light absorption cross-section. When the size of a NC is smaller than its Bohr radius, quantum confinement effect will appear, inducing an increasing of the bandgap. This kind of NCs is also called as quantum dots (QDs). Normally NCs can be synthesized through a solution method and dispersed in a solvent, and thus be easily solution-processed, as shown in Figure 4.5. QDs in solution are known as colloidal quantum dots (CQDs) as well. CQDs possess a tunable bandgap, which depends on the particle size. Together with the abundant inorganic semiconductor types, the CQDs material covers a very wide bandgap range. As a result, different PDs based on CQDs have been reported with sensing wavelength from ultraviolet (UV) to MIR. Nevertheless, CQD-based PDs suffer from the limited carrier mobility in the CQD film as the discrete dispersion of the particles in the film. In 2006, Gerasimos Konstantatos et al. reported a PD based on a film of PbS QDs. Ligand exchange was conducted on these QDs to change the original long oleic acid ligand to a short butylamine ligand. This exchange decreased the distance between QDs and largely improved the conductivity of the QD film. As a result, the device demonstrated a

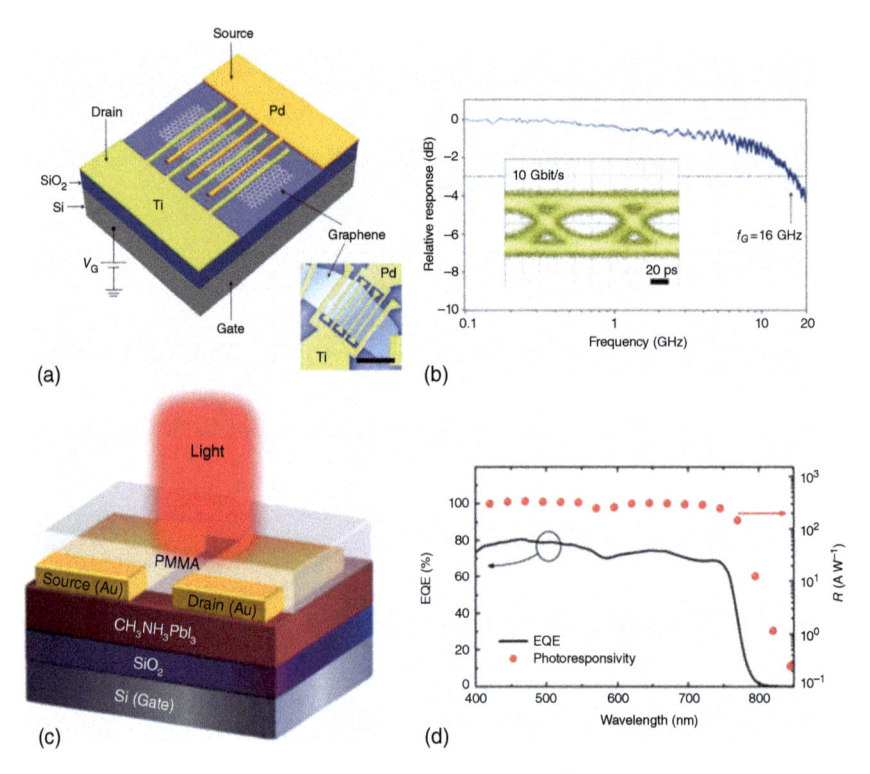

84 *4 Polymer Nanocomposites for Photodetectors*

Figure 4.4 (a) Schematic of a PD based on graphene. (b) Responsivity of the graphene-based PD along with light modulation frequency. Source: Mueller et al. [13] (c) Schematic of a PD based on MHP film. (d) Spectral response of the MHP-based PD. Source: Li et al. [14]. Licensed under CC-BY-4.0.

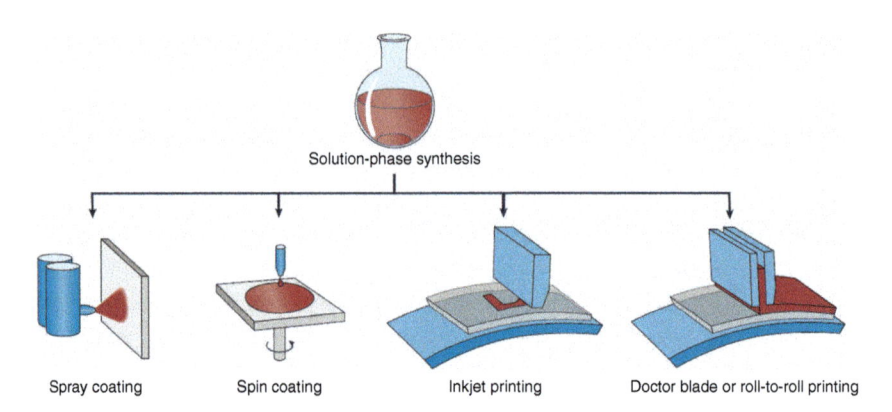

Figure 4.5 Illustration of the solution synthesis and processability of colloidal nanocrystals. Source: de Arquer et al. [1].

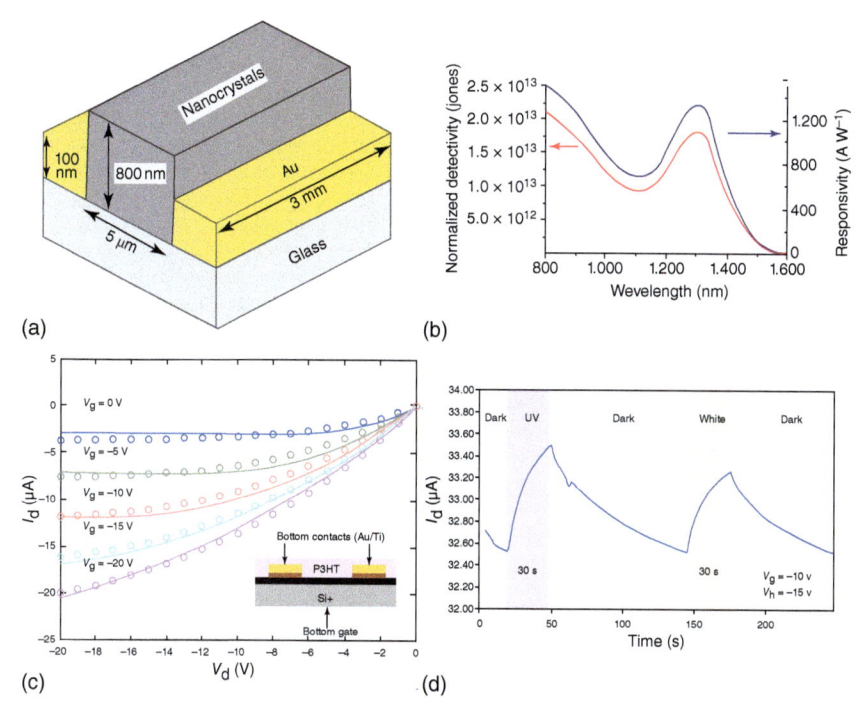

Figure 4.6 (a) Schematic of a photoconductor based a PbS QDs film. (b) Spectral response of the photoconductor. Source: Konstantatos et al. [19]. (c) Output characteristic of a P3HT phototransistor. The inset is the schematic of the device structure. (d) Instant response of the device to light illumination. Source: Scarpa et al. [20].

high D^* of 1.8×10^{13} Jones at infrared region of 1.3 μm, which is even higher than the PD based on epitaxially grown InGaAs film, as shown in Figure 4.6a,b [19]. Taking advantages of the emergence of ligand exchange technique as well as the development of solution-synthesis techniques of CQDs, PDs based on different NCs, including CdTe [21], CdS [22], Bi_2S_3 [23], PbSe [24], HgTe [25], etc., have been reported, demonstrating exciting device performance. Nevertheless, the discrete feature of the QDs, which is the cost of solution-synthesis and processability, consistently hinders the transport of charge carriers in a QD film.

Polymer, which is a kind of organic, has been long regarded insulative, like plastic and rubber. Since 1963 [26], it is realized that the polymer can be conductive, which is current known as "conductive polymer." The conductive polymer normally has a conjugated molecular structure [27]. By using different conjugated structures and branch chains, many kinds of conductive polymers have been synthesized and studied for its application in PDs [28]. Most conductive polymers behave like semiconductor, with their highest occupied molecular orbit (HOMO) and lowest unoccupied molecular orbit (LUMO) corresponding to the valence band and conduction band of semiconductor, respectively. Conductive polymer possesses merits including solution-synthesis, solution-processability, low cost, and flexible, and has been involved in the research of organic electronics for device like organic

Table 4.2 Features of different novel materials in different regards.

Materials	Subclass	Solution-processability	Sensing wavelength	Light harvesting ability	Flexibility	Carrier mobility
Inorganic	Nanocrystals	Good	Wide and tunable	Strong	Good	Low
	2D materials	No	Fixed	Weak	Good	High
Organic	Polymers	Good	Fixed	Moderate	Good	Moderate
	Small molecules	Moderate	Fixed	Moderate	Good	Moderate
Perovskite	All inorganic	Good	Visible	Strong	Weak	High
	Inorganic–organic hybrid	Good	Visible	Strong	Weak	Moderate

solar cells, organic transistors, organic light-emitting diode, and organic PDs. For example, P3HT, which is a well-known conductive polymer, has been applied in a photoconductor as the active layer, as shown in Figure 4.6c,d [20]. Nevertheless, there are some weaknesses of conductive polymer for their application in optoelectronic devices: (i) The synthesis of polymer is less controllable compared with NCs; (ii) The exciton binding energy in polymer is high, inducing a difficult dissociation of exciton; (iii) The bandgap of conductive polymer is less tunable compared with CQDs, and mostly lies in visible region [1, 27, 28].

The PDs based on novel semiconductors demonstrate exciting or interesting performance in different aspects. There are many review papers can be referred to [1]. Regarding the performance, the PDs based on novel materials have surpassed the traditional devices in some aspects. However, due to their different material features, as shown in Table 4.2, there is no perfect single material that can fulfill the demands to the next generation PDs in all aspects including all performance parameters, cost, flexibility, transparency, etc. Furthermore, it is worthy of noting that the strengths of the novel materials are pretty complementary. Hybridization of different materials may produce a better material system for a PD with improved performance in more aspects. This idea was raised a long times ago [29], based in which massive works have been conducted [9, 30]. Figure 4.7 shows some representative works based on different hybrid material systems, including 2D–2D, 2D–MPH, 2D–CQD, polymer–small molecular organic (SMO), and polymer–CQD. Among them, PoNa is of particular significance. The good dissolvability of polymer and the uniformity of the polymer solution and film make it an excellent matrix for a hybrid material system. The large interface area in a nanocomposite is a golden site for the dissociation of photo-induced excitons. The flexibility of polymer matrix, as well as the optoelectronic properties of all components, can be well preserved in the hybrid film.

Dependent on the counterpart materials, PoNas are applied in different devices. For instance, the nanocomposite of P3HT and SMO, PCBM, is widely used in organic

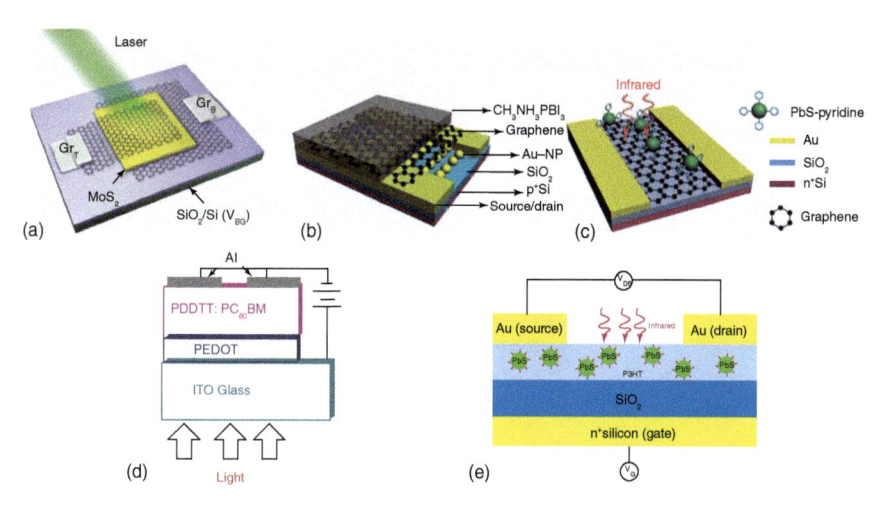

Figure 4.7 Schematic of a PD based the hybrid of graphene-MoS$_2$ (a), graphene-CH$_3$NH$_3$PbI$_3$ (b) graphene-PbS QDs (c) P3HT-PCBM (d), and P3HT-PbS QDs (e). Source: (a) Yu et al. [31]; (b) Sun et al. [32]; (c) Sun et al. [33]; (d) Gong et al. [34]; (e) Sun et al. [12].

solar cells [10]. P3HT and PCBM are p-type and n-type organic semiconductors, respectively. Their nanocomposite forms a bulk heterojunction, as illustrated in Figure 4.8a. It works as a p–n junction but with much larger interface area compared with conventional planar structure. The photo-induced excitons can dissociate to holes and electrons efficiently in the interface, which are then extracted to electrodes through P3HT and PCBM, respectively. Obviously, this kind of device can be utilized as a PDi as well [28]. If the counterpart in a PoNa is inorganic material, normally the inorganic counterpart would disperse discretely, as shown in Figure 4.8b. The inorganic counterpart can be NCs, CQDs, MHP QDs, 2D material QDs, nanopatterns, etc. The composite film is uniform with polymer as the matrix. There is no pathway between different nanoparticles for the extraction of the charge carriers in the inorganic counterpart to the electrode. Therefore, the photodetection device based on this kind of composite is usually photoconductor. Figure 4.8 demonstrates two typical device structures used for different PoNas. However, there are many exceptions need to be addressed carefully. The PoNas that have been reported for their application in PD are discussed in the below Section 4.4 specifically regarding the material composition.

4.4 Photodetectors Based on Polymer Nanocomposites

A PoNa comprises at least one polymer as the matrix. The other counterpart is soluble organic and/or inorganic. The organic counterpart can be polymer and/or small molecular. The inorganic counterpart normally is colloidal NCs. Divided by the component number, there are binary and ternary PoNas reported in the photodetection applications. Specifically, the binary PoNa includes compositions of polymer–polymer, polymer–small molecule, and polymer–inorganic NCs.

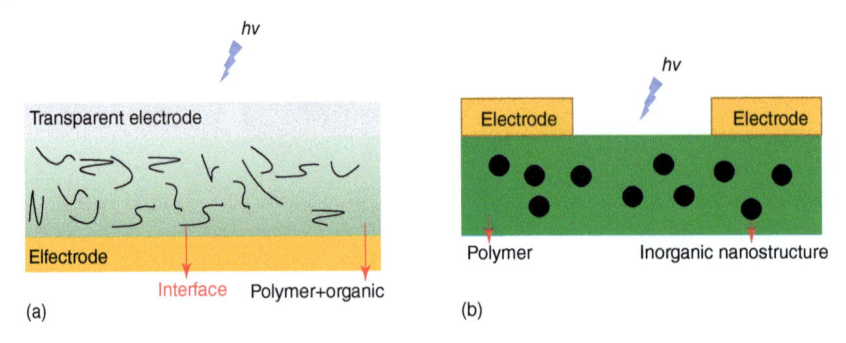

Figure 4.8 (a) Schematic of a typical PDi based on the hybrid of polymer–organic; (b) schematic of a typical photoconductor based on polymer-nanostructure.

The ternary PoNa includes composition of polymer–polymer–small molecule, polymer–small molecule–small molecule, and polymer–small molecule–inorganic NCs. Photodetectors based on different types of PoNa will be separately discussed in this Section 4.4.

4.4.1 Polymer–Polymer Nanocomposite

A film of polymer–polymer binary nanocomposite is deposited from a mixture solution of the two polymers. Therefore, their molecules must have similar polarity to dissolve in a same solvent. The two polymers form bulk heterojunction in the composite, making the composite film suitable for both PDi and PC. The composite film has excellent flexibility, which is inherited from the two constitutes. There are a lot of reports on PDs based on different polymer–PoNa. Table 4.3 lists the material composition, device performance, and year of the work of those reports collected by the authors.

In 1995, Yu and Heeger reported [35] the photovoltaic conversion and charge separation in the spin coated nanocomposite of poly[2-methoxy-5-(2-ethylhexyloxy)-1,4-phenylenevinylene] (MEH-PPV, Figure 4.9a) and poly(2,5-di(hexyloxy)cyanoterephthalylidene) (CN-PPV, Figure 4.9b) with the MEH-PPV as a donor and CN-PPV as an acceptor. The device demonstrated a carrier collection efficiency of 5% electron/photon and an energy conversion efficiency of 0.9%, 20 times larger than in diodes made with pure MEH-PPV and 100 times larger than in diodes made with CN-PPV. In the sensing wavelength 350–690, the R and EQE of the device increased with increasing reverse bias voltage, reaching 0.3 A W^{-1} and 80% at −10 V, respectively, as shown in Figure 4.9d. Not long after that in the same year 1995, J. Halls et al. in University of Cambridge [36] demonstrated the interpenetrating network formed from the phase segregated mixture of MEH-PPV and CN-PPV (Figure 4.9c), which provided both the spatially distributed interface necessary for efficient charge photogeneration, and the means for separately collecting the electrons and holes. Under reverse bias, the EQE of the nanocomposite in a sensing wavelength 300–700 nm raised rapidly, reaching 15% and 40% at a reverse bias of 3.5 and 10 V, and considerably higher values under forward bias. These performance

Table 4.3 Summary of the reported photodetectors based on polymer–polymer nanocomposite.

Polymer 1	Polymer 2	Device	R (A W^{-1})	EQE (%)	D* (Jones)	λ (nm)	Year	References
MEH-PPV	CN-PPV	PDi	0.3	80	—	350–690	1995	[35]
MEH-PPV	CN-PPV	PDi	—	40	—	300–700	1995	[36]
F8BT	PFB	PDi	—	0.55	—	310–620	2000	[37]
P3HT	F8TBT	PDi	0.1	25	—	400–600	2009	[38]
F8BT	MDMO-PPV	PDi	—	2.4	—	470	2011	[39]
P3HT	P1/P2	PDi	—	—	1.61×10^{10}	300–900	2015	[40]
P3HT	PMMA	PDi	166.45 m	—	—	350–650	2015	[41]
PTB7-Th	PIIG-NDI	PDi	0.12	2.68	1.2×10^{12}	400–900	2016	[42]
PDTP-DPP	PNDI	PDi	36.0	—	7.3×10^{11}	300–1100	2017	[43]
PDTP-DPP	PNDI-DPP10	PDi	49.1	—	2.4×10^{12}	300–1100	2017	[43]
PDTP-DPP	PNDI-DPP30	PDi	29.2	—	7.2×10^{11}	300–1100	2017	[43]
PDTP-DPP	PNDI-DPP50	PDi	14.9	—	3.5×10^{10}	300–1100	2017	[43]
PolyD	PolyA/A'	PDi	93.5	12.9	4.7×10^{12}	300–1000	2017	[44]
P3HT	PIDT-2TPD	PDi	0.16	30	1.1×10^{12}	400–700	2018	[45]
PTB7-Th	PNDI-2OD/5DD/POP	PDi	—	30.6	3.0×0^{13}	300–800	2018	[46]
PBDB-T	PNDI-FT10	PDi	0.34	70	1.0×10^{12}	350–850	2018	[47]
PBnDT-FTAZ	P(NDI2OD-T2)	PDi	0.21	—	—	400–800	2019	[48]
P3HT	PZ1	PDi	8	31 700	6.1×10^{12}	300–800	2019	[50]
PTB7-Th	TQ1	PDi	40	35	10^{11}	300–900	2020	[51]

figures of the MEH-PPV:CN-PPV PDi were much better than those for similar devices made with aluminum electrodes and either MEH-PPV or CN-PPV alone.

In 2000, Jonathan et al. reported a PDi based on the nanocomposite of poly(9,9-dioctylfluorene-*alt*-benzothiadiazole) (F8PT, Figure 4.10a) and dioctyl fluorene–triarylamine copolymer PFB (Figure 4.10a) [37]. The device demonstrated that at 410 nm, the photocurrent in the blend PDi is some 880 times higher than in the F8BT diode and 14 times greater than in the PFB device. The spectral response of the blend PDi broadly follows the quantity of light absorbed in the polymer layer, indicating that absorption of light throughout the polymer bulk contributes to the photocurrent. At a sensing wavelength of 300–650 nm, the spin-coated F8BT:PFB PDi film demonstrated the highest amount of scattering, lowest EQE (0.25% at 500 nm, Figure 4.10b), and dark current 4.0×10^{-9} A cm^{-2}. In 2009, Keivanidis et al. reported [38] that the PDis deposited via spin-coating the nanocomposites of poly([9,9-dioctylfluorene]-2,7-diyl-*alt*-[4,7-bis(3-hexylthien-5-yl)-2,1,3-benzothiadiazole]-2,2,-diyl) (F8TBT) and P3HT (Figure 4.10c). Different polymers including PFB and poly[2,7-(9,9-di-*n*-octylfluorene)-*co*]-1,4-phenylene[(4-*sec*-butylphenyl)

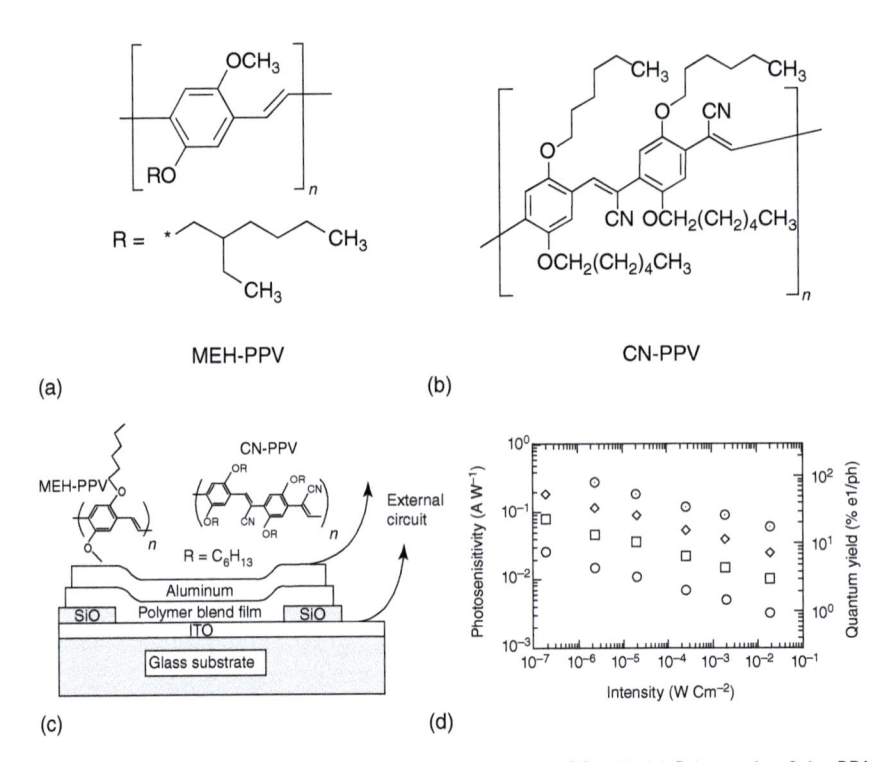

Figure 4.9 Molecular structure of MEH-PPV (a) and CN-PPV (b). (c) Schematic of the PDi fabricated based on the composite of MEH-PPV and CN-PPV. Source: Halls et al. [36]. (d) Performance of a PDi based on MEH-PPV-CN-PPV nanocomposite. Source: Yu and Heeger [35].

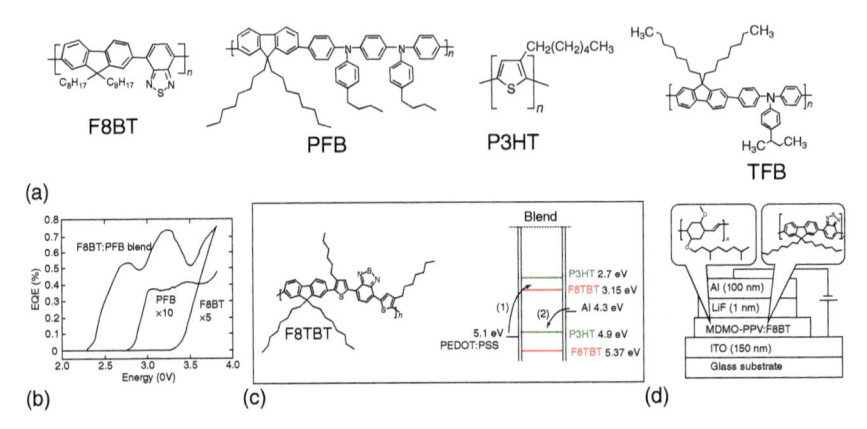

Figure 4.10 (a) Molecular structure of several conductive polymers. (b) EQE spectrum of the PDi based on nanocomposite of F8BT and PFB. Source: Halls et al. [37]. (c) Molecular structure of F8TBT and energy diagrams of PDi based on nanocomposites of F8TBT and P3HT. Source: Keivanidis et al. [38]. (d) Schematic of the PDi fabricated based on the composite of F8BT and MDMO-PPV. Source: Fukuda et al. [39].

imino]-1,4-phenylene)] (TFB, Figure 4.10a) were applied as the electron-blocking layer in the PDi. The effect of F8TBT as a hole-blocking layer in the device was also studied. In these devices, dark current can be reduced when both an electron and a hole-blocking layer were used but the EQE falls significantly. The dark current density values as low as 40 pA mm^{-2} are achieved with a corresponding EQE of 20% when an electron-blocking layer was used. Furthermore, the blend of F8BT with poly(2-methoxy-5-(3′,7′-dimethyloctyloxy)-1,4-phenylenevinylene) (MDMO-PPV, Figure 4.10a) demonstrated an improved photoconductive property, based on which a PDi was obtained via spin-coating process in 2011 by Fukuda et al. (Figure 4.10d) [39]. At 470 nm, the device displayed maximum EQE of 2.4% at 15 MV m^{-1} when the doping ratio of MDMO-PPV-doped F8BT was 50 wt%. This EQE is approximately 5.1 times higher than that (0.47%) of a reference device with undoped F8BT. In addition, the signal-to-noise ratio (SNR) (i.e. photocurrent/dark current) was improved by doping MDMO-PPV because of the reduced dark current density. The highest SNR was found to be 447.

In Keivanidis' work [38], P3HT was involved in the PDi as a polymer matrix. This may not be the first report on all polymer PD based on P3HT but definitely not the final one. P3HT is a well investigated p-type conductive polymer applied in organic optoelectronic devices. Different polymer counterparts have been used to form a bulk heterojunction with P3HT as the active layer in PD. For example, in 2015, Wang et al. synthesized two n-type polymer semiconductors based on ladder conjugated polypyrrones, which were named P1 and P2 [40]. The blend of P3HT and either P1 or P2 was used as the bulk heterojunction active layer in PDi (Figure 4.11a), with the electron-deficient polypyrrone as an acceptor and P3HT as a donor. The PD reported shows spectral response from 300 to 900 nm with the maximal D^* up to 10^{11} Jones and a D^* in an order of 10^{10} Jones at 800 nm. The relatively low EQE values of 22% at −5 V may be due to the large domain size of the active layer and thus insufficient exciton dissociation. Therefore, the device performance was expected to improve after the miscibility and aggregation of the active layers were carefully optimized. It was found that insulative polymer can be used to improve the crystallinity of the conductive polymer. In 2015, Li Zhang et al. reported that by blending poly(methacrylic acid methyl ester) (PMMA, Figure 4.11b), which is a dielectric polymer, into P3HT, the crystallinity of the P3HT film can be improved [41]. The improvement reached to the summit with PMMA of 20 wt%. However, the PDi based on blend of P3HT and PMMA showed a best photo-response performance when the content of PMMA is 50 wt%. The device demonstrated a maximum photoresponsivity of 166.45 mA W^{-1} under the illumination of 600 nm with intensity of 65 μW cm^{-2}. In 2018, Murto et al. [45] utilized the spin-coated nanocomposite of P3HT and PIDT-2TPD (1 : 1, 1 : 2, and 2 : 2) as the active layer in PDs. With a blend of P3HT and PIDT-2TPD in 2 : 1 ratio, the PD displayed a strong R reaching 0.16 A W^{-1} at 680 nm and EQE broadly above 30% and maximum D^* of 1.1×10^{12} Jones at 610 nm, indicating the PIDT-2TPD (Figure 4.11c) a highly functional new type of acceptor. Recently in 2019, a new acceptor PZ1 (Figure 4.11d) was applied in the P3HT-based PDis [50]. Kaixuan Yang et al. reported a PDi using the blend of P3HT and PZ1 sensing light with wavelength

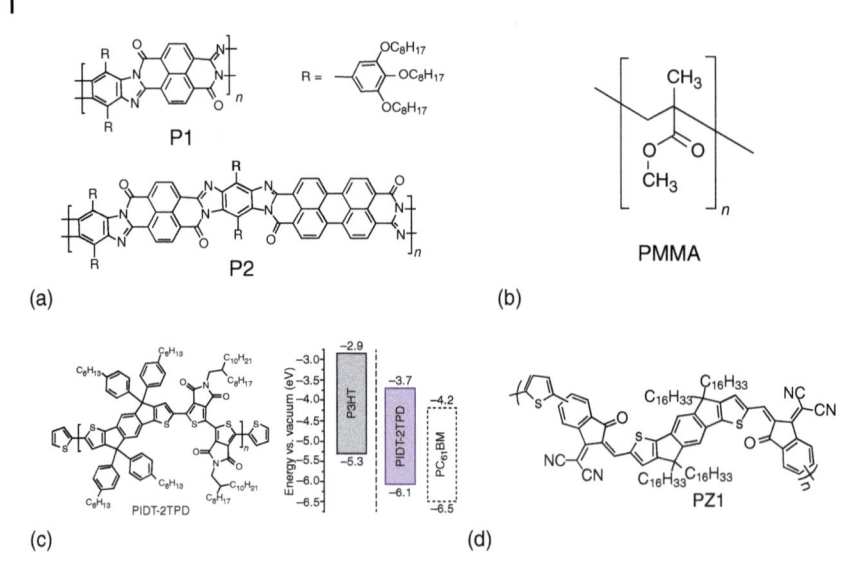

92 | *4 Polymer Nanocomposites for Photodetectors*

Figure 4.11 (a) Molecular structure of P1 and P2. Source: Wang et al. [40]. (b) Molecular structure of PMMA. (c) Molecular structure of P2TD-2TPD and its band structure. Source: Murto et al. [45]. Reproduced with permission. Copyright 2018, American Chemical Society. (d) Molecular structure of PZ1. Source: Zhang et al. [49].

of 300–800 nm. The device demonstrated an EQE of 46 700% at 375 nm and 31 700% at 615 nm under a −20 V bias. The corresponding D^* and R were 5.6×10^{12} Jones at 375 nm and 6.1×10^{12} Jones at 615 nm, respectively. These reports imply that the P3HT is a promising building block for assembly of low-cost, flexible, fast, and high-performance PDs.

In 2016, Xiaofen Wang et al. constructed acceptor–acceptor conjugated acceptor polymer including PIIG-NDI(OD), PIIG-PDI(EH), and PIIG-PDI(OD) (Figure 4.12c) [42]. PTB7-Th (Figure 4.12c) was used as the donor polymer to be blended with different acceptor polymer forming bulk heterojunction for different PDis curve. The highest R of PIIG-PDI(OD) and PIIG-PDI(EH)-based all-polymer PDs reached 0.12 A W^{-1}, which is comparable with inorganic PDis and the fullerene based OPDs. The PIIG-PDI(OD) PD showed the overall best performance among these different devices, with a high D^* of about 10^{12} Jones (Figure 4.12d). The work proves the feasibility of acceptor–acceptor type conjugated polymer as the acceptor in a bulk heterojunction-based PD.

Liuyong Hu and his coworkers had made consistent contribution to the all-polymer PDs. In 2017, Hu et al. [43] assembled the two organic units forming a polymer, PNDI-DPP, which is a kind of non-fullerene acceptor (Figure 4.12a). Dependent on the ratio of the two units, three polymers named PNDI-DPP10, PNDIDPP30, and PNDI-DPP50 were presented. A battery of PDis (Figure 4.12b) was fabricated using PDTP-DPP as the donor and different PNDI-DPPs as acceptors. The PD based on PNDI-DPP acceptor demonstrated an improved performance compared with PNDI when the content of the DPP is appropriate. Specifically, the PNDI-DPP10 device showed the best performance with an R of 30.4 mA W^{-1}

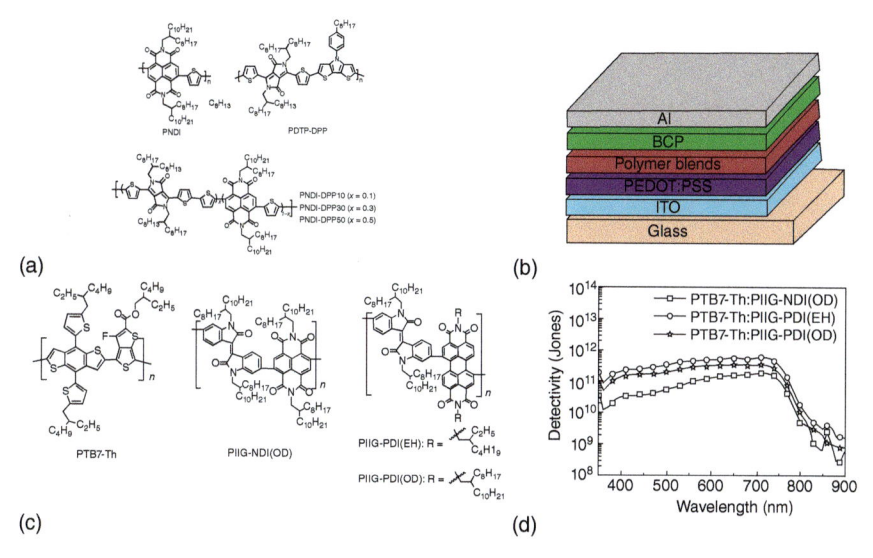

4.4 Photodetectors Based on Polymer Nanocomposites

Figure 4.12 (a) Molecular structure of PNDI, PNDI-DPP10, PNDI-DPP30, PNDI-DPP50, and PDTP-DPP. (b) Schematic of the PDi based on the bend of PDTP-DPP and PNDI or PNDI-DPP. (c) Molecular structure of PTB7-Th, PIIG-NDI(OD), PIIG-PDI(EH), and PIIG-PDI(OD). (d) Spectral detectivity of the PDi based on the blend of PTB7-Th and PIIG-NDI. Source: Wang et al. [42].

at 400 nm and 49.1 mA W^{-1} at 900 nm, and D^* of 1.5×10^{12} Jones at 400 nm and 2.4×10^{12} Jones at 900 nm. The PD based on PNDI-DPP30 had a lower dark current density 5.4×10^{-9} A cm^{-2}, giving rise to a similar D^* value of 7.2×10^{11} Jones at 900 nm. However, due to the low R of 10.3 mA W^{-1} at 400 nm and 14.9 mA W^{-1} at 900 nm, along with a dark current density 5.5×10^{-9} A cm^{-2}, the PD based on PNDI-DPP50 had the lowest specific D^* of 2.4×10^{11} Jones at 400 nm and 3.5×10^{10} Jones at 900 nm.

In the same year, Hu et al. [44] continued on the acceptor–acceptor strategy in organic PD. They used PDTP-DPP as the donor, which they named after polyD in the report. PDNI and its derivative, which are named after polyA′ and polyA, respectively, were polymerized together forming polyAA′. The polymerization was conducted with the polyA′ weight ratio of 25%, 50%, and 75%, which are coded with polyAA′25, polyAA′50, and polyAA′75, respectively. Besides, the blends of the polyA and polyA′ with the same component ratios were also prepared for comparison, which were polyA/A′25, polyA/A′50, and polyA/A′75, respectively. A series of PDis were fabricated and characterized using the blending of polyD and different acceptors. The photocurrent density was obtained with 3.9×10^{-3} A cm^{-2} for polyAA′50 vs. 8.9×10^{-4} A cm^{-2} for polyA/A′50. The EQE values of the representative devices under −0.1 V bias were found to be 6.3%, 7.3%, and 7.1% at 900 nm for polyA/A′25, polyA/A′50, and polyA/A′75 based PDs, respectively, Surprisingly, the EQE values are 12.9%, 12.4%, and 10.7% at 900 nm under −0.1 V bias for the devices based on polyAA′25, polyAA′50, and polyAA′75, respectively. In addition, the relatively best D^* values (4.7×10^{12}, 4.7×10^{12}, and 1.6×10^{12} Jones at 900 nm) were obtained for

94 | *4 Polymer Nanocomposites for Photodetectors*

polyAA′25, polyAA′50, and polyAA′75, respectively. All the results demonstrated that the polymerized acceptor–acceptor polymer gives better PDs than each single acceptor component as well as the blend of the two acceptor components. This work explicitly showed the validity of the polymerization strategy for different acceptor polymers.

In the next year, again, Hu et al. used the PTB7-Th as the donor in the binary polymer system, but polymer PNDI with different side chain as the acceptor this time. The polymer acceptors applied in their study are shown in Figure 4.13a. Dependent on the side chain, they are PNDI-5DD, PNDI-2OD, and PNDI-POD. PDis with device structure shown in Figure 4.13b were fabricated based on different donor–acceptor systems. Through a battery of characterization, it is revealed that the side chains of NDI-based acceptor polymers can significantly affect the electron mobility, blend

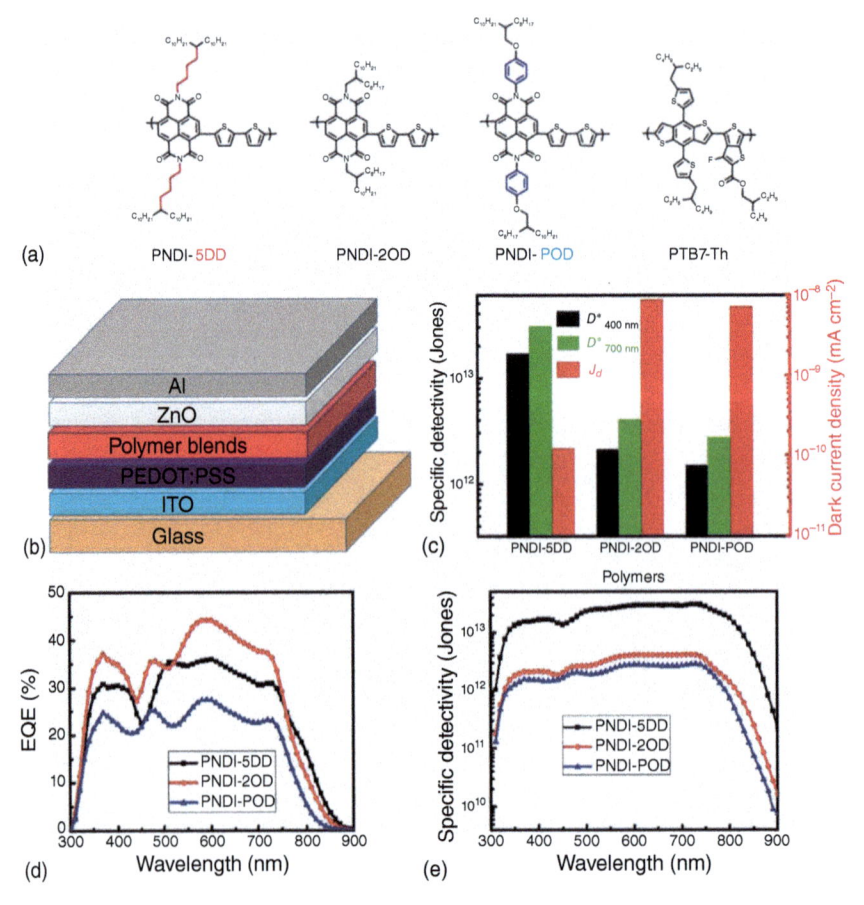

Figure 4.13 (a) Molecular structure of the acceptor polymers of PNDI-5DD, PNDI-2OD, and PNDI-POD and the donor polymer of PTB7-Th used in this work. (b) The structure of the PDi used in this work; (c) dark current density and specific detectivity (D^*) at 400 and 700 nm of the device. (d, e) The EQE spectra and detectivity of the device under −0.1 V bias. Source: Hu et al. [46].

film morphology, and then device performance. The device based on PNDI-5DD, which had a less steric hindrance group in the side chain, showed the relatively high EQE and lowest dark current density, inducing a D^* of over 10^{13} Jones in the spectral region of 300–800 nm under −0.1 V bias (Figure 4.13c–e). This better performance benefited from the smoother surface and preferred molecular stacking induced by the side chain. This work shed a light on the relationship between the device performance and the molecular structure in an all-polymer PDi [46].

In 2018, Xiaofeng Xu et al. reported the utilization of another copolymer PNDI-FT10 as the acceptor and poly[(2,6-(4,8-bis(5-(2-ethylhexyl)thiophen-2-yl)-benzo[1,2-b:4,5-b′]dithiophene))-alt-(5,5-(1′,3′-di-2-thienyl-5′,7′-bis(2-ethylhexyl) benzo[1′,2′-c:4′,5′-c′]dithiophene-4,8-dione))] (PBDB-T, Figure 4.14a) as the donor in an all-polymer PD [47]. The blend of PNDI-FT10 (Figure 4.14a) and PBDB-T forming a bulk heterojunction was applied as the active layer. They have conducted a lot of studies on the PDi based on this PoNa system, including different acceptor/donor ratio, inverted device with P3HT as buffer layer (Figure 4.14c, f), inverted device with PBDB-T as buffer layer (Figure 4.14d), semitransparent device using transparent electrodes (Figure 4.14e), and roll-to-roll fabrication of a flexible device (Figure 4.14b). The inverted PDis with small active areas showed good dark current in an order of 10^{-7} A cm^{-2} and one of the highest EQE over 70% among devices, especially the high D^* over 10^{12} Jones with planar photo-response from 450 to 740 nm. Electron blocking layers (P3HT and PBDB-T) were successfully formed on the active layers by using transfer-printing technique. It significantly lowered the dark current to 1.1×10^{-8} A cm^{-2} without limiting the outstanding EQE. Higher D^* over 3.0×10^{12} Jones in the entire visible region was therefore achieved with the maximal value of 5.8×10^{12} Jones at 650 nm, which is among the best-performing all-polymer PDs in that time. The large-area and semitransparent device was successfully fabricated and characterized on either inflexible glass/indium tin

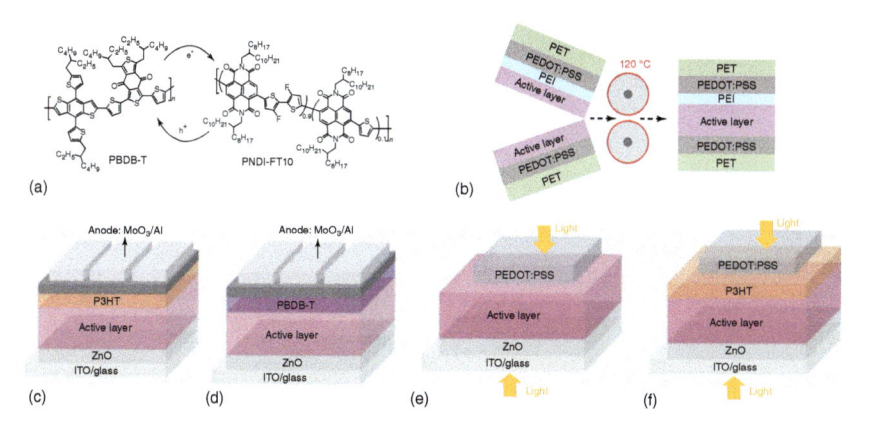

Figure 4.14 (a) Molecular structure of PBDB-T and PNDI-FT10; (b) schematic of roll-to-roll lamination fabrication using the active layer as the adhesive on flexible PET substrate; (c) device structure of the inverted PDi with a P3HT buffer layer; (d) device structure of the inverted PDi with a PBDB-T buffer layer; (e) device structure of the semitransparent PDi; (f) device structure of the semitransparent PDi with a P3HT buffer layer. Source: Xu et al. [47].

96 | *4 Polymer Nanocomposites for Photodetectors*

oxide (ITO) or flexible PET/PEDOT:PSS substrates, which were realized for the first time using a lamination method. The semitransparent devices could sense light from both sides with nearly identical D^* over 1.0×10^{11} Jones. This work clearly demonstrated the advantages of PoNa toward flexible, transparent, and high-performing PDs.

The roll-to-roll lamination fabrication method present by Xiaofeng Xu et al. is a very suitable method for flexible device, and worthy of further development. Recently in 2020, Yuxin Xia, Xiaofeng Xu, and their coworkers reported a PDi using all polymer active layer with PNDI-T10 (Figure 4.15a) as the acceptor, and either PTB7-Th or TQ-1 (Figure 4.15a) as the donor [51]. Furthery, both acceptor–donor bilayer structure and bulk heterojunction structure were studied and compared. Among them, the PTB7-Th bulk heterojunction device had the highest D^* up to 10^{11} Jones. This device also showed bandwidth of 320 kHz bandwidth under 0 V and 470 kHz under −4 V, which was believed to be the highest reported for all-polymer OPDs in that time. A transparent and flexible device was fabricated using the roll-to-roll lamination method, which perform well as a heart rate sensor, as shown in Figure 4.15c. This work is of significance as it demonstrated a real application of the all-polymer PDi.

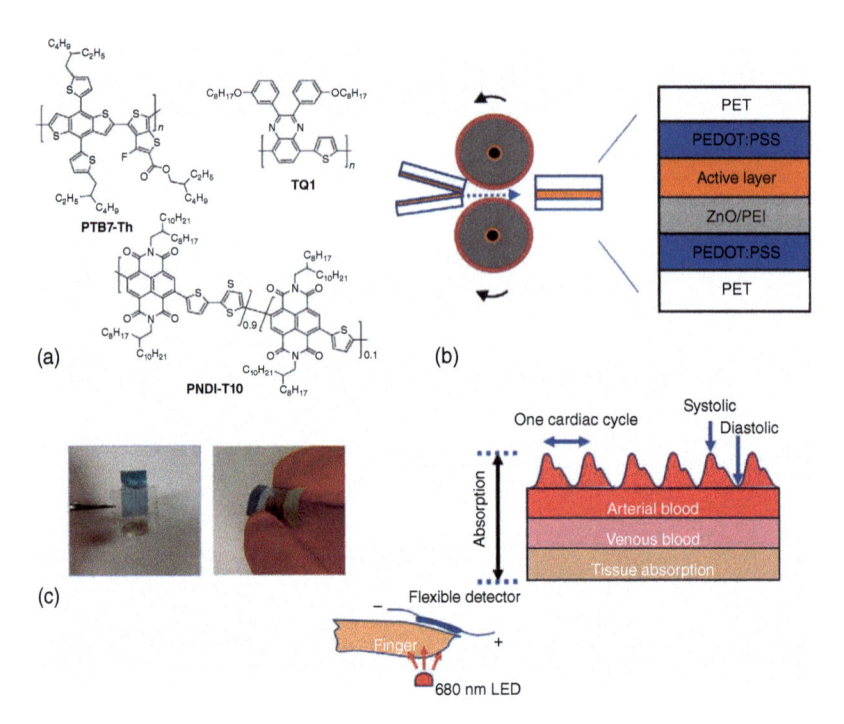

Figure 4.15 (a) Molecular structure of TQ1, PTB7-Th, and PNDI-T10; (b) schematic of roll-to-roll lamination fabrication in this work; (c) diagram of heart rate sensing by putting our detector on one side of fingertip while LED on the other side. Source: Xia et al. [51]. Reproduced with permission. Copyright 2020, Wiley-VCH Verlag GmbH & Co. KGaA, Weinheim.

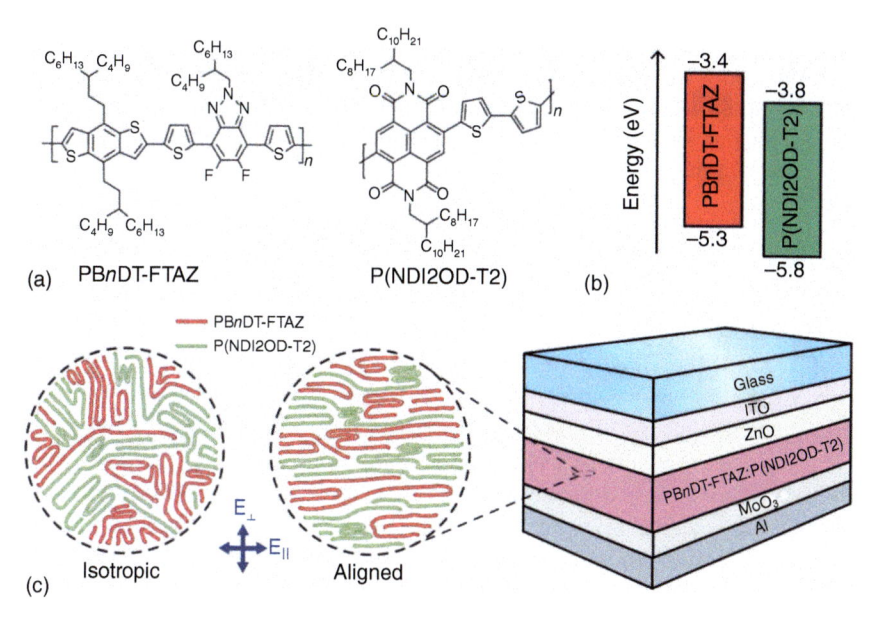

4.4 Photodetectors Based on Polymer Nanocomposites | 97

Figure 4.16 (a) Molecular structure of PBnDT-FTAZ and P(NDI2OD-T2); (b) HOMO and LUMO energy levels for PBnDT-FTAZ and P(NDI2OD-T2); (c) top-view schematic of polymer chain organization in the case of spun-cast isotropic film and a strain-oriented polarization-sensitive film. Schematic includes relative orientation of the polarized light field (⊥ and ∥). Right, the organic PD device stack. Source: Sen et al. [48].

There is other report on all-polymer PDs feature differently that is worthy of mention. In 2018, Sen et al. [48] demonstrated the high-performance all-polymer PDi based on bulk heterojunction composed of an electron donor polymer PBnDT-FTAZ and acceptor polymer P(NDI2OD-T2), which is sensitive to linearly polarized light throughout the visible spectrum (Figure 4.16). Thanks to the complementary absorption spectra of the two polymers, the device demonstrated a wide response spectrum from 350 to 800 nm. The blend film exhibits good ductility with the ability to accommodate large strains of over 60% without fracture. Therefore, a large uniaxial strain can be applied to the film resulting in in-plane alignment. Both polymers showed in-plane backbone alignment and maintain packing order after being strained. The strain-oriented detectors have maximum photocurrent anisotropies of 1.4 under transverse polarized light while maintaining peak R of 0.21 A W^{-1} and a bandwidth of about 1 kHz. The sensitivity of the device to polarized light should be attributed to the aligned molecular structure. This intrinsic polarization sensitivity removes the need for polarizers placed on the input side of a conventional detector, simplifying polarized light detectors. And the device apparently shows promising potential for flexible electronics [48].

The PDs base on the nanocomposite of polymer and polymer is also known as all polymer PDsPDNCPD. The kind of PD demonstrates a wide range of advantages including flexibility, transparency, polarized light sensitive, high D^*, etc., which are mostly inherited from the all polymer feature. Nevertheless, polymer has the disadvantages including high difficulty in synthesis, less controllability on polymerization process, narrow raw material selection range, and so on. Especially, it is worthy of noting that most all polymer PDs can only detect visible. Therefore, other materials beyond polymer should be considered as a counterpart in PoNa to further elevate and broaden the performance of PDs.

4.4.2 Polymer–Small Molecular Organic Nanocomposite

A SMO can be well mixed with a polymer for a nanocomposite since the solution stage due to their solubility in organic solvent. The SMO applied in organic PDs refers to fullerene (C_{60}, Figure 4.17b) or its derivative mostly. Fullerene-based SMOs are n-type semiconductors with low-lying LUMO energy levels, fast photo-induced electron-transfer, and high intrinsic electron mobility [55]. By attaching different functional group to the fullerene molecular, the electronic and optical properties of the SMOs can be adjusted. Generally, the SMO is normally used as an electron acceptor, replacing the acceptor polymer in the bulk heterojunction discussed earlier. A donor polymer should be applied as the counterpart for a PDi. P3HT and MEH-PPV (Figure 4.17a) are two mostly frequently used in this system. In the following text, the research on polymer–SMO nanocomposite using P3HT and MEH-PPV as the polymer component for photodetection application is reviewed and discussed for an in-depth understanding in this topic.

4.4.2.1 MEH-PPV–Small Molecular Organic Nanocomposite

The material system of MEH-PPV—SMO was first universally investigated in 1990s (Table 4.4). Professor Alan J. Heeger in University of California, Santa Barbara,

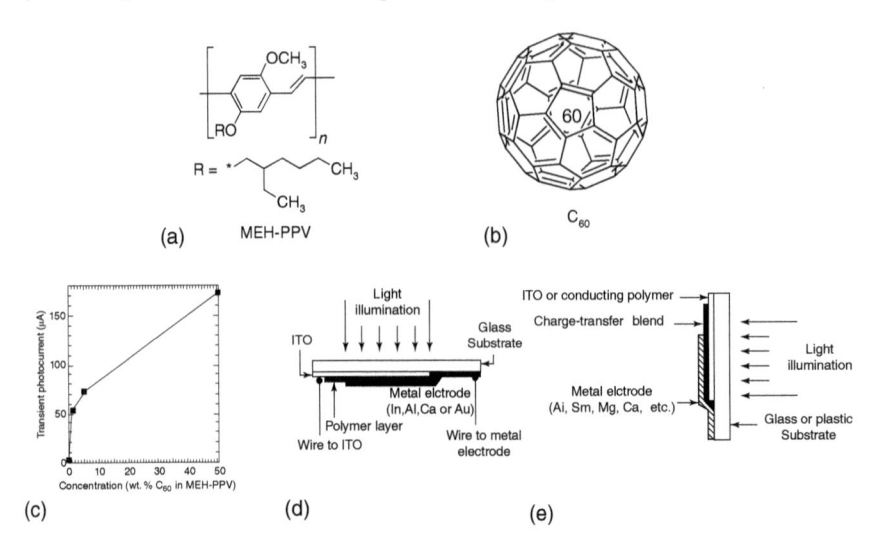

Figure 4.17 Molecular structure of MEH-PPV (a) and C_{60} (b). (c) Transient photocurrent in a PD based on (MEH-PPV)-C_{60} blend with different weight ratio. Source: Lee et al. [29]. (d) Device structure of a PD based on the blend of (MEH-PPV)-C_{60}. Source: Yu et al. [53]. (e) Schematic of the PDi based on blend of (MEH-PPV)-C_{60}. Source: Yu et al. [54]. © 1997, SPIE.

and his coworkers have made profound contribution to the organic photodetectors based on MEH-PPV-SMO blend since the very beginning. In 1993, they first studied the carrier generation and transfer process in the MEH-PPV-C$_{60}$ blend [29]. They revealed that by introducing some C$_{60}$ into the MEH-PPV, the lifetime of the photo-induced carrier got prolonged, and the early time recombination got inhibited. They attributed this change to the enhanced spatial separation of the electron and hole due to the existence of C$_{60}$ as well as the interface. Also, they concluded that thiophene based conductive polymer like P3OT had a less early time recombination of photo-induced carrier. The photocurrent in the blend polymer was 2 orders of magnitude larger than the pure MEH-PPV film (Figure 4.17c). This work demonstrated the promising potential of the blend of MEH-PPV-C$_{60}$ in photodetection application. In the next year, they successfully fabricated a PDi using the blend (Figure 4.17d) [53]. The device showed a R of 0.41 A W^{-1} at 370 nm and an EQE of 135%. They speculated that an avalanche mechanism may exist in the device for the EQE beyond one. Nevertheless, we think there is a photoconductor gain process due to the trapping of electrons in the C$_{60}$. After all, the device showed a R of 0.3 A W^{-1} at 550 nm, which is higher than the commercial UV-enhanced Si PDi at that time.

Later in 1997, Heeger and his coworkers reported photovoltaic and photodetection device based on the blend of MEH-PPV-C$_{60}$ (Figure 4.17e) [54]. They dig deeply into the mechanism and nailed down the electron donor and acceptor roles of MEH-PPV and C$_{60}$. Large size PD and image sensor were presented in that work, clearly demonstrating the potential of this material system for real application. In this work the thiophene-based conductive polymer, P3OT, and the derivative of C$_{60}$, was also applied as an electron donor and acceptor, respectively. They demonstrated even better performance, resulting in an intense study in the following two decades on the thiophene-based conductor polymer (P3HT) and derivatives of C$_{60}$ (PCBM, PC$_{70}$BM).

4.4.2.2 P3HT-Small Molecular Organic Nanocomposite

Table 4.5 shows some reported PDs using P3HT-SMO nanocomposite in the active layer. From the table it can be found the fullerene-based SMOs are the mostly used counterpart, including PCBM, and PC$_{71}$BM. The molecular structure of the SMOs related in this table is shown in Figure 4.18. The devices presented here are all based on PDi using the bulk heterojunction PoNa. Due to that the P3HT-PCBM system has

Table 4.4 Summary of some reported PDs based on MEH-PPV–small molecular organic nanocomposite.

SMO	Device	Fabrication	EQE (%)	D* (Jones)	λ (nm)	R (A W^{-1})	Band width (Hz)	Year	References
C$_{60}$	PDi	Drop casting	—	—	230–1220	—	—	1993	[56]
C$_{60}$	PDi	Spin casting	135	—	300–700	0.41	—	1994	[53]
C$_{60}$/PCBM	PDi	Spin casting	75	—	350–550	0.26	100	1997	[54]

Table 4.5 Summary of some reported photodetectors based on P3HT – small molecular organic nanocomposite.

SMO	Fabrication	Device	EQE (%)	D* (Jones)	λ (nm)	Band width (Hz)	R (A W^{-1})	τ_{rise}/τ_{decay}	Year	References
PCBM	Spin coating	PD	—	—	300–800	—	—	—	2008	[57]
PCBM	Spin coating	PD	80	10^{13}	350–750	—	—	—	2012	[58]
PCBM	Spray coating	PD	2	—	660	15k	—	—	2013	[59]
PCBM	Spin coating	PD	39	1.2×10^{11}	350–700	—	0.13	—	2013	[60]
PCBM	Spin coating	PD	35	3.15×10^{11}	400–700	790 k	0.18	—	2013	[61]
PCBM	Spin coating	PC	—	—	520	—	—	1.32/3.73 s	2014	[62]
PC$_{71}$BM	Spin coating	PC	16 700	—	300–700	—	51.7	—	2015	[63]
PCBM	Spin coating	PD	64	2.8×10^{12}	300–700	1	0.188	—	2015	[64]
PC$_{71}$BM	Spin coating	PC	6830	1.32×10^{14}	300–650	—	347.2	—	2015	[65]
PCBM	Spray coating	PD	38	1.33×10^{12}	300–800	—	0.16	—	2015	[66]
PCBM	Spin coating	PC	37 500	—	300–800	—	—	—	2015	[67]
DC-IDT2T	Spin-coating	PC	4000	2.60×10^{13}	300–800	—	131.4	—	2016	[68]
PCBM	Spin coating	PC	3250	1.04×10^{12}	350–800	11.4	14.25	18/20 μs	2017	[71]
PCBM	Spin coating	PC	208.11	9.10×10^{12}	350–750	10 M	921.30	—	2017	[75]
PC$_{71}$BM	Spin coating	PC	53 500	—	300–800	—	278	0.74/43 s	2017	[69]
PCBM	Spin coating	PC	8180	7.75×10^{11}	400–800	—	—	—	2017	[70]
PCBM	Spin coating	PD	25.3	8×10^{11}	400–700	—	0.086	—	2018	[76]
PCBM	Spin coating	PC	12 000	10^{12}	350–750	—	54	24/78 μs	2018	[72]
PCBM	Spray coating	PD	—	—	300–1000	—	—	—	2018	[80]
PCBM	Spin coating	PC	—	1.48×10^{14}	300–800	—	>40	null/0.42 s	2018	[73]
PC$_{71}$BM	Spin coating	PC	12 000	6.8×10^{-12}	400–1100	—	8.8	—	2019	[74]
PCBM	Spin coating	PD	0.44	8.8×10^{9}	300–600	—	—	0.35/0.49 s	2019	[78]
PCBM	Spin coating	PD	—	4.8×10^{14}	300–600	—	2.63×10^{2}	10/10.5 μs	2019	[77]
PC$_{71}$BM	Spin coating	PD	4767	10^{13}	200–800	—	14.44	—	2020	[79]

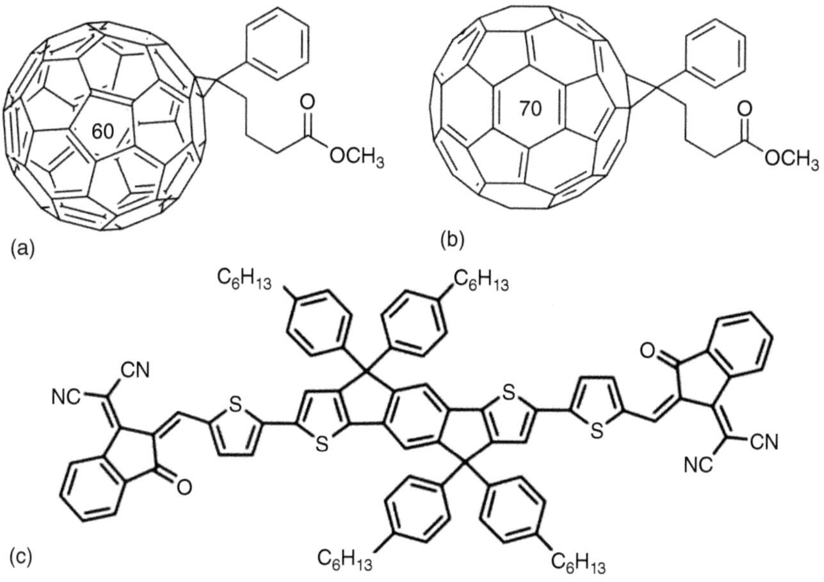

Figure 4.18 Molecular structure of PCBM (a), PC$_{71}$BM (b), and DC-IDT2T (c).

been well studied in organic solar cells [10], the researches on PDs listed in this table are about material treatment, device structure optimization, and others.

In 2012, Liu et al. showed that novel treatment on the material film can furtherly promote the device performance. They demonstrated a treatment method on the P3HT-PCBM polymer film with solvent vapor annealing and post production annealing (Figure 4.19a) [58]. This method was proved benefit to the crystallinity of this film. The device with the treatment showed a double performance in EQE and D^*, which specifically are EQE approximately 80%, D^* of over 1×10^{13} Jones, the dark current magnitude of nA cm^{-2} and over 120 dB of linear dynamic range (Figure 4.19b). In the same year, Binda et al. [59] fabricated a PD by consecutively depositing solution-processed transparent anode, the spray coated active layer, and the evaporated aluminum cathode onto the fiber (Figure 4.19c). The anode layer was deposited both onto the fiber facet and sideways on the fiber cladding, where the electrical contact to the readout electronics was taken. The cathode contact was taken on axis with a soft spring contact. It was presented that the device showed an almost constant external EQE net of about 2%, dark current density ~10 nA cm^{-2}, and a bandwidth around 15 kHz. Even though the performance of the device needs to be further improved, this work demonstrated the applicability of the PoNa film in a curved device, and realized the fabrication of the device by the spray the active material, which is suitable for massive production. In 2013, Arredondo et al. fabricated a P3HT-PCBM based PDi with an inverted device structure [60]. The inverted device has a structure of Cr/Al/Cr/P3HT-PCBM/PEDOT:PSS/Ag, as shown in Figure 4.19d. The object of developing the inverted structure is to enhance device stability by suppressing the interaction in the ITO/PEDOT interface, and avoid the absorption by ITO electrode through a top-absorbing design. Compared

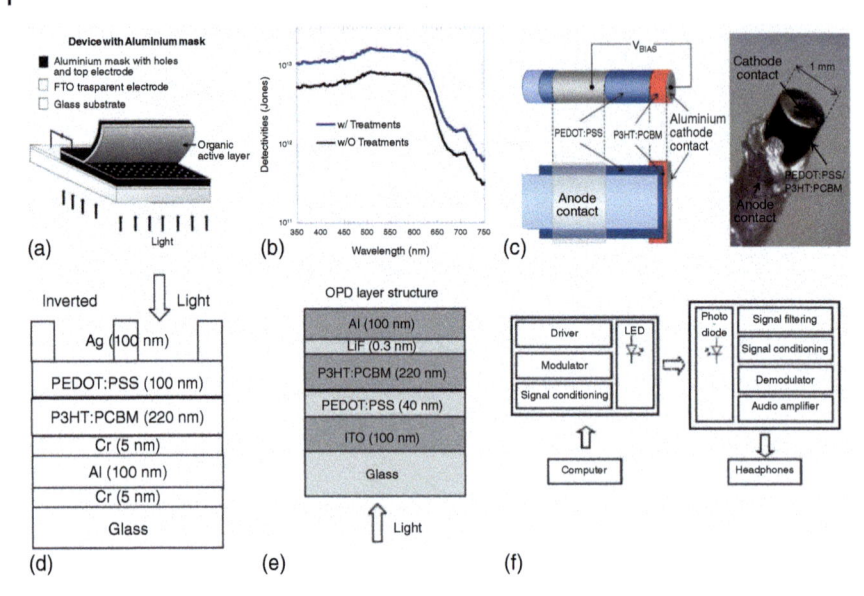

Figure 4.19 (a) Schematic representation of the PDi with the indented mask between the P3HT-PCBM active layer and the FTO. Source: Macedo et al. [57]. (b) D^* at bias of -0.5 V for polymer PDs with and without combined treatments. Source: Liu et al. [58]. (c) Schematic of the PDi fabricated on a fiber (left) and the picture (right) of the integrated fiber-PDi prototype. Source: Binda et al. [59]. Reproduced with permission. Copyright 2013, Wiley-VCH Verlag GmbH & Co. KGaA, Weinheim. (d) P3HT-PCBM based PDi with an inverted structure. Source: Arredondo et al. [60]. (e) P3HT-PCBM based PDi with a standard structure; (f) block diagram of the visible light communication system using the standard device. Source: Arredondo et al. [61]. Licensed under CC-BY-4.0.

with standard device (Figure 4.19e), the inverted PDs show slightly higher EQE and R, which are 39% and 0.13 A W^{-1} at 530 nm, respectively. The bandwidth of the inverted device is 700 kHz at -3 V bias, which is a little lower than the 780 kHz of the standard device. In another report in the same year by Arredondo et al. [61], they successfully applied the standard device, which has a large bandwidth of 790 kHz at -6 V bias, in an optical communication system (Figure 4.19f). These works demonstrate the potential of the PDi based on P3HT-PCBM nanocomposite in a real application.

In 2014, Shuming Li et al. studied the photoconductor based on the blend of P3HT and PCBM. The blend film in this device is regarded as a uniform conductive layer for charge transport [62]. To improve the photo-response, a photonic crystal structure was integrated underneath the conductive layer, as shown in Figure 4.20a,b. This strategy induced a 1.6 times enhancement of the on/off ratio of light/dark current, but a slower response speed. This work provided a feasible way to increase the photocurrent of a photoconductor. In the same year, Lingliang Li et al. furthery investigated the photoconductor based on the blend of a donor polymer and an acceptor SMO. They used the PC$_{71}$BM as the acceptor SMO in their work instead of PCBM, and P3HT as the donor polymer. Their device demonstrated a maximum EQE of 16 700% and R of 51 700 mA W^{-1} for the 1% of PC$_{71}$BM doping weight ratio

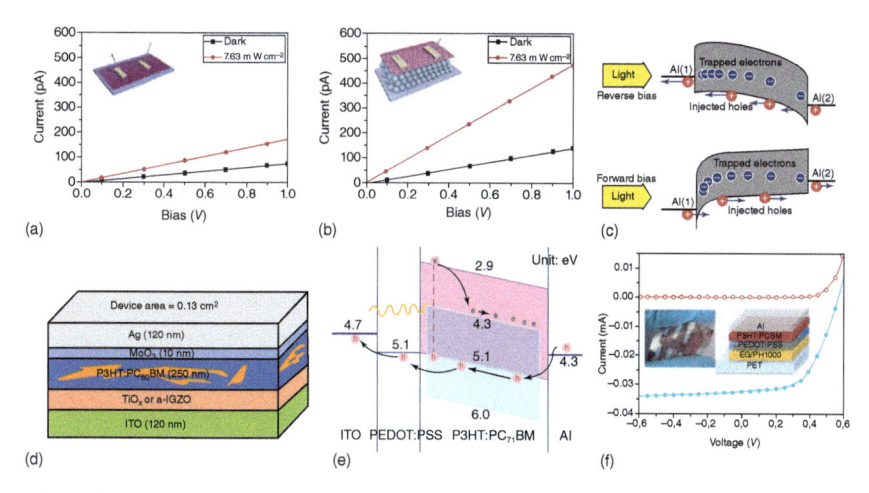

Figure 4.20 *I–V* curve of the control photoconductor (a) built on a bare glass substrate and the device built on the PC film with a stopband (b) at 520 nm. Source: Li et al. [62]. (c) The spatial band diagrams of the PDi under reverse and forward bias, respectively. Source: Li et al. [63]. Licensed under CC-BY-4.0. (d) Schematic of the PDi. Source: Arora et al. [64]. (e) The energy level alignment in the device. Source: Wang et al. [65]. (f) *I–V* characteristics of the PDs fabricated on regular PET (inset shows the fabricated device and schematic device structures of the PDi). Source: Liu et al. [66]. Reproduced with permission. Copyright 2014, Wiley-VCH Verlag GmbH & Co. KGaA, Weinheim.

[63]. This work reveal that an internal gain can be obtained in a photoconductor, as shown in Figure 4.20c, which can induce a very large EQE as well as the *R*. In the next year, Wenbin Wang, Lingliang Li, and their coworkers reported another PD based on P3HT-PC$_{71}$BM nanocomposite. The highest EQE values at 390 and 610 nm of 110 700% and 115 800% under −19 V bias along with the corresponding *R* and *D** values arrived to 568.6 A W^{-1} and 2.17 × 10^{14} Jones under 610 nm light illumination and −19 V bias were reported using this device [65]. They describe the internal gain as "photomultiplication," which is actually due to the long time trapping of electron in the isolated PC$_{71}$BM (Figure 4.20e). They have also fabricated similar photoconductors using the blend of P3HT and PCBM in 2015 [67], and blend of P3HT and fullerene-free SMO, DC-IDT2T in 2016 [68]. These devices all demonstrated very large EQE. Especially in 2017, Wenbin Wang, Lingliang Li, and their coworkers again reported the photomultiplication effect in a PD based on P3HT and PC$_{71}$BM. When their weight ratio was 100 : 1, the device demonstrated a narrowband response with full width half maximum (FWHM) less than 30 nm. This small FWHM should be attributed to the sharp absorption edge of active layer with small amount of PC$_{71}$BM. Besides, the device showed a higher EQE, *R*, and *D** of 53 500%, 278 A W^{-1}, 1.3 × 10^{11} Jones, respectively [69]. In the same year, another similar device with PCBM instead of PC$_{71}$BM was reported, demonstrating a narrowband photo-response with FWHM of 30 nm, and a maximum EQE of 8180% for light of 650 nm at 60 V [70].

Also in 2015, Arora et al. [64] proposed another way to improve the performance based on the P3HT-PCBM blend. They fabricated a PDi with a layer of indium–gallium–zinc-oxide (α-IGZO) or TiO$_x$ added as an electron transport layer, as shown in Figure 4.20d. It was found that the EQE at the wavelength of 530 nm for the P3HT-PCBM device without electron transport layer was 56%, which was enhanced to 64% and 62% with a-IGZO and TiO$_x$, respectively. It was revealed that the thickness of a-IGZO has major impact on device performance. The optimized thickness of 7.5 nm produced efficient P3HT-PCBM-based PDs with a dark current of 3.6×10^{-8} A cm^{-2} and D^* of 2.8×10^{12} Jones at 550 nm. Furthermore, stability and reproducibility were also enhanced with a-IGZO, facilitating fabrication of large area P3HT-PCBM PDs. On the other hand, furthermore, graphene was introduced into the P3HT-PCBM based PDi for flexible device. In 2015, Zhaoyang Liu et al. reported a PDi with configuration of device configuration of exfoliated graphene (EG)/PH1000/PEDOT:PSS/P3HT-PCBM/Al (Figure 4.20f). This device showed the highest R, D^* and short circuit current density of 0.16 A W^{-1}, 1.33×10^{12} Jones and 0.61 mA cm^{-2}, respectively [66]. The device was stable during folding and unfolding for over 100 cycles. The pronounced stretching and bending stability of P3HT-PCBM hybrid films enable a corresponding flexible PD.

Photomultiplication is very useful in PD for high EQE and R. There is a consistent research in scientific community on this topic. In 2017, Yue Wang et al. reported on a PDi based on the spin coated P3HT-PCBM blend film with polyethylenimine ethoxylated (PEIE) modified ITO electrode (Figure 4.21a) [71]. The device demonstrated a maximum EQE of 3250% under 370 nm (3.07 μW cm^{-2}) at −1 V, a SNR 2.48×10^4, a bandwidth of 11.4 MHz, a rise time of 18 μs and a fall time of 20 μs. The R and D^* of the device is 5.61 A W^{-1} and 1.42×10^{12} Jones, respectively [71]. With an EQE beyond 100%, this device demonstrated an apparent photomultiplication effect. The transparent PEIE applied here can efficiently blocks the unnecessary electronic charge injection between the active film and the electrode, which dramatically decrease the dark current. Under illumination, the photoexcited charges accumulated in the PEIE modified ITO region finally can tunnel through the barrier with the help of the applied reverse bias, leading to a large photocurrent. Later in 2018, Yue Wang et al. demonstrated a similar device, which has an active layer of P3HT-PCBM blend and a PEIE modified ITO electrode (Figure 4.21b). In visible light range, this device showed an EQE of 1000–12 000%, a R from 4 to 54 A W^{-1}, a D^* up to 10^{12} Jones. The optimized device had a fast rise time of 24 μs. It is worthy of noting that a photomultiplication PD normally has a slow response speed. Yue Wang et al. reported a relatively fast response speed with rise time of tens of μs [72].

In 2019, Yi-Lin Wu et al. applied the photomultiplication PD in an image sensor array [73]. The PD they used has a structure as shown in Figure 4.21c. The blend of P3HT-PCBM with weight ratio of 100 : 1 was used as the active layer. Electrode was optimized regarding carrier injection, and stacked rectifying layers were used. Finally, the device demonstrated a high $R > 40$ A W^{-1} and a D^* of 1.48×10^{14}. The rise and fall time of the device were 5 and 1.14 seconds, respectively, which are long due to the photomultiplication effect. The organic image sensor based on this device exhibited a high pixel photo-response and a weak-light imaging capability even at 1 μW cm^{-2}. Based on the aforementioned discussed device architecture with

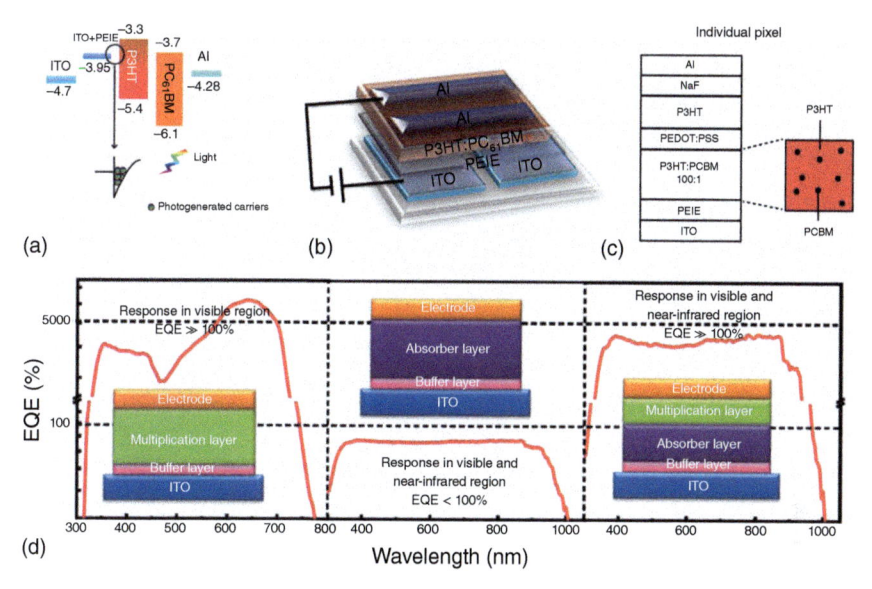

4.4 Photodetectors Based on Polymer Nanocomposites

Figure 4.21 (a) The energy level diagram of the PD. Source: Wang et al. [71]. © 2017, OSA Publishing (b) Schematic of the organic photomultiplication PD. Source: Wang et al. [72]. (c) Schematic of the organic image sensor. Source: Wu et al. [73]. (d) The EQE spectra and device structure schematics of the single-layered photomultiplication organic PD (left), organic PDi (center), and double layer photomultiplication-PDi organic PD. Source: Zhao et al. [74].

a photomultiplication effect, in 2019, Zijin Zhao et al. added an extra light absorber layer to the device, and thus largely enhanced the device performance, as shown in Figure 4.21d [74]. Specifically, the blend of P3HT-PC$_{71}$BM was used as the multiplication layer, and the blend of PM6-Y6 was used as the absorber layer. The response range of the PD is primarily determined by materials in the absorber layer, and the EQE of the PD is mainly controlled by the multiplication layer. The optimal PD exhibited a broad response covering 350–950 nm, a large EQE value of ~1200%, and a D^* of ~6.8×10^{-12} cm Hz$^{1/2}$ W^{-1} under a 10 V bias.

To improve the device performance and facility for fabrication of PD based on polymer–SMO nanocomposite, research on device structure design, film treatment, and fabrication method has been continuously proceed to. In 2017, Tiening Wang et al. reported a PD using blend of P3HT-PCBM as the active layer. Additionally, a poly[(9,9-bis(30-(N,N-dimethylamino)propyl)-2,7-fluorene)-alt-2,7-(9,9-dioctylfluorene)] (PFN) layer was inserted between the anode and active layer as a dark current inhibition layer (Figure 4.22a) [75]. As a result, the dark current density of the device was effectively reduced from 0.07 mA cm^{-2} to 1.92×10^{-5} mA cm^{-2} under a −0.5 V bias. The resulting device also showed a SNR of 1.93×10^5, a bandwidth of 10 MHz, and a D^* of 9.10×10^{12} Jones at a low reverse bias of −0.5 V and light of 550 nm. In 2018, Sung Hyun Park et al. used the 3D printing technique to produce organic PDs (Figure 4.22b) [76]. The printed P3HT-PCBM-based PD achieved an EQE of 25.3%, a D^* of 8×10^{11} Jones, and a LDR of 100 dB, which is comparable with that of micro-fabricated counterparts and yet created solely via a one-pot custom

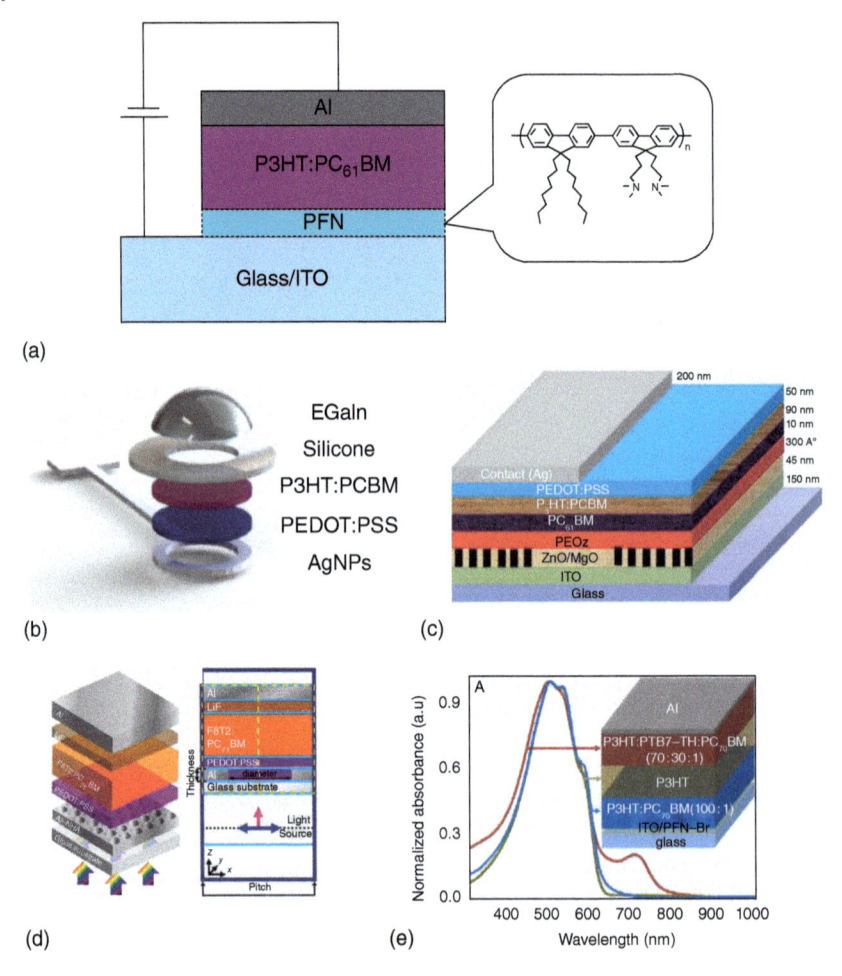

Figure 4.22 (a) Device structure of the PD and the molecular structure of PFN. Source: Wang et al. [75]. Licensed under CC-BY-3.0. (b) 3D-printed bulk-heterojunction PDs. Source: Park et al. [76]. (c) Schematic of the PD with the optimized layers thickness specified. Source: Maity et al. [77]. (d) Schematic of the PD with plasmonic aluminum nanohole arrays integrated. Source: Esopi et al. [78]. (e) Normalized absorption spectra measured for a binary blend P3HT-PC70BM (100 : 1) layer, a pristine P3HT layer, and a ternary blend P3HT-(PTB7-Th)-PC$_{70}$BM (70 : 30 : 1) layer. Inset: Schematic diagram illustrating the cross-sectional view of the PD. Source: Lan et al. [79]. Licensed under CC-BY-4.0.

built 3D-printing tool housed under ambient conditions. In the same year, Chia-Te Yen et al. reported an organic photodetector using a spray-coated P3HT-PCBM blend layer [80]. Even though the performance of the device was not specified, the facile way for film deposition has an advantage for massive production. In 2019, Santanu Maity and Tiju Thomas reported an improved organic photodetector with glass/ITO/ZnO/ZnMgO/PEOz/PCBM/P3HT:PCBM/CH$_3$-SAM/PEDOT:PSS/Ag structure (Figure 4.22c) [77]. The novelty is that a layer of ZnMgO nanorods with mostly textured (with (002) orientation) and hexagonal morphology are applied in the device using a post-thin-film-deposition chemo-thermal process. This process helped in passivation of trap states in ZnO, and thus improved charge collection. With ZnO layer, the device obtained a promising broadband UV photo-response. The optimized device showed a D^* of 4.8×10^{14} Hz$^{1/2}$ W^{-1} and a R of 2.63×10^2 A W^{-1}. Plasmonic effect, which induced the near-field enhancement, can be used to improve the performance of PD. In 2019, Monica R. Esopi and Qiuming Yu reported an organic PD with plasmonic Al nanohole arrays (Al-NHAs) integrated as transparent electrodes (Figure 4.22d) [78]. The Al-NHAs can produce strong UV absorbance in the active layer and enhanced internal electric field intensity. The device with Al-NHAs showed two narrow response peaks under reverse bias and one broad response peak under forward bias. The bias-dependent photo-response switching achieved by the Al-NHA-based devices is attributed to the plasmonic-enhanced internal electric field, which acts as a small forward bias and increases the driving force for hole diffusion. Plasmonic Al-NHA electrodes provide an approach for making ITO-free, flexible organic UV PDs with improved photo-response, and bias-dependent response tunability.

Recently, Zhaojue Lan et al. reported another kind of organic photodetector, which have two different light absorber layers, with one for NIR and the other for visible (Figure 4.22e) [79]. The blend of P3HT-PC$_{71}$BM was used as the visible absorber layer. Another electron donor material, PTB7-Th, was added to this blend forming a NIR absorber layer by extending absorption edge from 650 to 800 nm. These two kinds of layers were connected in series through a P3HT layer. These three layers played the role of active layer in the presented PD. In the presence of NIR light, photocurrent was produced in the NIR absorber layer due to the trap-assisted charge injection at the organic/cathode interface at a reverse bias. In the presence of visible light, photocurrent was produced in the visible absorber layer, enabled by the trap-assisted charge injection at the anode/organic interface at a forward bias. A high R of >10 A W^{-1} was obtained in both short and long wavelengths. The dual-mode OPD exhibits an NIR light response operated at a reverse bias and a visible light response operated at a forward bias, with a high D^* of $\sim 10^{13}$ Jones in both NIR and visible light ranges. This work provides a new way to enhance the performance and extend the response spectrum of an organic photodetector, which is to add polymer component into the polymer–SMO nanocomposite system.

4.4.3 Polymer–Polymer–Small Molecular Organic Nanocomposite

A ternary blend formed by adding the third polymer into the binary blend of polymer–SMO has at least three advantages: (i) The third polymer may have a different absorption spectrum, which can extend the photo-response spectrum of the PD; (ii) By adding new component, the interface which can facilitate the exciton dissociation is larger; (iii) There is a Förster transfer process between the two polymers, which is that the photo-induced exciton may transfer from one polymer to the other, and then get dissociated in a preferred interface [81]. On the other hand, there are also some disadvantages for adding a new polymer into the binary system. (i) The functionality of two polymers would be different. By mixing them

Table 4.6 Summary of some reported PDs based on polymer–polymer–small molecular organic nanocomposite.

Polymer 1	Polymer 2	SMO	EQE (%)	D* (Jones)	λ (nm)	R (A W^{-1})	τ_{rise}/τ_{decay}	Year	References
PTOPT	POMeOPT	C$_{60}$	60	—	350–700	—	—	1997	[82]
P3HT	PTB7-Th	PC$_{71}$BM	38 000	1.91×10^{13}	300–800	229.5	—	2015	[83]
MEH-PPV	PFO-DBT	PC$_{71}$BM	—	—	400–1000	4 m	—	2015	[84]
MEH-PPV	PFO-DBT	PC$_{71}$BM	0.37	—	450–750	3.9 m	800/500 ms	2015	[85]
P3HT	PBDT-TS1	PC$_{71}$BM	830@390 nm 720@625 nm 330@760 nm	10^{12}	300–820	3	—	2017	[86]
P3HT	PTB7-Th	PCBM	100	2.4×10^{10}	300–900	2.5	—	2018	[87]

4.4 Photodetectors Based on Polymer Nanocomposites | 109

together, the inferior one replaces some volume of the superior one; (ii) Mixing different materials may deteriorate the crystallinity of the components. Therefore, the weight ratio of the ternary material system must be optimized carefully to produce a well-performing PD.

Table 4.6 lists some reports on PDis using ternary blend of polymer–polymer–SMO as the active layer. All the devices were prepared via spinning–coating process. Early in 1997, L. S. Roman et al. reported a PDi using ternary blend of poly(3-(4-octylphenyl)-2,2′-bithiophene) (PTOPT), poly(3-(2′-methoxy-5′-octylphenyl)thiophene) (POMeOPT), and C_{60} (Figure 4.23a) [82]. C_{60} was used as the electron acceptor and the other two polymers were donors. It was believed that C_{60} acts as a "compatibilizer" in the parlance of polymer technology, inducing a more efficient blending. POMeOPT showed a better interaction to C_{60}, making it necessary in this system. The device showed a wide photo-response spectrum from 350 to 700. The EQE reach up to 60% under a reverse bias of 2 V. In 2015,

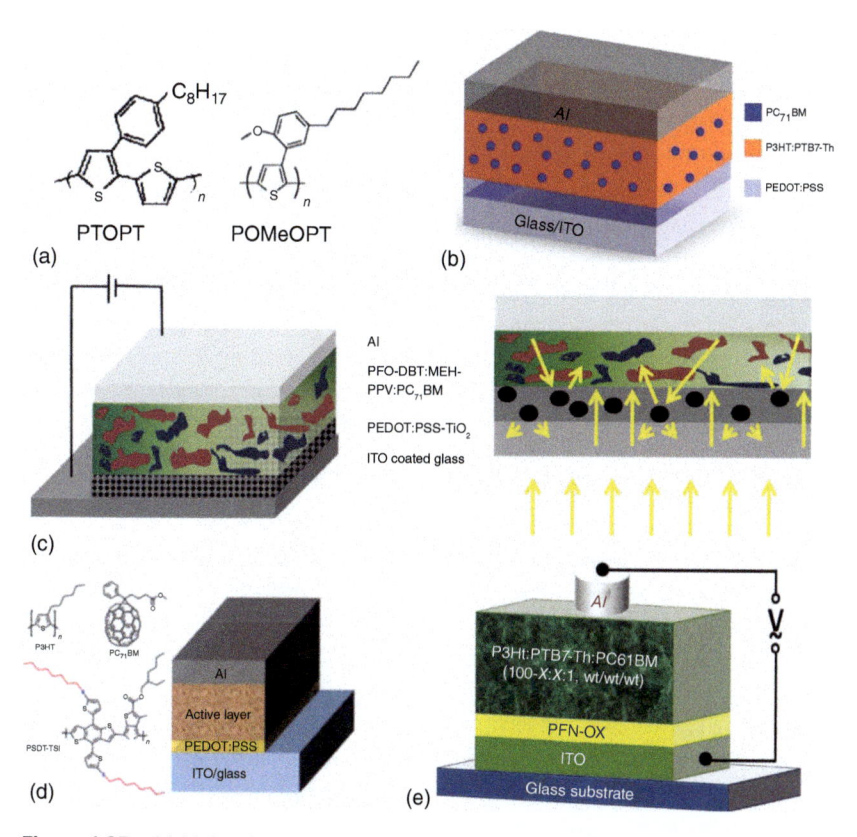

Figure 4.23 (a) Molecular structure of PTOPT and POMeOPT. (b) Device structure of the PD based on ternary blend of P3HT-(PTB7-Th)-PC$_{71}$BM. Source: Wang et al. [83]. (c) Device structure of the PD based on ternary blend of (PFO-DBT)-(MEHPPV)-PC71BM. Source: Zafar et al. [84]. (d) Device structure of the PD based on ternary blend of P3HT-(PBDT-TS1)-PC$_{71}$BM. Source: Gao et al. [86]. © 2017, IOP Publishing (e) Device structure of the PD based on ternary blend of P3HT-(PTB7-Th)-PCBM. Source: Miao et al. [87].

Wenbin Wang et al. presented a PDi with structure shown in Figure 4.23b [83]. Besides the traditional P3HT-PC$_{71}$BM system, another widely used donor polymer PTB7-Th was added to the active layer. Due to the narrow bandgap feature of PTB7-Th, the device showed a photo-response in NIR region up to 800 nm. The aforementioned photomultiplication effect was observed in device. When the weight ratio of P3HT:PTB7-Th:PC$_{71}$BM is 50 : 50 : 1, the device showed the highest EQE of about 38 000% in the spectral range from 625 to 750 nm under a −25 V bias. This work demonstrated the spectral response enhancement effect of the ternary blend system. In the same year, Qayyum Zafar et al. successively reported two kinds of PDis using ternary blend of (PFO-DBT)-(MEH-PPV)-PC$_{71}$BM. The structure of one of the devices is shown in Figure 4.23c [84]. Although they focused on the study on electrode, the device demonstrated a wide response spectrum from 400 to 1000 nm. In the other report [85], a layer of LiF was used as the hole-blocking layer. The device showed a similar performance with R of about 3.9 mA W^{-1}. In 2016, Mile Gao et al. reported a PDi based on blend of P3HT-(PBDT-TS1)-PC$_{71}$BM, as shown in Figure 4.23d [86]. The device showed EQE of 830%, 720%, and 330% under 390, 625, and 760 nm light illumination, respectively, under −10 V bias. What's more important, the response spectrum extended to 820 from 720 nm by introducing PBDT-TS1 into the active layer. In 2018, Jianli Miao et al. demonstrated a PD based on ternary blend of P3HT-(PTB7-Th)-PCBM (Figure 4.23e) [87]. The device showed tunable spectral response dependent on the bias polarity, which is a broadband spectral response from 350 to 800 nm under forward bias and a narrowband spectral response from 750 to 850 nm with a FWHM of about 40 nm under reverse bias. Similar to the predecessor [83], the response spectrum of the device increased from 650 to 800 nm by introducing PTB7-Th into the active layer. The photomultiplication effect was found in this device, leading to an EQE above 400% for broadband response, and up to 200% for narrow band response.

4.4.4 Polymer–Small Molecular Organic–Small Molecular Organic Nanocomposite

Due to the success of polymer–polymer–SMO ternary blend system, which is actually the donor–donor–acceptor system, the donor–acceptor–acceptor system, which is normally realized in polymer–SMO–SMO ternary blend, come to be the research interest for PD. For instance, in 2010, Fang-Chung Chen et al. reported a broadband PD by introducing a dye, Ir-125 (Figure 4.24a), into the blend of P3HT and PCBM [91]. The response spectrum of the device got broadened from 700 to 1100 nm, which is of particular significance due to the realization of NIR detection. Besides, the dye successful enhanced, if not created, the photomultiplication effect in the device, inducing an EQE > 7000%, and R of 32.4 A W^{-1} at a low bias of −1.5 V. Some reported PDis based on polymer–SMO–SMO system are summarized in Table 4.7 for reference. In 2012, Shao-Tang Chuang, Fang-Chung Chen, and their coworkers added another NIR sensitive dye into the P3HT-PCBM-(Ir-125) system (Figure 4.24b,c) [88]. By using two organic NIR dyes in the active layer, a high EQE (>5500%) and a high R (23.0 A W^{-1}) under a low reverse bias was obtained.

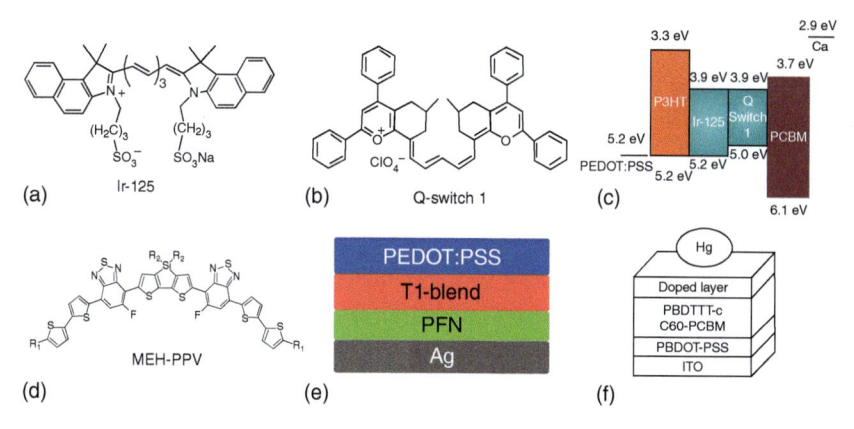

Figure 4.24 Molecular structure of Ir-125 (a) and Q-switch 1 (b); (c) energy level diagram for P3HT, Ir-125, Q-switch 1, and PCBM. Source: Chuang et al. [88]. (d) Molecular structure of T1; (e) device structure of the PD based on ternary blend of P3HT-T1-PC$_{71}$BM. Source: Pace et al. [89]. (f) Device structure of the PD based on ternary blend of PBDTT-C$_{60}$-PCBM. Source: Herrbach et al. [90].

Table 4.7 Summary of some reported photodetectors based on polymer–small molecular organic–small molecular organic nanocomposite.

Polymer	SMO$_1$	SMO$_2$	EQE (%)	D* (Jones)	λ (nm)	R (A W^{-1})	Year	References
P3HT	PCBM	Ir-125	7200	—	300–1050	32.4	2010	[91]
P3HT	PCBM	Ir-125/ Q-switch 1	8000	—	700–1100	23	2012	[88]
P3HT	PC$_{71}$BM	T1	35	—	300–800	—	2014	[89]
PBDTTT-c	PCBM	C$_{60}$	72	1.93 × 10^{13}	400–900	0.26	2016	[90]

Meanwhile, substantially improved EQEs higher than 200% at the NIR region have been demonstrated. These works showed the advantage of this strategy for PDs with broad spectral responses and high quantum efficiencies simultaneously. In 2014, Giuseppina Pace et al. reported their work on integrating barrow bandgap conjugated molecule T1 (Figure 4.24d) into the P3HT-PC$_{71}$BM blend to fabricate a PD (Figure 4.24e) [89]. The response spectrum extended to 800 nm due to the absorption of T1. They finally presented a semitransparent, all-printed detectors with broadband wavelength response obtained when light impinges both from the top and from the bottom side. In 2016, J. Herrbach et al. reported a PDi using the blend of donor polymer PBDTTT-c and acceptor SMOs PCBM and C$_{60}$ (Figure 4.24f) [90]. Their interest in this work is to use a doped PBDTT-c film to replace the traditional PEDOT:PSS layer. The device demonstrated a wide response spectrum from 400 to 900 nm, an EQE of 73%, and a D* up to 1.93 × 10^{13} Jones. This excellent performance benefited largely from the ternary blend active layer.

4.4.5 Polymer–Inorganic Nanocrystals Nanocomposite

Blending polymer and inorganic NCs can combine the desirable electronic and optoelectronic properties within a single composite, such as tunable bandgap and strong light absorption of the NCs component, and the easy processing of the organic component. By applied NCs with proper bandgap, the response spectrum of the PD based on blend can be easily extended to UV or IR region. The nanocomposite of conductive polymer and inorganic NCs is prepared by dissolve two components in a solvent or two compatible solvents. A uniform composite film can be obtained by spin-coating the solution on arbitrary substrate. To avoid aggregation of a single component in the film, the surface property of the NCs, amount ratio, and process condition must be taken care of seriously. The progress on the blend materials obviously depends on the development on the synthesis of polymer and NCs. Therefore, the mostly studied benzene-based polymer MEH-PPV and thiophene-based polymer P3HT are also two main polymers used for this blend. This section will discuss the works of PDs fabricated using (MEH-PPV)–inorganic NCs and P3HT–inorganic NCs blend as the active layer.

4.4.5.1 MEH-PPV–Inorganic Nanocrystals Nanocomposite

Table 4.8 lists some reported on PDs using the nanocomposite of MEH-PPV and inorganic NCs as the active layer. As early as 1996, Greenham et al. [92] studied the charge separation and transport in spin-coated composite materials formed by blending conjugated polymer MEH-PPV with the cadmium selenide (CdSe) or cadmium sulfide (CdS) NCs. The nanocrystals may either be in direct contact with the polymer, or, alternatively, the surface of each nanocrystal may be coated with a surfactant molecule, trioctylphosphine oxide (TOPO), which forms a barrier of 11 Å thickness between the nanocrystal core and the polymer (Figure 4.25a). Significant quenching of the luminescence was found when the nanocrystal surface is not coated with TOPO, indicating that charge transfer occurs at the polymer/nanocrystal interface. For a pure polymer device, the EQE was only 0.014%, consistent with results reported previously for this kind of device. Adding

Table 4.8 Summary of some reported photodetectors based on MEH-PPV–inorganic nanocrystals nanocomposite.

NCs	Device	EQE (%)	λ (nm)	R (A W^{-1})	τ_{rise}/τ_{decay}	Year	References
CdSe/CdS	PC	71	420–700	—	—	1996	[92]
CdSe	PC	71	420–650	—	—	1997	[93]
TiO$_2$	PD		310–630	50 m	100 ns/10 μs	1999	[94]
PbS	PC	10^{-5}	1100–1400	—	—	2004	[95]
PbSe	PC	150	400–800	—	—	2005	[96]
PbS	PC	0.38	650–1200	3.1 × 10^{-3}	—	2005	[97]
PbSe	PDi	10 000	400–600	—	—	2007	[98]

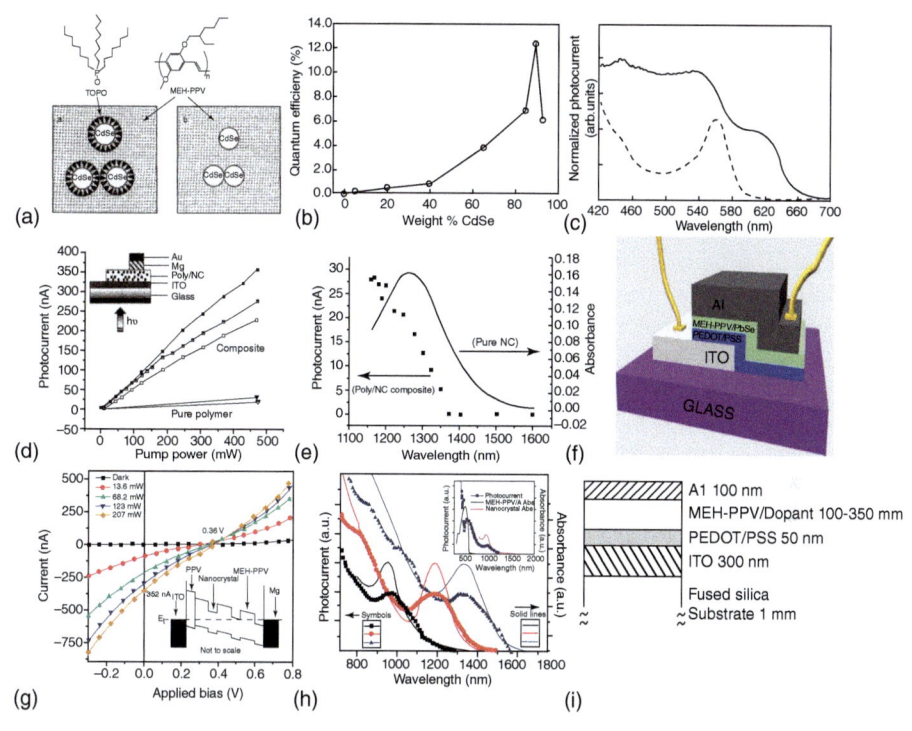

Figure 4.25 (a) Schematic diagram of MEH-PPV/nanocrystal composites, showing the chemical structure of MEH-PPV and trioctylphosphineoxide (TOPO). (left) CdSe nanocrystals with surfaces coated by (TOPO); (right) CdSe nanocrystals with naked surfaces; (b) Short-circuit quantum efficiency for devices containing 5-nm-diameter CdSe nanocrystals, as a function of CdSe concentration. Excitation was at 514 nm; (c) Spectral response of the short circuit current for a device containing 90 wt% 5-nm-diameter CdSe nanocrystals (solid line), and for a pure MEH-PPV device (dashed line). Source: Greenham et al. [92]. (d) Photocurrent as a function of pump power at 5 V sample bias for polymer/NC blends (solid lines) and pure polymer (dashed lines). Three separate polymer/NC blends with the same NC concentration are shown to indicate the repeatability of the result. The pump wavelength is 975 nm allowing selective excitation of the NCs. The inset shows the device. (e) Photocurrent spectral response of the polymer/NC blend (dots) compared to the absorption of the NCs in solution (line). Source: McDonald et al. [95]. (f) The schematic layout of the ITO/PEDOT:PSS/MEHPPV: PbSe/Al photodetector on a glass substrate. Source: Qi et al. [96]. (g) Dark current and photocurrent vs. of the photodetector based on (MEH-PPV)-PbS NCs nanocomposite. Inset: Proposed simplified band diagram depicting the relative energy alignments; (h) photocurrent spectral responses and absorption spectra. Source: McDonald et al. [97]. (i) Schematic device structure of the photodetector based on (MEH-PPV)-PbSe NCs nanocomposite. Source: Campbell et al. [98]. Reproduced with permission. Copyright 2007, AIP publishing.

5 wt% of nanocrystal gave a factor of 6 improvements in efficiency. Increasing the nanocrystal concentration up to 90 wt% further improved the quantum efficiency to a value of 12%. At 23 and 13 V, this device gave quantum efficiencies of 52% and 71%, respectively. The photo-response spectrum of the composite device showed the response in the region of 600–660 nm where there is no absorption in the polymer (Figure 4.25b,c). This response corresponds to light that was absorbed in

the nanocrystals, with subsequent hole transfer onto the polymer. At wavelengths below 600 nm, the response is due to a combination of absorption in the polymer and in the nanocrystals. Due to the poor electron transport in the polymer, only those electrons generated close to the electrode would contribute to the photocurrent by reaching the electrode without recombination. The peak photocurrent therefore occurs where the optical density is low and light is absorbed throughout the thickness of the device, rather than at high optical densities where almost all the light is absorbed close to the electrode. Furtherly in 1997, Greenham et al. did another research on the blend of MEH-PPV and CdSe NCs. It demonstrated that the composite material of conjugated polymer and semiconductor nanocrystal is an ideal material to achieve charge separation. And at high nanocrystal concentration, both kinds of carriers can be transmitted through the path way form by each component, thereby greatly improving the photodetection devices efficiency. Nevertheless, it implies the opposite condition, which is the NCs dispersed in the matrix discretely, forming trapping centers of minority carriers instead of pathway. Then there will be only transport of majority carrier in the matrix for photocurrent. If the trapping is long enough, a photoconductive gain may be available, inducing a high EQE. In 1999, Narayan and Singh reported a PDi with structure of ITO/MEHPPV-TiO$_2$ NCs/Al [94]. Due to the wide bandgap of the TiO$_2$, the device can sense short wavelength light at 310 nm. The spectral response is sensitive to the magnitude and the polarity of the bias, showing a R of ~50 mA W^{-1} at 29 V. The device in the reverse bias responded to the 50 µs wide pulse with a rise time of <100 ns and a decay time of <10 µs. The persistence photocurrent effect was observed in the device, which could be attributed to the existence of the TiO$_2$ NCs.

With the development of NCs, CQDs of narrow bandgap semiconductors like PbS CQDs were brought out for infrared photodetection. In 2004, McDonald et al. reported [95] that the photo-response at infrared wavelengths of 975–1300 nm in a PD using (MEH-PPV)–PbS CQDs blend as the active layer material (Figure 4.25d). The photocurrent was attributed to absorption in the nanocrystals with subsequent hole transfer to the polymer. Nevertheless, the quantum efficiency calculated from the photocurrent was small, with a value at 10^{-5}, although it was larger than the value of device of pure polymer (Figure 4.25d,e). In the next year, the same team successfully largely improved the device performance by optimizing the device structure (Figure 4.25g) [97]. The quantum efficiency under same bias condition increased up to 3% in the NIR region. A R of 3.1×10^{-3} A W^{-1} was achieved under illumination of 975 nm. These works reveal the potential of narrow bandgap CQDs for solution-processed PDs with sensitivity far beyond 800 nm.

Successively in 2015 and 2017, Difei Qi et al. [96] and I. H. Campbell et al. [98] reported PDs based on the blend of MEH-PPV and PbSe CQDs (Figure 4.25f,i). In their works, instead of spectral response, they focused more on the internal gain, which induced an EQE larger than one. The former attributed the gain to the sum of the carrier multiplication in PbSe nanocrystal quantum dots via multiple exciton generation, and the efficient charge conduction through the host polymer material (photoconductive gain). The latter attributed the gain to only the photoconductive gain, which is the circulation of hole carriers through the device in response to electrons trapped in the polymer layer. Either way, it was proved that a considerable gain can be expected in a polymer–inorganic NCs blend-based PD (Table 4.9).

4.4.5.2 P3HT–Inorganic Nanocrystals Nanocomposite

NCs of oxide semiconductor like TiO_2 and ZnO have a wide bandgap corresponding to wavelength of about 320 nm, making them suitable for UV sensing. In 2009, Feng Yan et al. reported a phototransistor using the composite of P3HT and TiO_2 NCs as the active layer material (Figure 4.26a) [99]. The transfer characteristic of the device showed a parallel shift under a UV light illumination of 370 nm. TiO_2 NCs with different shapes were studied in the device. This work deeply investigated the photo-response mechanism of the phototransistor, and proposed a photo-gating effect instead of the photo-induced carrier injection, to be responsible for the photo-response process. It said that the minority would be trapped in the TiO_2 NCs, which would apply an extra bias to the conductive P3HT matrix, and thus induce a change to the conductive condition (Figure 4.26b). This mechanism would be adopted widely in the following similar work. In 2011, Zhenhua Sun et al. reported another phototransistor based on P3HT and TiO_2 NCs blend [101]. This work furtherly verified the photo-gating mechanism. The device showed well response to UV of 370 nm with $R > 1\,A\,W^{-1}$ (Figure 4.26c). What's more, they showed that adding NCs into P3HT not certainly deteriorate the quality of the polymer matrix. NCs with proper shape could improve the P3HT crystallinity, and thus increase the mobility. The two phototransistors discussed earlier have a structure of thin film transistor. The gate electrode would be used to adjust the photo-response through a field-effect effect. Regardless of it, the device is actually a photoconductor, which may have a photoconductive gain inside for a large EQE and R. In 2012, Fawen Guo et al. reported their PD using the blend of P3HT and ZnO NCs (Figure 4.26d) [103]. The device was realized, transitioning from a PDi with a rectifying Schottky contact in the dark to a photoconductor with an Ohmic contact under illumination. Therefore, the device combined the low dark current of a PDi and the high R of a photoconductor at illumination of 360 nm. At a bias of −9 V, the external EQE and R reach up to 340 600% and $1001\,A\,W^{-1}$, respectively (Figure 4.26e), which overwhelmed pure inorganic device. A very high D^* of 2.5×10^{14} Jones was presented benefited from the low dark current. The transient response result showed a rise time (output signal changing from 10% to 90% of the peak output value) of 25 μs, which was limited by the rising edge of the light pulse from the optical chopper. The decay of the photocurrent after switching off the ultraviolet pulse has a fast component of 142 μs and a slow component of 558 μs, which indicates the existence of two channels for the recombination of holes (Figure 4.26f).

CQDs of lead chalcogenide narrow bandgap semiconductors like PbS and PbSe NCs are of intense interest for their application in IR region. Their combination with P3HT produced PDs with better performance than MEH-PPV. In 2011, Dorota Jarzab et al. studied the charge transfer between PbS NCs and P3HT using time-resolved photoluminescence method [100]. They confirmed the hole transfer from PbS to P3HT when mixing them together, as shown in Figure 4.27a. This work

Table 4.9 Summary of some reported photodetectors based on P3HT–inorganic nanocrystals nanocomposite.

NCs	Device	Fabrication	EQE (%)	D^* (Jones)	λ (nm)	R (AW^{-1})	Speed	Year	References
TiO$_2$	PC	Spin coating	—	—	—	—	—	2009	[99]
PbS	PC	Drop casting	—	—	550–900	—	—	2011	[100]
TiO$_2$	PC	Spin coating	—	—	350–900	>1	—	2011	[101]
PbS	PC	Spin coating	—	—	895	2×10^4	—	2012	[12]
CdSe	PC	Spin coating	—	—	300–700	—	$\tau_{rise} < 0.1$ s $\tau_{decay} < 0.1$ s	2012	[102]
ZnO	PC	Hydrolysis	340 600	2.5×10^{14}	300–475	1001	BW = 9.4 kHz	2012	[103]
PbS	PC	Direct writing	—	2.1×10^{12}	400–1000	—	$\tau_{rise} \leq 0.16$ s $\tau_{decay} \leq 0.12$ s	2015	[104]
CdTe	PC	Spin coating	4300	4.51×10^{12}	300–800	13.6	$\tau_{rise} = 2$ μs	2015	[105]
PbSe	PC	Spin coating	—	5.02×10^{12}	300–800	500	—	2015	[106]
PbS$_x$Se$_{1-x}$	PC	Spin coating	—	1.77×10^9	800–1600	15.4	—	2016	[107]
Cu$_2$Te	PD	Spin coating	—	—	300–1100	70	—	2016	[108]
PbS	PC	Spin coating	—	5.0×10^{10}	400–1600	—	—	2017	[109]

4.4 Photodetectors Based on Polymer Nanocomposites

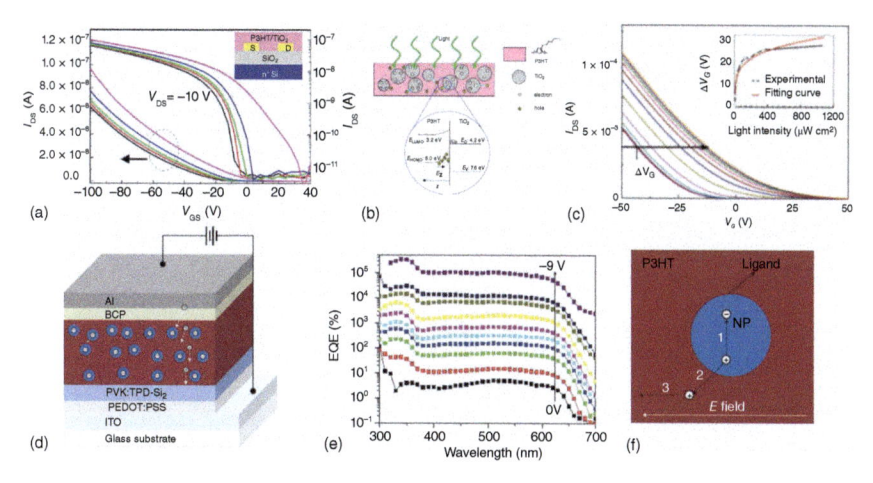

Figure 4.26 (a) Transfer characteristics of a PD based on a composite of P3HT and TiO$_2$ nanoparticles. The inset is the cross-section of the device; (b) schematic diagram showing charge accumulation in a P3HT/TiO$_2$ composite film under light illumination. Source: Yan et al. [99]. (c) Transfer characteristics of a PD based on a TiO$_2$-P3HT composite under 370 nm UV illumination with different intensity. Inset: Shift of gate voltage vs. light intensity *I*. The fitting curve is shown in red. Source: Sun et al. [101]. (d) Schematic layout of the PD based on polymer-ZnO composite; (e) EQEs of the P3HT–ZnO device under different reverse bias; (f) illustration of electron–hole pair generation (1), splitting (2), hole transport and electron trapping process (3) in the nanocomposite. Source: Guo et al. [103]. Reproduced with permission. Copyright 2012, Macmillan Publishers Limited.

can be regarded as the physics fundamental of relevant works. In 2012, Zhenhua Sun et al. reported a phototransistor based on P3HT-PbS QDs nanocomposite (Figure 4.27b) [12]. This work compared the effect of PbS QDs with different surface ligands. It revealed that a short ligand is necessary for efficient charge transfer between PbS and P3HT. The device using PbS QDs with pyridine ligand shows a high R of 2×10^4 A W^{-1} under illumination of 895 nm, which implied an apparent photoconductive gain. In 2015, Jewon Yoo et al. reported a novel device fabrication technique, which is meniscus-guided, direct-writing technique, as shown in Figure 4.27c [104]. The technique produced nanowire (NW) of blend of P3HT and PbS QDs between Au and Al electrodes. The device demonstrated a high D^* of 2.1×10^{12} Jones in the UV–vis range. A stretchable UV–vis–NIR (from 365 to 940 nm) PD based on NW array (3×3) was fabricated and characterized systematically, showing a nearly identical photo-response under extreme (up to 100%) and repeated stretching (up to 100 cycles), indicating their excellent mechanical and photoelectrical stability. In 2015, Haowei Wang et al. demonstrated their phototransistor of P3HT-PbSe QDs blend (Figure 4.27d) [106]. By introducing PbSe into P3HT matrix, the device was able to sense NIR light with wavelength up to 1600 nm. A R and D^* of 500 A W^{-1} and 5.02×10^{12} Jones were achieved, respectively, at $V_{DS} = -40$ V and $V_G = -40$ V with 40 mW cm^{-2} of 980 nm laser illumination. The device had a top gate structure, which makes it stable in air as the dielectric layer could play a role of encapsulation. In the next year, Taojian Song et al. reported a PD with the same device structure as Haowei Wang et al. (Figure 4.27e), but with

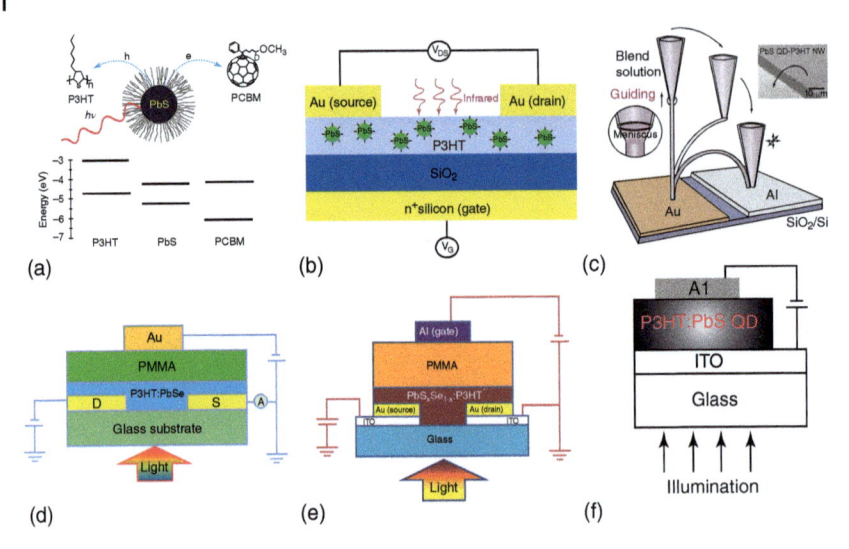

Figure 4.27 (a) Chemical structure of P3HT and PCBM, and cartoon of the PbS-NCs with ligands; lower part: schematic energy-level diagram of P3HT, PCBM, and PbS-NCs. Source: Jarzab et al. [100]. (b) Schematic diagram of P3HT-PbS QDs hybrid phototransistor. Source: Sun et al. [12]. (c) Scheme of direct writing of a PbS QD-P3HT hybrid NW arch by the meniscus-guided approach. Inset: SEM image of a single hybrid NW arch on an Au–Al electrode. Source: Yoo et al. [104]. Reproduced with permission. Copyright 2015, Wiley-VCH Verlag GmbH & Co. KGaA, Weinheim. (d) Schematic of PD based on P3HT-PbSe QDs blend. Source: Wang et al. [106]. (e) The cross-section diagram of the infrared PD based on P3HT-PbS$_x$Se$_{1-x}$ QDs blend. Source: Song et al. [107]. (f) Schematic of the PD based on P3HT-PbS QDs blend. Source: Xu et al. [109] 2017, Royal Society of Chemistry. Licensed under CCBY 3.0.

PbS$_x$Se$_{1-x}$ QDs instead of PbSe QDs in the active layer [107]. The optimized device demonstrated a R and D^* of 55.98 mA W^{-1} and 1.02×10^{10} Jones, respectively, at low $V_{DS} = -10$ V and $V_G = 3$ V under 980 nm laser with an illumination intensity of 0.1 mW cm^{-2}. Likely, the device has the ability to sense NIR light. The sensing spectrum would change according to the different x value in PbS$_x$Se$_{1-x}$. A maximum wavelength up to 1600 nm can be detected by this device. In 2017, Wenzhan Xu et al. presented another photoconductor based on blend of P3HT and PbS QDs (Figure 4.27f), which has a wide response spectrum from 300 to 1600 nm [109]. The device exhibited a D^* greater than 10^{10} Jones and an EQE of over 80% in the visible region and 10% in the infrared region.

There are other chalcogenide semiconductor QDs widely applied in blend-based PDs. In 2012, Xianfu Wang et al. reported a photodetector based on the hybrid of P3HT and CdSe nanowire heterojunction (Figure 4.28a), which exhibited excellent mechanical flexibility and stability [102]. Figure 4.28b shows the photocurrent of the device increased and decreased as a response to the on/off states of the light with a power of 140 mW cm^{-2} at a bias of 3.0 V. The switching in the two states was very fast and reversible, allowing the device to act as a high-quality photosensitive switch (Figure 4.28a). Flexible photodetectors with excellent were fabricated on both poly(ethylene terephthalate) (PET) substrates and printing

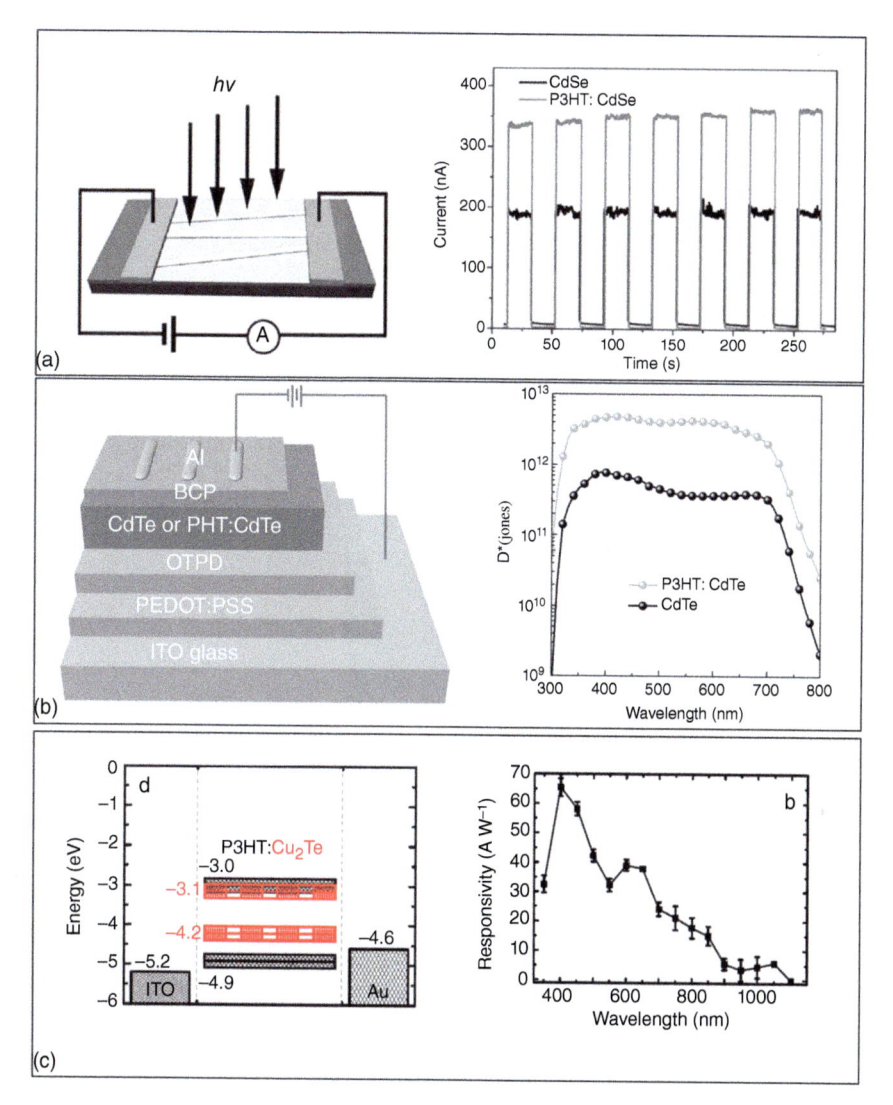

Figure 4.28 (a) Schematic illustration (left) and reproducible on/off switching (right) of the P3HT-CdSe NW hybrid. Source: Wang et al. [102]. (b) The device structure of CdTe or P3HT-CdTe nanocomposite PDs (left) and the spectral specific detectivity of CdTe and P3HT:CdTe devices under −5 V bias. Source: Wei et al. [105]. (c) Energy diagram of the PD based on P3HT–Cu$_2$Te blend before contact and without bias applied (left). Photocurrent spectrum of the device (right). Source: Arciniegas et al. [108].

paper. The flexible devices were successfully operated under bending up to almost 180° and showed an extremely high on/off switching ratio (larger than 500), a fast time response (about 10 ms), and excellent wavelength-dependence, which are very desirable properties for its practical application in high-frequency or high-speed flexible electronic devices. In 2015, Haotong Wei et al. reported a photodetector using P3HT and CdTe QDs with device structure shown in Figure 4.28b [105].

By comparison, it was revealed that the P3HT matrix surrounding CdTe QDs can passivate the deep traps in the CdTe surface. The shallow traps remained could trigger a gain of 50, and a R of 10.8 A W^{-1} at 350 nm. The hybrid device showed a universal performance improvement across the whole response spectrum from 300 to 800 nm (Figure 4.28b). A D^* of 4.51×10^{12} Jones and response time of 2 μs were presented. NCs with novel shapes have been applied in PD also and demonstrated interesting effect. In 2016, Milena P. Arciniegas et al. reported a PD with structure of Au/P3HT-Cu$_2$Te nanodisks/ITO (Figure 4.28c) [108]. Unlike the device discussed earlier, this device has a bulk heterojunction configuration like the P3HT-PCBM system, as there is a Cu$_2$Te network formed in the active layer. The hybrid device showed a performance enhancement compared with the pure P3HT device, which can be attributed to the interface between organic and inorganic components facilitating the dissociation of exciton. The device demonstrated a broadband response from 300 to 1100 nm, with a highest R of about 70 A W^{-1}. This work shows that the bulk heterojunction concept is also applicable in organic–inorganic hybrid system.

4.4.6 Polymer–Small Molecular Organic–Inorganic Nanocrystals Nanocomposite

Since that a blend is able to inherit the virtues of all the components, it is easy to rise the idea to combine polymer, SMO, and inorganic NCs for a hybrid material, which is supposed to manifest the most excellent properties. This idea has been practiced, with outstanding PDs produced. Most of them are based on the classical polymer–SMO system of P3HT and PCBM. The follows are the discussion on them.

Table 4.10 lists seven representative works of photodetector based on P3HT-PCBM–inorganic NCs blend. In 2008, Hsiang-Yu Chen et al. added CdTe QDs into the P3HT-PCBM hybrid, and then applied this ternary hybrid film to a PD, with device structure shown in Figure 4.29a [11]. The CdTe has a benign surface ligand, PMDTC, making the high ratio of CdTe in the hybrid possible. The CdTe QDs would help the dissociation of exciton and then trapped the electron for a long time. These trapped electrons lowered the hole injection barrier from the electrode to P3HT. In other words, a photoconductive gain would be formed with the hole circulated in the circuit. Therefore, the device demonstrated huge EQE of as about 8000% at 350 nm and about 600% at 700 nm were reached at voltages of −4.5 V. This work is one of the earliest reports regarding the photoconductive gain induced by NCs. In the next year, Tobias Rauch et al. applied PbS QDs to sensitize the P3HT-PCBM layer of a PDi, to enable the NIR photodetection function to this device [110]. The PD finally demonstrated a good sensitivity to light from 1000 to 1900 nm, with an EQE of 51%, D^* of 2.3×10^9 Jones, and a bandwidth of 2.5 kHz. What's more, based on the PDi, they successfully fabricated a full functional practical NIR imager (Figure 4.29b). The charge transfer process in a P3HT-PCBM-PbS QDs blend was clarified in 2011 by Dorota Jarzab et al. through a time-resolved photoluminescence technique (Figure 4.27a). It was proved that the inorganic NCs play an important

Table 4.10 Summary of some reported photodetectors based on P3HT-PCBM–inorganic nanocrystals nanocomposite.

Inorganic nanocrystals	Device	Fabrication	EQE (%)	D^* (Jones)	λ (nm)	R (A W^{-1})	Speed	Year	References
CdSe	PC	Spin coating	8000	—	350–750	—	—	2008	[11]
CdSe	PDi	Spin coating	51	2.3×10^9	1000–1900	0.5	BW = 2.5 kHz	2009	[110]
PbS	PC	Drop casting	—	—	550–900	—	—	2011	[100]
PbS/ZnO	PC	Spin coating	1624	1.01×10^{12}	300–1050	5.60	$\tau_{rise} = 160$ μs $\tau_{decay} = 80$ μs	2014	[111]
Au	PC	Sputtering and spin coating	1500	—	400–800	—	—	2014	[112]
CdTe	PC	Spin coating	200	7.3×10^{11}	400–900	—	BW = 900 Hz	2016	[113]
PbS	PD	Spray coating	—	—	X-ray	—	$\tau_{rise} < 100$ ms $\tau_{decay} < 100$ ms	2016	[114]

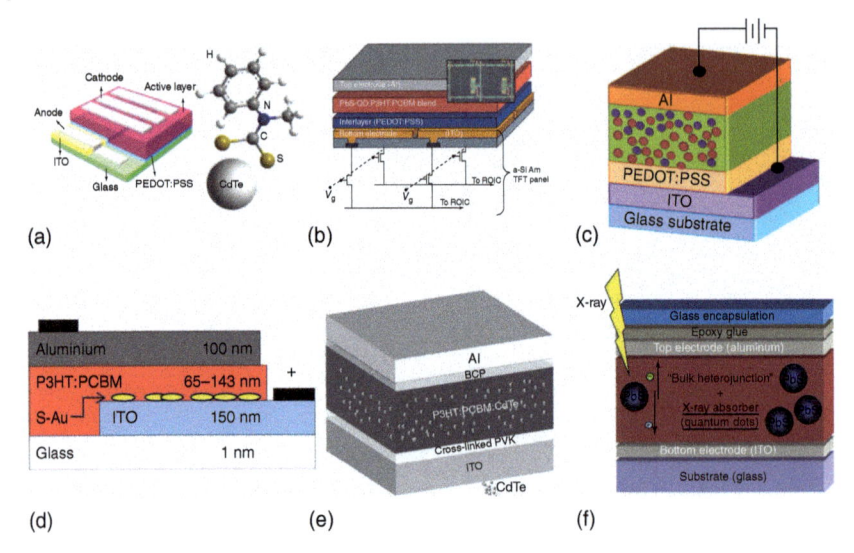

Figure 4.29 (a) Device structure of the PDi based on P3HT-PCBM-CdTe QDs blend and chemical structure of the PMDTC ligand (not to scale). Source: Chen et al. [11]. (b) Schematic of the PD based on P3HT-PCBM-PbS QDs blend with an a-Si AM backplane. The inset shows an optical micrograph of two active matrix pixels with a pixel pitch of 154 mm. Source: Rauch et al. [110]. (c) Device structure of the PD based on P3HT, PCBM, PbS QDs and/or ZnO NCs. Source: Dong et al. [111]. (d) Schematic of PD based on P3HT-PCBM-(S-Au). Source: Melancon et al. [112]. (e) Schematic of the PD based on P3HT-PCBM-CdTe blend. Source: Shen et al. [113]. (f) Schematic illustration of a hybrid organic PD based on a P3HT-PCBM-PbS QDs blend. Source: Ankah et al. [114].

role of light absorber, and the photo-induced holes and electrons transfer to P3HT and PCBM, respectively, and then they would be extracted to the corresponding electrodes through the P3HT or PCBM pathway (if there is a PCBM pathway). In 2014, Rui Dong et al. added ZnO NCs into the P3HT-PCBM-PbS QDs blend, making quaternary hybrid (Figure 4.29c). For comparison, the P3HT-PCBM-ZnO NCs device showed a larger EQE in UV range, but smaller EQE than P3HT-PCBM-PbS QDs device. The P3HT-PCBM-ZnO-PbS QDs device showed a university high EQE across the whole spectrum from 300 to 1050 nm. The EQE was 1624%, 1391%, and 166% at 350, 500, and 930 nm, respectively. The corresponding R was 4.58, 5.60, and 1.24 A W^{-1}. This work revealed the importance of the balanced harvesting of the light in UV and NIR region for a PD. The plasmonic effect has also been used in this kind of device. In 2014, Justin M. Melancon and S. R. Živanović fabricated a photoconductor with structure shown in Figure 4.29d [112]. A layer of Au nanostructure (S-Au) was firstly sputtered on the substrate. The S-Au layer has a plasmonic resonance peak at about 580 nm. The plasmonic effect induced the near-field enhancement in the surface of S–Au, increasing the light absorption cross-section of the P3HT-PCBM surrounding. The performance of the device got improved largely resulting to a broadband gain. This work actually provided a very inspiring strategy to improve the performance of PDs. In 2016, the P3HT-PCBM-CdTe QDs system was applied in PD again as shown in Figure 4.29e by Liang Shen et al. [113].

In this work, they explicitly manifest their intention to make photoconductive gain through introducing CdTe as the electron container. By this way, they successfully realized a 20 folder increase of the EQE compared with the device without QDs. The device showed outstanding performance with the parameters listed in Table 4.10. In the same year, G.N. Ankah et al. reported a PD based on the P3HT-PCBM-PbS QDs, but for the detection of X-ray (Figure 4.29f) [114]. It was demonstrated the sensing ability of this kind of device, and did optimization on the film thickness. This is a novel but significant application raised in this research field, which deserves more attention in scientific community.

4.5 Outlook

Due to the rapid development in material science, the solution-processabilities of organics and inorganics are becoming similar more and more, which makes their nanocomposites abundant enough to seek for candidates toward various high-performing PDs. Along with more organic and inorganic NCs prepared, the research on this topic will be pushed ahead endlessly. Nevertheless, the research should not be skimming on the surface but either shed light on the internal physical process or produce practical device. To do the former, more completed performance parameters of the device should be acquire in the report, including R, D^*, bandwidth, NEP, etc. To do the latter, the performance stability, ambient endurance, flexibility of the device should be concerned more.

List of Abbreviations

B	bandwidth
CQDs	colloidal quantum dots
D^*	specific detectivity
EQE	external quantum efficiency
G	gain
HOMO	highest occupied molecular orbit
IC	integrated circuit
LUMO	lowest unoccupied molecular orbit
MHP	metal-halide perovskites
MIR	mid-infrared
NC	nanocrystal
NIR	near-infrared
PD	photodetector
P_{in}	incident optical power
PC	photoconductor
PDi	photodiode
QDs	quantum dots
R	responsivity

References

1 de Arquer, F.P.G., Armin, A., Meredith, P., and Sargent, E.H. (2017). Solution-processed semiconductors for next-generation photodetectors. *Nat. Rev. Mater.* 2: 16100.
2 Jansen-van Vuuren, R.D., Armin, A., Pandey, A.K. et al. (2016). Organic photodiodes: the future of full color detection and image sensing. *Adv. Mater.* 28: 4766–4802.
3 Chen, H.Y., Liu, H., Zhang, Z.M. et al. (2016). Nanostructured photodetectors: from ultraviolet to terahertz. *Adv. Mater.* 28: 403–433.
4 Konstantatos, G. and Sargent, E.H. (2010). Nanostructured materials for photon detection. *Nat. Nanotechnol.* 5: 391–400.
5 Li, N., Lan, Z.J., Cai, L.F., and Zhu, F.R. (2019). Advances in solution-processable near-infrared phototransistors. *J. Mater. Chem. C* 7: 3711–3729.
6 Huo, N.J. and Konstantatos, G. (2018). Recent progress and future prospects of 2D-based photodetectors. *Adv. Mater.* 30: 1801164.
7 Li, J.H., Niu, L.Y., Zheng, Z.J., and Yan, F. (2014). Photosensitive graphene transistors. *Adv. Mater.* 26: 5239–5273.
8 Xie, C. and Yan, F. (2017). Flexible photodetectors based on novel functional materials. *Small* 13: 2201–2227.
9 Li, C., Huang, W.C., Gao, L.F. et al. (2020). Recent advances in solution-processed photodetectors based on inorganic and hybrid photo-active materials. *Nanoscale* 12: 2201–2227.
10 Heeger, A.J. (2014). 25th anniversary article: bulk heterojunction solar cells: understanding the mechanism of operation. *Adv. Mater.* 26: 10–28.
11 Chen, H.Y., Lo, M.K.F., Yang, G.W. et al. (2008). Nanoparticle-assisted high photoconductive gain in composites of polymer and fullerene. *Nat. Nanotechnol.* 3: 543–547.
12 Sun, Z.H., Li, J.H., and Yan, F. (2012). Highly sensitive organic near-infrared phototransistors based on poly(3-hexylthiophene) and PbS quantum dots. *J. Mater. Chem.* 22: 21673–21678.
13 Mueller, T., Xia, F.N.A., and Avouris, P. (2010). Graphene photodetectors for high-speed optical communications. *Nat. Photonics* 4: 297–301.
14 Li, F., Ma, C., Wang, H. et al. (2015). Ambipolar solution-processed hybrid perovskite phototransistors. *Nat. Commun.* 6: 8238.
15 Green, M.A., Ho-Baillie, A., and Snaith, H.J. (2014). The emergence of perovskite solar cells. *Nat. Photonics* 8: 506–514.
16 Dou, L.T., Yang, Y., You, J.B. et al. (2014). Solution-processed hybrid perovskite photodetectors with high detectivity. *Nat. Commun.* 5: 5404.
17 Deng, H., Yang, X.K., Dong, D.D. et al. (2015). Flexible and semitransparent organolead triiodide perovskite network photodetector arrays with high stability. *Nano Lett.* 15: 7963–7969.
18 Hu, X., Zhang, X.D., Liang, L. et al. (2014). High-performance flexible broadband photodetector based on organolead halide perovskite. *Adv. Funct. Mater.* 24: 7373–7380.

19 Konstantatos, G., Howard, I., Fischer, A. et al. (2006). Ultrasensitive solution-cast quantum dot photodetectors. *Nature* 442: 180–183.

20 Scarpa, G., Martin, E., Locci, S. et al. (2009). Organic thin-film phototransistors based on poly(3-hexylthiophene). *J. Phys. Conf. Ser.* 193: 4380–4386.

21 Zhang, Y.J., Hellebusch, D.J., Bronstein, N.D. et al. (2016). Ultrasensitive photodetectors exploiting electrostatic trapping and percolation transport. *Nat. Commun.* 7: 11924.

22 Dai, Y.J., Wang, X.F., Peng, W.B. et al. (2017). Largely improved near-infrared silicon-photosensing by the piezo-phototronic effect. *ACS Nano* 11: 7118–7125.

23 Konstantatos, G., Levina, L., Tang, J., and Sargent, E.H. (2008). Sensitive solution-processed Bi_2S_3 nanocrystalline photodetectors. *Nano Lett.* 8: 4002–4006.

24 Sarasqueta, G., Choudhury, K.R., and So, F. (2010). Effect of solvent treatment on solution-processed colloidal PbSe nanocrystal infrared photodetectors. *Chem. Mater.* 22: 3496–3501.

25 Tang, X., Ackerman, M.M., Chen, M.L., and Guyot-Sionnest, P. (2019). Dual-band infrared imaging using stacked colloidal quantum dot photodiodes. *Nat. Photonics* 13: 277.

26 Bolto, B.A., Mcneill, R., and Weiss, D.E. (1963). Electronic conduction in polymers. III. Electronic properties of polypyrrole. *Aust. J. Chem.* 16: 1090–1103.

27 Facchetti, A. (2011). π-Conjugated polymers for organic electronics and photovoltaic cell applications. *Chem. Mater.* 23: 733–758.

28 Baeg, K.J., Binda, M., Natali, D. et al. (2013). Organic light detectors: photodiodes and phototransistors. *Adv. Mater.* 25: 4267–4295.

29 Lee, C.H., Yu, G., Moses, D. et al. (1993). Sensitization of the photoconductivity of conducting polymers by C-60 – photoinduced electron-transfer. *Phys. Rev. B* 48: 15425–15433.

30 Tian, W., Liu, D., Cao, F.R., and Li, L. (2017). Hybrid nanostructures for photodetectors. *Adv. Opt. Mater.* 5: 1600468.

31 Yu, W.J., Liu, Y., Zhou, H.L. et al. (2013). Highly efficient gate-tunable photocurrent generation in vertical heterostructures of layered materials. *Nat. Nanotechnol.* 8: 952–958.

32 Sun, Z.H., Aigouy, L., and Chen, Z.Y. (2016). Plasmonic-enhanced perovskite-graphene hybrid photodetectors. *Nanoscale* 8: 7377–7383.

33 Sun, Z.H., Liu, Z.K., Li, J.H. et al. (2012). Infrared photodetectors based on CVD-grown graphene and PbS quantum dots with ultrahigh responsivity. *Adv. Mater.* 24: 5878–5883.

34 Gong, X., Tong, M.H., Xia, Y.J. et al. (2009). High-detectivity polymer photodetectors with spectral response from 300 nm to 1450 nm. *Science* 325: 1665–1667.

35 Yu, G. and Heeger, A.J. (1995). Charge separation and photovoltaic conversion in polymer composites with internal donor/acceptor heterojunctions. *J. Appl. Phys.* 78: 4510–4515.

36 Halls, J., Walsh, C., Greenham, N.C. et al. (1995). Efficient photodiodes from interpenetrating polymer networks. *Nature* 376: 498–500.

37 Halls, J.J., Arias, A.C., MacKenzie, J.D. et al. (2000). Photodiodes based on polyfluorene composites: influence of morphology. *Adv. Mater.* 12: 498–502.

38 Keivanidis, P.E., Khong, S.-H., Ho, P.K. et al. (2009). All-solution based device engineering of multilayer polymeric photodiodes: minimizing dark current. *Appl. Phys. Lett.* 94: 123.

39 Fukuda, T., Kimura, S., Honda, Z., and Kamata, N. (2011). Blue-sensitive organic photoconductive device with MDMO-PPV doped F8BT layer. *Mol. Cryst. Liq. Cryst.* 539: 202/[542]–209/[549].

40 Wang, Q., Qi, J., Qiao, W., and Wang, Z.Y. (2015). Soluble ladder conjugated polypyrrones: synthesis, characterization and application in photodetectors. *Dyes Pigm.* 113: 160–164.

41 Zhang, L., Yang, D., Wang, Y. et al. (2015). Performance enhancement of FET-based photodetector by blending P3HT with PMMA. *IEEE Photonics Technol. Lett.* 27: 1535–1538.

42 Wang, X., Lv, L., Li, L. et al. (2016). High-performance all-polymer photoresponse devices based on acceptor–acceptor conjugated polymers. *Adv. Funct. Mater.* 26: 6306–6315.

43 Hu, L., Qiao, W., Han, J. et al. (2017). Naphthalene diimide–diketopyrrolopyrrole copolymers as non-fullerene acceptors for use in bulk-heterojunction all-polymer UV–NIR photodetectors. *Polym. Chem.* 8: 528–536.

44 Hu, L., Qiao, W., Zhou, X. et al. (2017). Effect of compositions of acceptor polymers on dark current and photocurrent of all-polymer bulk-heterojunction photodetectors. *Polymer* 114: 173–179.

45 Murto, P., Genene, Z., Benavides, C.M. et al. (2018). High performance all-polymer photodetector comprising a donor–acceptor–acceptor structured indacenodithiophene–bithieno [3,4-c] pyrroletetrone copolymer. *ACS Macro Lett.* 7: 395–400.

46 Hu, L., Han, J., Qiao, W. et al. (2018). Side-chain engineering in naphthalenediimide-based n-type polymers for high-performance all-polymer photodetectors. *Polym. Chem.* 9: 327–334.

47 Xu, X., Zhou, X., Zhou, K. et al. (2018). Large-area, semitransparent, and flexible all-polymer photodetectors. *Adv. Funct. Mater.* 28: 1805570.

48 Sen, P., Yang, R., Rech, J.J. et al. (2019). Panchromatic all-polymer photodetector with tunable polarization sensitivity. *Adv. Opt. Mater.* 7: 1801346.

49 Zhang, Z.-G., Yang, Y., Yao, J. et al. (2017). *Angew. Chem. Int. Ed.* 56: 13503. doi.10.1002/anie.201707678.

50 Yang, K., Wang, J., Miao, J. et al. (2019). All-polymer photodetectors with photomultiplication. *J. Mater. Chem. C* 7: 9633–9640.

51 Xia, Y., Aguirre, L.E., Xu, X., and Inganäs, O. (2020). All-polymer high-performance photodetector through lamination. *Adv. Electron. Mater.* 6: 1901017.

52 Qiu, L.-Z., Wei, S.-Y., Xu, H.-S. et al. (2019). Ultrathin polymer nanofibrils for solar-blind deep ultraviolet light photodetectors application. *Nano Lett.* 20: 644–651.

53 Yu, G., Pakbaz, K., and Heeger, A. (1994). Semiconducting polymer diodes: large size, low cost photodetectors with excellent visible–ultraviolet sensitivity. *Appl. Phys. Lett.* 64: 3422–3424.

54 Yu, G., Gao, J., Yang, C., and Heeger, A.J. (1997). *High-sensitivity photodetectors made with charge-transfer polymer blends.* In: *Photodetectors: Materials and Devices II*, 306–314. San Jose, CA, United States: International Society for Optics and Photonics.

55 Lai, Y.Y., Cheng, Y.J., and Hsu, C.S. (2014). Applications of functional fullerene materials in polymer solar cells. *Energy Environ. Sci.* 7: 1866–1883.

56 Lee, C., Yu, G., Moses, D. et al. (1993). Sensitization of the photoconductivity of conducting polymers by C_{60}: photoinduced electron transfer. *Phys. Rev. B* 48: 15425.

57 Macedo, A., Zanetti, F., Mikowski, A. et al. (2008). Improving light harvesting in polymer photodetector devices through nanoindented metal mask films. *J. Appl. Phys.* 104: 033714.

58 Liu, X., Wang, H., Yang, T. et al. (2012). Solution-processed ultrasensitive polymer photodetectors with high external quantum efficiency and detectivity. *ACS Appl. Mater. Interfaces* 4: 3701–3705.

59 Binda, M., Natali, D., Iacchetti, A., and Sampietro, M. (2013). Integration of an organic photodetector onto a plastic optical fiber by means of spray coating technique. *Adv. Mater.* 25: 4335–4339.

60 Arredondo, B., De Dios, C., Vergaz, R. et al. (2013). Performance of ITO-free inverted organic bulk heterojunction photodetectors: comparison with standard device architecture. *Org. Electron.* 14: 2484–2490.

61 Arredondo, B., Romero, B., Pena, J.M.S. et al. (2013). Visible light communication system using an organic bulk heterojunction photodetector. *Sensors* 13: 12266–12276.

62 Li, S., Xue, D., Xu, W. et al. (2014). Improving the photo current of the [60] PCBM/P3HT photodetector device by using wavelength-matched photonic crystals. *J. Mater. Chem. C* 2: 1500–1504.

63 Li, L., Zhang, F., Wang, J. et al. (2015). Achieving EQE of 16,700% in P3HT:PC 71 BM based photodetectors by trap-assisted photomultiplication. *Sci. Rep.* 5: 1–7.

64 Arora, H., Malinowski, P., Chasin, A. et al. (2015). Amorphous indium-gallium-zinc-oxide as electron transport layer in organic photodetectors. *Appl. Phys. Lett.* 106 (14): 143301.

65 Wang, W., Zhang, F., Li, L. et al. (2015). Improved performance of photomultiplication polymer photodetectors by adjustment of P3HT molecular arrangement. *ACS Appl. Mater. Interfaces* 7: 22660–22668.

66 Liu, Z., Parvez, K., Li, R. et al. (2015). Transparent conductive electrodes from graphene/PEDOT:PSS hybrid inks for ultrathin organic photodetectors. *Adv. Mater.* 27: 669–675.

67 Li, L., Zhang, F., Wang, W. et al. (2015). Trap-assisted photomultiplication polymer photodetectors obtaining an external quantum efficiency of 37 500%. *ACS Appl. Mater. Interfaces* 7: 5890–5897.

68 Wang, W., Zhang, F., Bai, H. et al. (2016). Photomultiplication photodetectors with P3HT: fullerene-free material as the active layers exhibiting a broad response. *Nanoscale* 8: 5578–5586.

69 Wang, W., Zhang, F., Du, M. et al. (2017). Highly narrowband photomultiplication type organic photodetectors. *Nano Lett.* 17: 1995–2002.

70 Miao, J., Zhang, F., Du, M. et al. (2017). Photomultiplication type narrowband organic photodetectors working at forward and reverse bias. *Phys. Chem. Chem. Phys.* 19: 14424–14430.

71 Wang, Y., Zhu, L., Hu, Y. et al. (2017). High sensitivity and fast response solution processed polymer photodetectors with polyethylenimine ethoxylated (PEIE) modified ITO electrode. *Opt. Express* 25: 7719–7729.

72 Wang, Y., Zhu, L., Wang, T. et al. (2018). Fast and sensitive polymer photodetectors with extra high external quantum efficiency and large linear dynamic range at low working voltage bias. *Org. Electron.* 62: 448–453.

73 Wu, Y.L., Fukuda, K., Yokota, T., and Someya, T. (2019). A highly responsive organic image sensor based on a two-terminal organic photodetector with photomultiplication. *Adv. Mater.* 31: 1903687.

74 Zhao, Z., Wang, J., Xu, C. et al. (2019). Photomultiplication type broad response organic photodetectors with one absorber layer and one multiplication layer. *J. Phys. Chem. Lett.* 11: 366–373.

75 Wang, T., Hu, Y., Deng, Z. et al. (2017). High sensitivity, fast response and low operating voltage organic photodetectors by incorporating a water/alcohol soluble conjugated polymer anode buffer layer. *RSC Adv.* 7: 1743–1748.

76 Park, S.H., Su, R., Jeong, J. et al. (2018). 3D printed polymer photodetectors. *Adv. Mater.* 30: 1803980.

77 Maity, S. and Thomas, T. (2019). Hybrid-organic-photodetector containing chemically treated ZnMgO layer with promising and reliable detectivity, responsivity and low dark current. *IEEE Trans. Device Mater. Reliab.* 19: 193–200.

78 Esopi, M.R. and Yu, Q. (2019). Plasmonic aluminum nanohole arrays as transparent conducting electrodes for organic ultraviolet photodetectors with bias-dependent photoresponse. *ACS Appl. Nano Mater.* 2: 4942–4953.

79 Lan, Z., Lei, Y., Chan, W.K.E. et al. (2020). Near-infrared and visible light dual-mode organic photodetectors. *Sci. Adv.* 6: eaaw8065.

80 Yen, C.-T., Huang, Y.-C., Yu, Z.-L. et al. (2018). Performance improvement and characterization of spray-coated organic photodetectors. *ACS Appl. Mater. Interfaces* 10: 33399–33406.

81 Chen, L., Roman, L.S., Johansson, D.M. et al. (2000). Excitation transfer in polymer photodiodes for enhanced quantum efficiency. *Adv. Mater.* 12: 1110–1114.

82 Roman, L.S., Andersson, M.R., Yohannes, T., and Inganäs, O. (1997). Photodiode performance and nanostructure of polythiophene/C_{60} blends. *Adv. Mater.* 9: 1164–1168.

83 Wang, W., Zhang, F., Li, L. et al. (2015). Highly sensitive polymer photodetectors with a broad spectral response range from UV light to the near infrared region. *J. Mater. Chem. C* 3: 7386–7393.

84 Zafar, Q., Najeeb, M.A., Ahmad, Z., and Sulaiman, K. (2015). Organic–inorganic hybrid nanocomposite for enhanced photo-sensing of PFO-DBT:MEH-PPV:PC$_{71}$BM blend-based photodetector. *J. Nanopart. Res.* 17: 372.

85 Zafar, Q., Ahmad, Z., and Sulaiman, K. (2015). PFO-DBT:MEH-PPV:PC$_{71}$BM ternary blend assisted platform as a photodetector. *Sensors* 15: 965–978.

86 Gao, M., Wang, W., Li, L. et al. (2017). Highly sensitive polymer photodetectors with a wide spectral response range. *Chin. Phys. B* 26: 018201.

87 Miao, J., Zhang, F., Du, M. et al. (2018). Photomultiplication type organic photodetectors with broadband and narrowband response ability. *Adv. Opt. Mater.* 6: 1800001.

88 Chuang, S.-T., Chien, S.-C., and Chen, F.-C. (2012). Extended spectral response in organic photomultiple photodetectors using multiple near-infrared dopants. *Appl. Phys. Lett.* 100: 9.

89 Pace, G., Grimoldi, A., Natali, D. et al. (2014). All-organic and fully-printed semitransparent photodetectors based on narrow bandgap conjugated molecules. *Adv. Mater.* 26: 6773–6777.

90 Herrbach, J., Revaux, A., Vuillaume, D., and Kahn, A. (2016). P-doped organic semiconductor: potential replacement for PEDOT:PSS in organic photodetectors. *Appl. Phys. Lett.* 109: 073301.

91 Chen, F.-C., Chien, S.-C., and Cious, G.-L. (2010). Highly sensitive, low-voltage, organic photomultiple photodetectors exhibiting broadband response. *Appl. Phys. Lett.* 97: 195.

92 Greenham, N.C., Peng, X., and Alivisatos, A.P. (1996). Charge separation and transport in conjugated-polymer/semiconductor-nanocrystal composites studied by photoluminescence quenching and photoconductivity. *Phys. Rev. B* 54: 17628.

93 Greenham, N., Peng, X., and Alivisatos, A. (1997). Charge separation and transport in conjugated polymer/cadmium selenide nanocrystal composites studied by photoluminescence quenching and photoconductivity. *Synth. Met.* 84: 545–546.

94 Narayan, K. and Singh, T.B. (1999). Nanocrystalline titanium dioxide-dispersed semiconducting polymer photodetectors. *Appl. Phys. Lett.* 74: 3456–3458.

95 McDonald, S., Cyr, P., Levina, L., and Sargent, E. (2004). Photoconductivity from PbS-nanocrystal/semiconducting polymer composites for solution-processible, quantum-size tunable infrared photodetectors. *Appl. Phys. Lett.* 85: 2089–2091.

96 Qi, D., Fischbein, M., Drndić, M., and Šelmić, S. (2005). Efficient polymer-nanocrystal quantum-dot photodetectors. *Appl. Phys. Lett.* 86: 093103.

97 McDonald, S.A., Konstantatos, G., Zhang, S. et al. (2005). Solution-processed PbS quantum dot infrared photodetectors and photovoltaics. *Nat. Mater.* 4: 138–142.

98 Campbell, I. and Crone, B. (2007). Bulk photoconductive gain in poly(phenylene vinylene) based diodes. *J. Appl. Phys.* 101: 024502.

99 Yan, F., Li, J., and Mok, S.M. (2009). Highly photosensitive thin film transistors based on a composite of poly(3-hexylthiophene) and titania nanoparticles. *J. Appl. Phys.* 106: 074501.

100 Jarzab, D., Szendrei, K., Yarema, M. et al. (2011). Charge-separation dynamics in inorganic–organic ternary blends for efficient infrared photodiodes. *Adv. Funct. Mater.* 21: 1988–1992.

101 Sun, Z., Li, J., Liu, C. et al. (2011). Enhancement of hole mobility of poly(3-hexylthiophene) induced by titania nanorods in composite films. *Adv. Mater.* 23: 3648–3652.

102 Wang, X., Song, W., Liu, B. et al. (2013). High-performance organic–inorganic hybrid photodetectors based on P3HT:CdSe nanowire heterojunctions on rigid and flexible substrates. *Adv. Funct. Mater.* 23: 1202–1209.

103 Guo, F., Yang, B., Yuan, Y. et al. (2012). A nanocomposite ultraviolet photodetector based on interfacial trap-controlled charge injection. *Nat. Nanotechnol.* 7: 798–802.

104 Yoo, J., Jeong, S., Kim, S., and Je, J.H. (2015). A stretchable nanowire UV–vis–NIR photodetector with high performance. *Adv. Mater.* 27: 1712–1717.

105 Wei, H., Fang, Y., Yuan, Y. et al. (2015). Trap engineering of CdTe nanoparticle for high gain, fast response, and low noise P3HT:CdTe nanocomposite photodetectors. *Adv. Mater.* 27: 4975–4981.

106 Wang, H., Li, Z., Fu, C. et al. (2015). Solution-processed PbSe colloidal quantum dot-based near-infrared photodetector. *IEEE Photonics Technol. Lett.* 27: 612–615.

107 Song, T., Cheng, H., Fu, C. et al. (2016). Influence of the active layer nanomorphology on device performance for ternary PbS$_x$Se$_{1-x}$ quantum dots based solution-processed infrared photodetector. *Nanotechnology* 27: 165202.

108 Arciniegas, M.P., Stasio, F.D., Li, H. et al. (2016). Self-assembled dense colloidal Cu$_2$Te nanodisk networks in P3HT thin films with enhanced photocurrent. *Adv. Funct. Mater.* 26: 4535–4542.

109 Xu, W., Peng, H., Zhu, T. et al. (2017). A solution-processed near-infrared polymer: PbS quantum dot photodetectors. *RSC Adv.* 7: 34633–34637.

110 Rauch, T., Böberl, M., Tedde, S.F. et al. (2009). Near-infrared imaging with quantum-dot-sensitized organic photodiodes. *Nat. Photonics* 3: 332.

111 Dong, R., Bi, C., Dong, Q. et al. (2014). An ultraviolet-to-NIR broad spectral nanocomposite photodetector with gain. *Adv. Opt. Mater.* 2: 549–554.

112 Melancon, J.M. and Živanović, S.R. (2014). Broadband gain in poly(3-hexylthiophene): phenyl-C$_{61}$-butyric-acid-methyl-ester photodetectors enabled by a semicontinuous gold interlayer. *Appl. Phys. Lett.* 105: 163301.

113 Shen, L., Fang, Y., Wei, H. et al. (2016). A highly sensitive narrowband nanocomposite photodetector with gain. *Adv. Mater.* 28: 2043–2048.

114 Ankah, G., Büchele, P., Poulsen, K. et al. (2016). PbS quantum dot based hybrid-organic photodetectors for X-ray sensing. *Org. Electron.* 33: 201–206.

5

Polymer Nanocomposites for Pressure Sensors

Qi-Jun Sun[1] and Xin-Hua Zhao[2]

[1] City University of Hong Kong, Department of Materials Science and Engineering, Tat Chee Avenue, Kowloon, Hong Kong, Hong Kong SAR 999077, China
[2] Southern University of Science and Technology, Department of Chemistry, Shenzhen, Guangdong 518055, China

5.1 Introduction

Pressure sensor is one type of electronic devices, which can transduce mechanical deformation into electrical signals. In the past decades, pressure sensor has obtained global research interest for their extensive application in touch screen, memories, wearable electronics, and robots [1–26]. Great achievement has been made for the pressure sensor devices based on microelectromechanical systems (MEMS) during the past 30 years; however, MEMS based pressure sensors are normally based on rigid silicon substrate, and their applications for flexible electronic skin are greatly limited. Alternatively, pressure sensors based on polymers or polymer nanocomposites are highly flexible and stretchable, indicating the tremendous potential for future wearable and robotic applications. Despite the great efforts that have been paid to the development of wearable and stretchable sensors, highly sensitive sensors with excellent stability for flexible displays, healthcare monitoring, and e-skin applications that are compatible with large-scale production remain challenging. Particularly, the tactile sensors with superior pressure sensitivity and a broad linear pressure range are highly desired for their potential application as components of various wearable electronic systems and devices.

Typically, according to the transduction principles, pressure sensing devices are mainly divided into the following categories including but not limited to capacitive [27–31], piezoresistive [32–39], piezoelectric [40–45], and triboelectric pressure sensors [46–49]. Capacitive pressure sensors based on MEMS have been commercialized and have much higher sensitivity than piezoresistive sensors. Nonetheless, complex circuit configurations are required to operate this type of pressure sensors. Piezoresistive pressure sensors have a simple device configuration and easy signal processing, and therefore, they have been widely investigated by researchers in recent years. On the other hand, piezoelectric and triboelectric

---

Polymer Nanocomposite Materials: Applications in Integrated Electronic Devices, First Edition.
Edited by Ye Zhou and Guanglong Ding.
© 2021 WILEY-VCH GmbH. Published 2021 by WILEY-VCH GmbH.

pressure sensors can be self-powered, i.e. no input current and voltage are required to drive these sensors. However, these sensors can show the reliable detection of dynamic pressures but the reliable static pressure detection is still challenging.

Thanks to the development of material science and engineering, polymer nanocomposites have been broadly employed in developing flexible electronics [50, 51]. Particularly, the conductive polymer nanocomposites have been extensively used as the electrode or sensing layer for flexible electronics and sensors. In order to achieve the conductive polymer nanocomposites, the strategies are to impart the conductive nanomaterials into the polymer matrix [52–54].

In this chapter, we begin with a brief introduction of the parameters, work mechanisms for capacitive-, piezoresistive-, piezoelectric-, and triboelectric-pressure sensors. Then, we introduce the technologies used in the development of polymer nanocomposite based materials for pressure sensors. After that, representative examples of polymer nanocomposite based pressure sensors will be introduced and discussed, but with more focus on capacitive and piezoresistive sensors and with brief discussion for piezoelectric and triboelectric pressure sensors. Additionally, we also present a more detailed overview of state-of-the-art piezoresistive pressure sensors, followed by a characterization of a specific piezoresistor based pressure sensor and its applications in healthcare monitoring, human motion detection, and robotic skin. In the end of this chapter, we also include a Table 5.1 to summarize and compare the performance of recently reported polymer nanocomposite-based pressure sensors and give conclusive remarks.

5.2 Parameters for Pressure Sensors

Pressure sensitivity and linear detection range are two of the most important parameters to estimate the performance of pressure sensors. The other important parameters such as response time, limit of detection (LOD), stability, and durability are also briefly introduced. All these parameters need to be considered in developing the polymer nanocomposite based pressure sensors.

5.2.1 Pressure Sensitivity

The pressure sensitivity (S) of tactile sensor is expressed as the relative output change to the applied pressure.

For piezoresistive pressure sensor, $S = (\Delta I/I_0)/\Delta P$

For capacitive pressure sensor, $S = (\Delta C/C_0)/\Delta P$

where ΔI, I_0, ΔC, C_0, and ΔP are the change of current, initial current, change of capacitance, initial capacitance, and change of applied pressure, respectively. The relative change of the electrical outputs (current, capacitance) as a function of the applied pressure usually plotted in a graph, and the slope of the curves implies the pressure sensitivity. A larger slope of the curve means higher pressure sensitivity of the pressure sensor.

Recently, highly sensitive pressure sensors with broad pressure range from below 1 kPa^{-1} to thousands of kPa^{-1} are reported. Although great progress has been made

Table 5.1 A summary of the performances of flexible and stretchable pressure sensors.

Materials	Sensing mechanism	Minimum detection	Maximum detection	Sensitivity	References
P(VDF-TrFE)/BrTiO$_3$-FET	Piezoelectric	0.1 MPa	0.5 MPa	—	[62]
P(VDF-TrFE)/PbTiO$_3$-FET	Piezoelectric	2 MPa	22 MPa	—	[63]
P(VDF-TrFE) nanofiber	Piezoelectric	0.1 Pa	12 Pa	0.41 V Pa^{-1}	[64]
LM-NP/PDMS composite	Triboelectric	—	200 kPa	2.52 V kPa^{-1}	[60]
Graphene/PU sponge	Piezoresistive	9 Pa	10 kPa	0.26 kPa^{-1}	[65]
Polypyrrole hollow-sphere	Piezoresistive	1 Pa	11 kPa	133 kPa^{-1}	[66]
Reduced graphene oxide (rGO) foam	Piezoresistive	163 Pa	49 kPa	15.2 kPa^{-1}	[67]
CNT/PDMS porous	Piezoresistive	0.25 kPa	100 kPa	—	[68]
PEDOT:PSS/PUD micropyramid	Piezoresistive	23 Pa	8 kPa	10.3 kPa^{-1}	[11]
CNTs/PDMS interlocked microdome	Piezoresistive	0.2 Pa	59 kPa	15.1 kPa^{-1}	[69]
rGO micropyramid	Piezoresistive	1.5 Pa	1.4 kPa	−5.5 kPa^{-1}	[54]
SWCNTs/PDMS microstructure	Piezoresistive	0.6 Pa	1.2 kPa	1.8 kPa^{-1}	[18]
Au-coated PDMS micropillar/polyaniline	Piezoresistive	15 Pa	3.5 kPa	2.0	[70]
CNT–PDMS composite with microdome arrays	Piezoresistive	0.2 Pa	59 kPa	15.1	[69]
Hierarchically structured graphene	Piezoresistive	1 Pa	12 kPa	8.5	[71]
Fingerprint-like 3D graphene film	Piezoresistive	0.2 Pa	75 kPa	110	[72]
CNT–graphene composite film	Piezoresistive	0.6 Pa	6 kPa	19.8	[73]
CNT-coated textile/Ni-coated textile	Piezoresistive	6 Pa	20 kPa	14.4	[74]
Porous multiwalled nanotube/PDMS/reverse micelle solution	Piezoresistive	—	—	≤0.025 kPa^{-1}	[68]
Carbonized silk nanofiber membrane	Piezoresistive	−0.8 Pa	5 kPa	34.47	[75]
Graphene oxide (GO)	Capacitive	0.24 Pa	1 kPa	0.8 kPa^{-1}	[76]
Fluorosilicone/air gap	Capacitive	0.5 kPa	190 kPa	0.91 kPa^{-1}	[77]
Silver NP–SBS composite/PDMS	Capacitive	—	20 kPa	0.21 kPa^{-1}	[78]

(Continued)

Table 5.1 (Continued)

Materials	Sensing mechanism	Minimum detection	Maximum detection	Sensitivity	References
Silver NWs/PDMS	Capacitive	—	1.4 MPa	1.62 MPa^{-1}	[79]
CNT–Ecoflex composite film/microporous ecoflex	Capacitive	0.1 Pa	130 kPa	0.601 kPa^{-1}	[80]
Porous PDMS/air gap	Capacitive	2.5 Pa	20 kPa	0.7 kPa^{-1}	[81]
PDMS microhairy	Capacitive	—	5 kPa	0.58 kPa^{-2}	[82]
PDMS microstructure OFET	Capacitive	3 Pa	20 kPa	0.55 kPa^{-1}	[83]
PDMS microstructure OFET	Capacitive	—	60 kPa	8.2 kPa^{-1}	[84]
Polystyrene microspheres	Capacitive	17.5 Pa	—	0.815 kPa^{-1}	[85]

in pressure sensitivity, it is still challenging to maintain high sensitivity over wide pressure range, which has recently attracted intensive research interests.

5.2.2 Linear Sensing Range

Linear relationship between the relative electrical outputs to external pressure is highly required, which can simplify the procedure for electrical signal collection and processing. Most of the reported pressure sensors show a sectional type of linear relationship, which is not usually a strict linear relationship. For practical applications, a linear relationship over the whole pressure sensing range is highly desired.

5.2.3 LOD and Response Speed

LOD and response speed are two of the critical parameters for pressure sensors. On the one hand, LOD can show the minimum force the sensor can detect, which is closely relevant to the detection accuracy of the sensors. The recently reported pressure sensing devices have demonstrated an LOD below 1 Pa, which can fulfill for the requirement in most of the real applications. Response time means the response speed of the sensor to the pressure. A high response speed is necessary in the real-time detection of external stimuli, such as the real-time detection of the wrist pulses, subtle human motions, and spatial pressure distribution.

5.2.4 Reliability

Excellent reliability is highly required for pressure sensors in practical applications. Reliability can be used to estimate the output consistence of a pressure sensor, which can be influenced by moisture, temperature, oxygen, baseline drift, and chemical stability of the sensing materials.

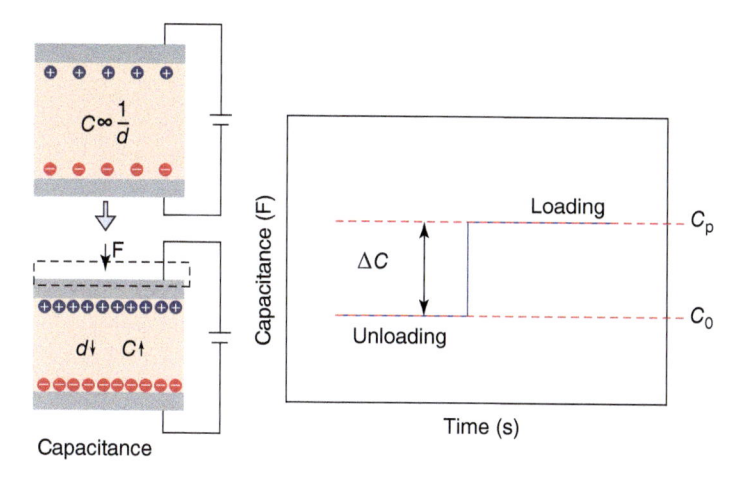

Figure 5.1 Sensing mechanism of a capacitive pressure sensor. Source: Zang et al. [21].

5.3 Working Principles and Examples of Polymer Nanocomposite Based Pressure Sensors

5.3.1 Capacitive Pressure Sensors

For capacitive pressure sensors, the capacitance of the capacitor is estimated by the following equation: $C = \varepsilon_0 \varepsilon_r A/d$, where ε_0, ε_r, A, and d are dielectric constant of the vacuum, relative dielectric constant of the dielectric layer, contact area between the dielectric layer and the electrodes, and dielectric thickness correspondingly, respectively. The external pressure can cause the capacitance change according to the contact area and the dielectric thickness.

Schematic illustration depicting the sensing mechanism of capacitive pressure sensors is depicted in Figure 5.1 [21]. Without any external pressure, the capacitor exhibits a very low capacitance (C_0) due to the relatively thick dielectric layer. With loaded external pressures, the capacitance increases to C_p because the compression force decreases the dielectric thickness, leading to the increment of capacitance by ΔC. Consequently, the external pressure could be detected by the capacitive sensor according to the change in capacitance.

The sensitivity of the capacitive pressure sensor can be enhanced by the employment of microstructured polymer nanocomposite or inside of the polymer composite [55, 56]. Regular microstructures such as pyramid, micro-pillars, hemisphere, and natural biomaterial inspired microstructures are widely employed for the development of capacitive sensing layer for pressure sensors. For example, capacitive pressure sensor based on the microstructured silver nanowires (AgNWs)-embedded polydimethylsiloxane (PDMS) electrode was fabricated by Shuai et al. showing high pressure sensitivity [30]. The schematic fabrication procedure of the top electrode, dielectric film, and microstructured AgNWs/PDMS nanocomposite based bottom electrode, and the assembling of the capacitive device is depicted in Figure 5.2.

136 | *5 Polymer Nanocomposites for Pressure Sensors*

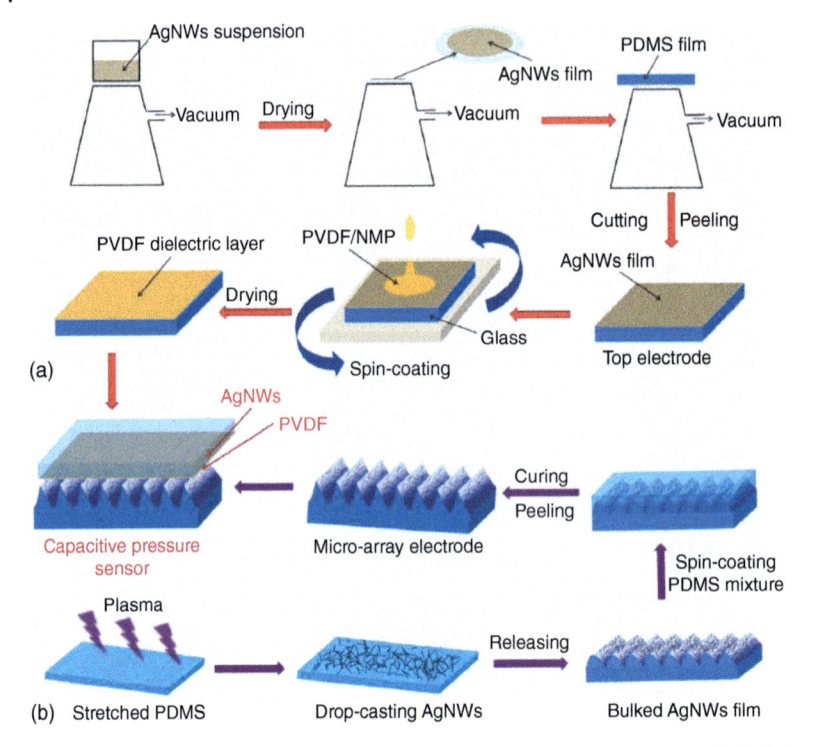

Figure 5.2 Schematic fabrication process of (a) the top electrodes and PVDF dielectrics, and (b) the capacitive pressure sensors. Source: Shuai et al. [30].

In detail, the top electrode was achieved via transferring the vacuum filtered AgNWs film on a rubber substrate. The poly(vinylidene fluoride) (PVDF) dielectric layer was prepared by spin-coating. The microstructured AgNWs-embedded flexible electrode is depicted in Figure 5.2b. First, the PDMS substrate on Si substrate was stretched and the AgNWs were then drop-casted on top of the PDMS substrate. The PDMS substrate was released and microstructured AgNWs/PDMS electrode was formed after drying the AgNW film. Finally, the upper electrode, PVDF dielectric layer, and bottom microstructured AgNWs/PDMS electrode were laminated layer-by-layer as shown in Figure 5.2b. The morphologies of the top electrode, PDMS mold, and AgNWs-embedded bottom electrodes were revealed by scanning electron microscope (SEM) (Figure 5.3). As indicated in Figure 5.3a,b, the AgNWs are embedded on PDMS surface uniformly and form interconnected networks, which enables the top electrode with good electrical conductivities. As shown in Figure 5.3c, the PVDF layer is uniformly formed, showing a thickness of 3.5 μm. As shown in Figure 5.3d, the PDMS mold shows a smooth and wave surface. In order to reveal the working principles of capacitive sensors based on the microstructured nanocomposite, they also fabricated the sensors based on the flat nanocomposite. The work principles of the capacitive pressure sensors with the flat and microstructured nanocomposite electrodes are schematically shown in Figure 5.4. The sensors fabricated with micro-array electrodes show enhanced

5.3 Working Principles and Examples of Polymer Nanocomposite Based Pressure Sensors | **137**

Figure 5.3 (a and b) SEM images of top electrodes. (c) Cross-sectional SEM image of top electrode with PVDF. (d–f) SEM images of PDMS mold, hybrid mold, and bottom electrode, respectively. Source: Reproduced with permission Shuai et al. [30]. Copyright 2013, American Chemical Society.

deformation under the same external pressure compared with that based on the flat electrodes, which is because the microstructured electrodes deform a lot compared with the flat ones. The relative capacitance responses to the related deformation behavior are shown in Figure 5.4c. The thickness of the PVDF dielectric film in the planar electrode based pressure sensor decreases under external pressures, which is the main reason for the capacitance variation. For the capacitive pressure sensor based on the microstructured AgNWs/PDMS electrode, external pressure changes the thickness of PVDF dielectric layer and the contact area between the microstructured AgNWs/PDMS electrode, simultaneously. Therefore, the capacitive sensors with the microstructured AgNWs/PDMS electrodes have higher pressure sensitivity than that of the planar electrode.

5.3.2 Piezoresistive Pressure Sensors

The piezoresistive mechanical sensors have been widely investigated because of their simple device configurations and easier signal processing. The work principle of piezoresistive pressure sensor is the resistance changes of the sensors to the external pressures as illustrated in Figure 5.5. Without external pressure, the polymer nanocomposite film in the pressure is in the high resistance state, and the sensor exhibits a low output current. Under external pressure, the piezoresistive pressure sensor shows an increased current output compared with the initial state at the same drive voltage. Consequently, the piezoresistive pressure sensors can measure the external pressures by the change of the resistance or the output current of the piezoresistive pressure sensor. Various polymer nanocomposites are used as the

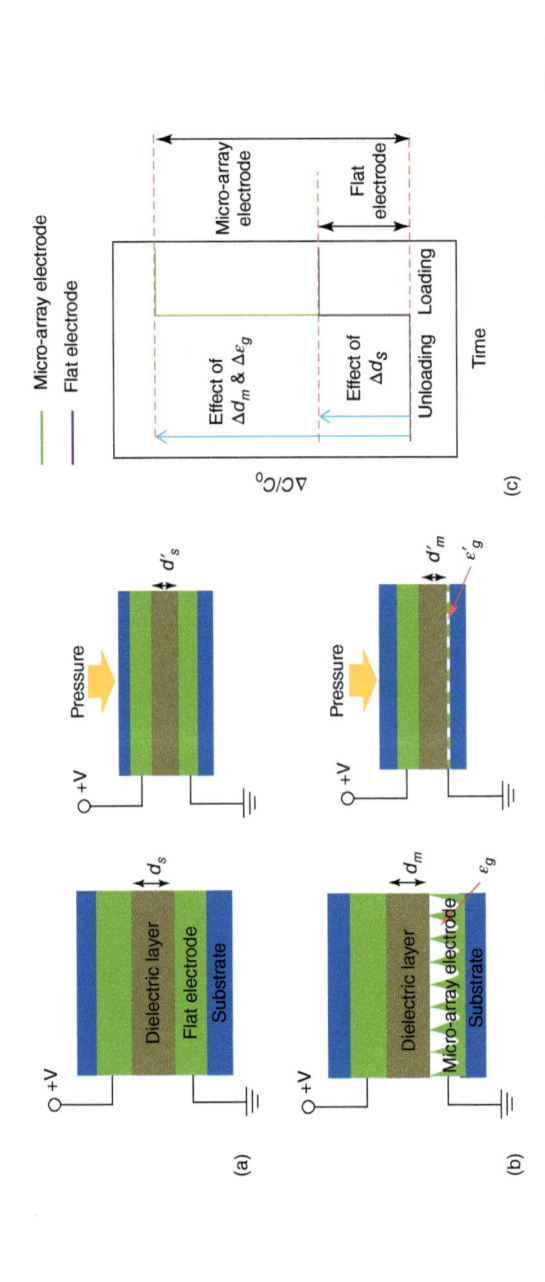

Figure 5.4 Work principles of the capacitive pressure sensors with the flat electrodes and micro-array electrodes respectively. Source: Shuai et al. [30].

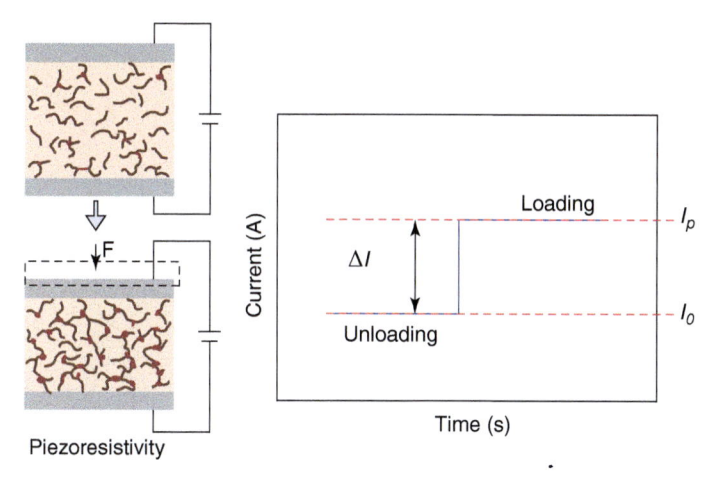

Figure 5.5 Work principle of piezoresistive pressure sensors. Source: Zang et al. [21].

pressure-sensitive layers for piezoresistive pressure sensors in the past decades [57, 58].

Lee et al. presented a highly sensitive piezoresistive pressure sensor based on a conductive polymer nanocomposite including polyurethane (PU) and sea-urchin shaped metal nanoparticles (SSNPs) with excellent transparency [57]. The piezoresistive pressure sensors are simply fabricated as depicted in Figure 5.6. First, the SSNPs were mixed with the PU matrix to obtain the conductive inks with various weight percentages. Secondly, the mixture as spin coated onto ITO/PET substrate to obtain the transparent piezoresistive film. After the composite inks solidified and covered with another piece of indium tin oxide (ITO)/poly(ethylene terephthalate) (PET) substrate on top as the upper electrode, the sensor fabrication was completed. The morphology of the SSNPs in the PU matrix was characterized as shown in Figure 5.6b, indicating that the microstructures of the metal particles could be kept intact in the PU matrices. Figure 5.6c shows the resistance responses of the sensor to external pressures from 0 to 18 kPa. The results demonstrate that the pressure sensor based on the SSNPs/PU nanocomposite exhibits large changes in resistance, enabling the feasible detection of tiny external pressures. As depicted in Figure 5.6d, the piezoresistive pressure sensor devices based on the nanocomposites with various SSNPs concentrations show excellent transparency, implying that the pressure sensors showed superior optical transparency and higher concentrations of the SSNPs can reduce the transparency of the polymer nanocomposites. To demonstrate their potential applications, the pressure sensors are fabricated into 6 × 6 array to detect the motion of human fingers (Figure 5.7).

The natural biological materials have inspired the researchers' ideas in material design. Recently, we have fabricated the bio-inspired graphite/PDMS polymer composite as pressure-sensitive layer for piezoresistive pressure sensors [58]. The bio-inspired structures on the surface of the polymer nanocomposite were duplicated from the sandpaper template. NaCl particles were used as sacrificial templates to realize hierarchical microstructures inside the polymer nanocomposite. The

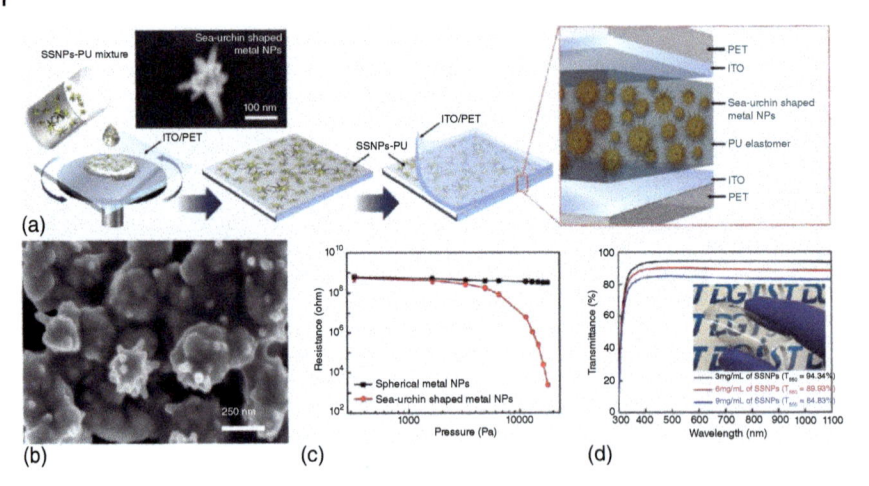

Figure 5.6 (a) Fabrication process of the piezoresistive skin sensor based on SSNPs/PU nanocomposite. (b) Morphology of SSNPs in PU matrix. (c and d) Resistance of the nanocomposite and optical transmittance of the sensor based on different SSNPs concentrations. Inset is the photograph of a transparent device. Source: (a) Reproduced with permission Lee et al. [57]. Copyright 2016, Wiley-VCH, (b–d) Lee et al. [57]. © 2016, John Wiley & Sons.

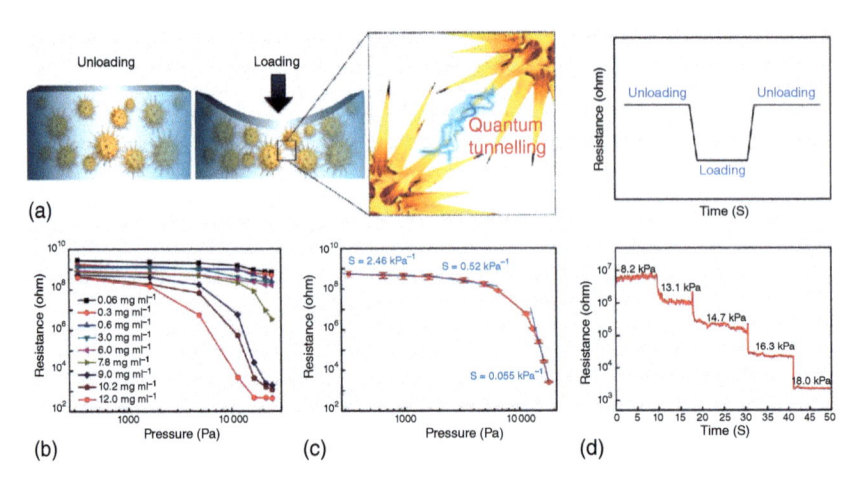

Figure 5.7 (a) Schematic illustrations showing work principles of the piezoresistive pressure sensors fabricated with the polymer nanocomposite. (b) Resistance changes to applied pressures. (c) Pressure sensitivity of the sensors. (d) Static pressure detection. Source: Lee et al. [57]. © 2016, John Wiley & Sons.

fabrication process of the polymer composite films and the pressure sensor is schematically shown in Figure 5.8. First, the graphite (G) powder, PDMS, and NaCl template were mixed to obtain G/PDMS/NaCl composite inks and the inks were bladed onto the sandpaper templates. Then another sandpaper template was covered on the top of the composite film. Thirdly, the microstructured polymer composite films were obtained after thermal annealing and removing the sandpapers. The G/PDMS porous polymer composite film with surface hierarchical

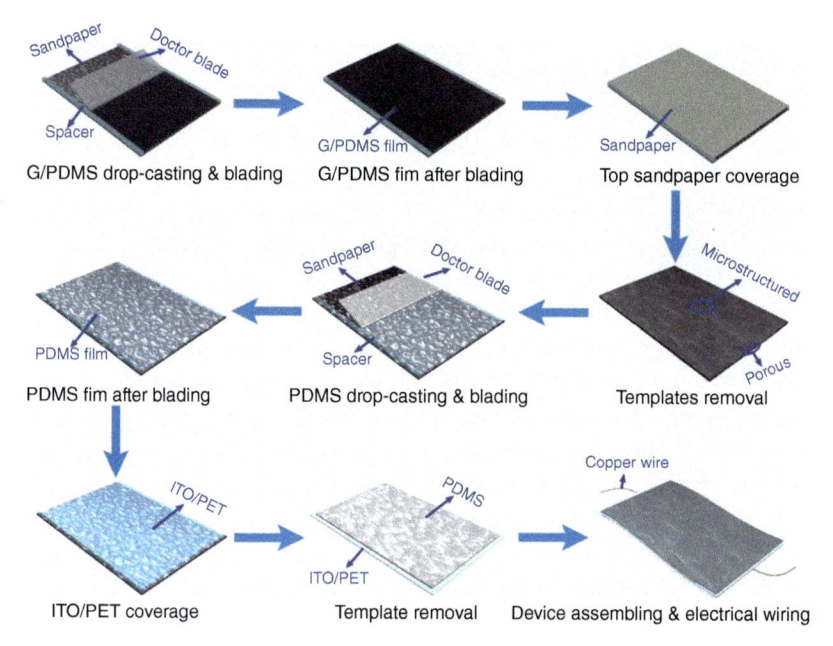

Figure 5.8 Schematic fabrication procedures of the piezoresistive pressure sensor based on the hierarchically microstructured G/PDMS polymer composite. Source: Sun et al. [58].

microstructures (denoted as porous@microstructured) was achieved by removing the NaCl templates in hot water. Adjusting the mass ratio between NaCl and PDMS can obtain composite films with different porosities. Planar polymer composite film with porous microstructures (denoted as porous@planar) and film with surface microstructures but no porous structures (denoted as microstructured only) were also fabricated as the active films for sensor construction. The preparation process of these composite films is similar with that of porous@microstructured composite film by choosing the template combination. Fourthly, microstructured PDMS substrate was fabricated by using the sandpaper template. Finally, microstructured composite film was sandwiched between ITO/PET and ITO/PET/PDMS flexible substrate to construct the pressure sensor devices.

The surface microstructures and sponge microstructures in the polymer nanocomposites play important roles in pressure sensing performance. The SEM image of the sandpaper template was characterized as shown in Figure 5.9a,b. It shows that the sandpaper template is with hierarchical microstructures on its surface. Figure 5.9d,e displays the morphology of G/PDMS composite film with different magnifications. As shown in Figure 5.9, randomly distributed ridges and hierarchical microstructures are obtained by using the sandpaper templates. The diameters are around 10 and 150 μm for the small and large holes, respectively, implying the hierarchical microstructures. The SEM images of NaCl template and cross-sectional composite film are shown in Figure 5.9f,g, respectively. It is observed that the diameters of porous structures inside the polymer composite are similar with the template particles, demonstrating a feasible approach to realize porous

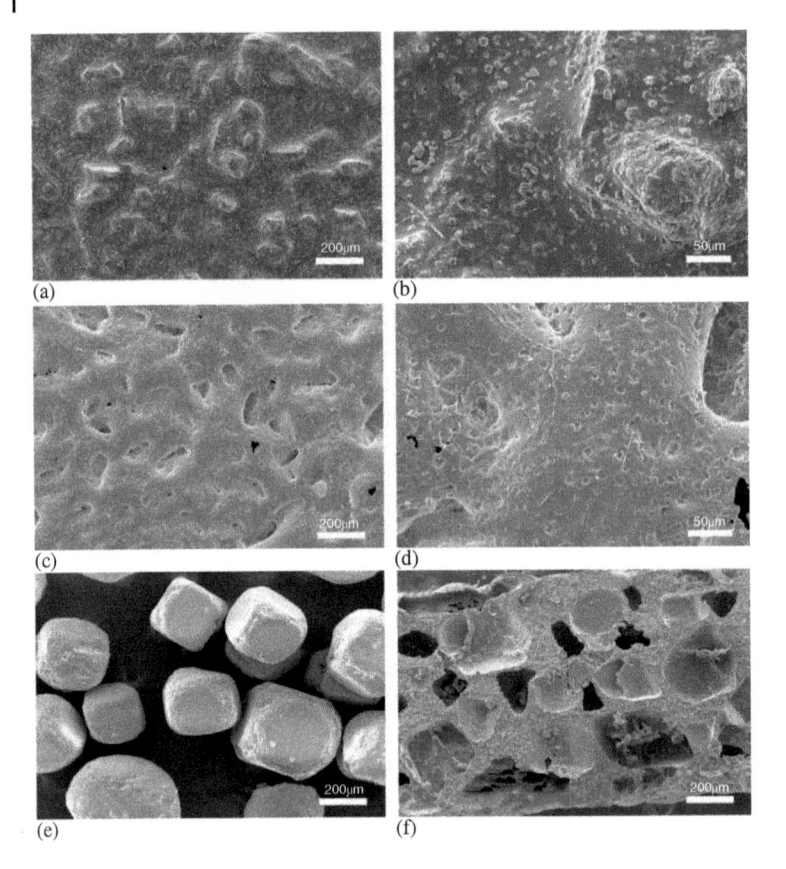

Figure 5.9 (a and b) SEM image of sandpaper template. (c and d) SEM images of the G/PDMS polymer composite. (e) SEM image of NaCl template. (f) Cross-sectional SEM image of the G/PDMS polymer composite. Source: Reproduced with permission Sun et al. [58]. Copyright 2019, Wiley-VCH.

polymer composite films. In order to investigate the sensitivity of the piezoresistive pressure based on the porous G/PDMS polymer composite, the sensors fabricated with three different polymer composite films were characterized in the pressure range from 0 to 150 kPa as shown in Figure 5.10a. The piezoresistive pressure sensor based on the porous@microstructured polymer composite shows much higher pressure sensitivity compared with those of the pressure sensors based on the porous@planar and microstructured only polymer composite films, which is because both the surface microstructures and inside microstructures of the polymer composite play important roles in the enhancement of the piezoresistive pressure sensors (Figure 5.10b). Furthermore, the piezoresistive pressure sensor was demonstrated with a LOD of 5 Pa, and a response time of 8 ms, which is comparable with that of human skin (Figure 5.10c–g). Additionally, the pressure sensor was test up to 25 000 loading/unloading cycles with negligible current decay, indicating an excellent long-term durability (Figure 5.10h). For practical application demonstrations, the piezoresistive pressure sensor was used to detect human wrist

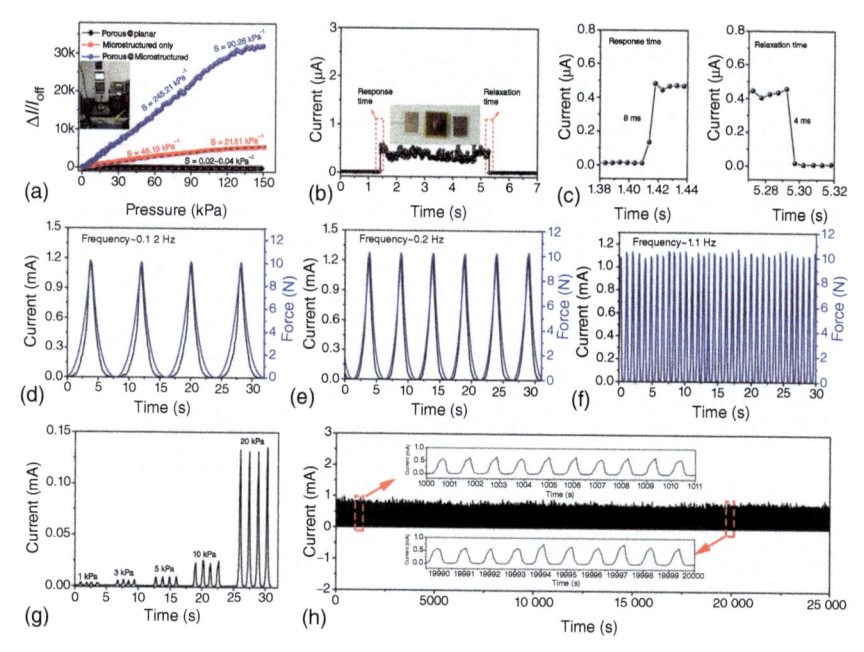

Figure 5.10 Characterization of the piezoresistive pressure sensor fabricated with porous G/PDMS polymer composite. (a) Pressure sensitivity of the pressure sensor in the pressure range from 0 to150 kPa. Inset showing the optical image of the force gauge. (b and c) Limit of detection and response time of the pressure sensor. (d–f) Current response under different frequencies. (g) Current response to various external pressures. (h) Durability measurement. Source: (a) Reproduced with permission Sun et al. [58]. Copyright 2019, Wiley-VCH, (b–h) Sun et al. [58].

pulse, subtle human body motions, and texture roughness of the objects, which will be discussed later in Section 5.4.1.

5.3.3 Piezoelectric and Triboelectric Tactile Sensors Based on Polymer Nanocomposites

Piezoelectric and triboelectric pressure sensors have obtained extensive attentions due to their application in wearable and healthcare monitoring electronics since Wang and Song reported ZnONW arrays for piezoelectric nanogenerator in 2006 [59]. These kinds of pressure sensors are self-powered devices and can detect external pressures without additional power supply. The work principle of piezoelectric pressure sensors is shown in Figure 5.11. These pressure sensors have been widely used to detect the dynamic pressures because of their merits of self-powering capability. However, it is difficult for the piezoelectric and triboelectric sensors to measure the static pressure although these sensors can detect the static pressure by the voltage change theoretically. Nevertheless, undesired voltage decay usually occurs when measuring the static pressures. Therefore, it is highly desired to develop the self-powered pressure sensors, which can detect both the static and dynamic pressures.

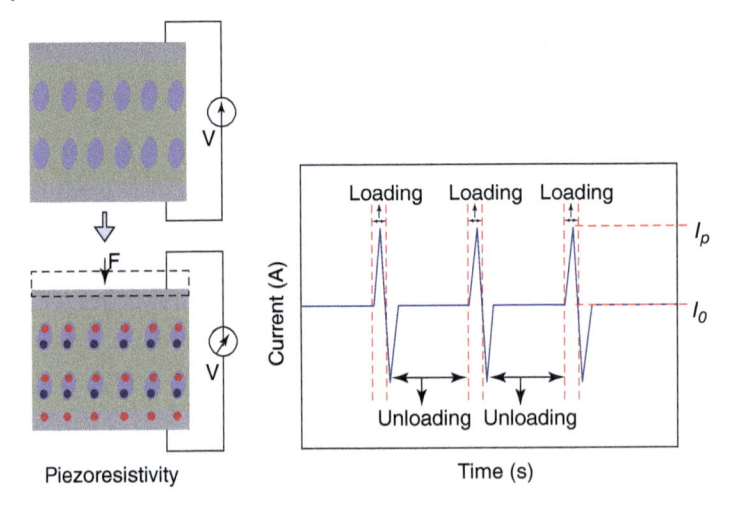

Figure 5.11 Piezoelectric pressure sensing mechanism. Source: Zang et al. [21].

Due to the excellent flexibility and electrical properties of polymer nanocomposites, they have been widely used in the piezoelectric and triboelectric pressure sensors. For example, Lee et al. reported a high-performance piezoelectric nanogenerator fabricated with a chemically-reinforced polymer nanocomposite [41]. They designed piezoelectric polymer nanocomposites to overcome the existing limitations for piezoelectric nanogenerators (PNGs). They achieved the chemical reinforcement of composite system for PNGs. The PNGs devices were realized by using the polymer nanoparticles composited of copolymers and maleic anhydride. Their composite systems show good dispersion of the piezoelectric ceramics (PZT) nanoparticles in the polymer matrices. The device structure of the PZT NP/elastomer composite based PNG is schematically shown in Figure 5.12a,b. Direct coupling the PZT NPs with the polymer matrix can enable the polymer nanocomposite with many advantages because the polymer networks can fix the PZT NPs as shown in Figure 5.12b. Integrating the PZT NPs into the composite film can improve the uniform distribution of PZT NPs, which can enhance the forces applied to the NPs during the film deformation. As depicted in Figure 5.12c, PZT-NH$_2$ is well distributed in the polymer matrix, which is because of the reaction of the amine functional group (–NH$_2$) of PZT with the maleic anhydride. The photograph in Figure 5.12d shows the excellent flexibility of the PNG devices.

The performance of the polymer nanocomposite based PNG is characterized as shown in Figure 5.13. To prove that the piezoelectricity is generated from the polymer nanocomposite, they carried out the switching polarity measurement by changing the connections between the device and the equipment. It is demonstrated that the voltage and current values of the PNG device under forward and reverse connections are the same just the differences in directions as shown in Figure 5.13a,b. A high stable durability is highly required for the practical applications. The durability test of the PNG device is characterized as shown in Figure 5.13c. It is observed that the PNG device exhibited an excellent

5.3 Working Principles and Examples of Polymer Nanocomposite Based Pressure Sensors | **145**

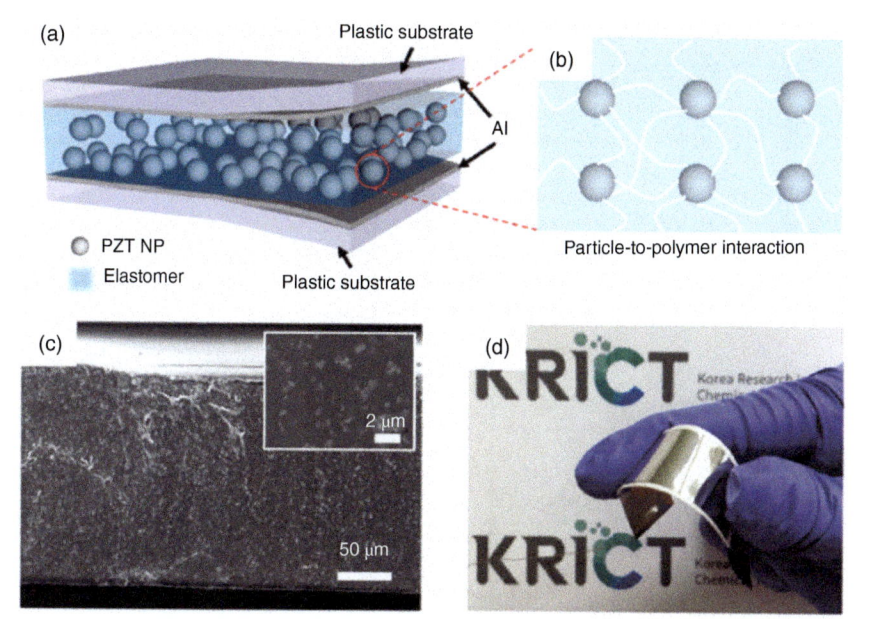

Figure 5.12 (a) Schematics device structure and (b) interconnections between the nanoparticle and the polymer matrix. (c) Cross-sectional SEM image of the polymer composite film. (d) Optical image of PNG device. Source: (a, b, d) Lee et al. [41], (c) Reproduced with permission Lee et al. [41]. Copyright 2018, Royal Society of Chemistry.

signal-to-noise ratio without noticeable voltage decay up to 5000 cycles, indicating excellent output voltage durability.

The polymer nanocomposites have also been extensively employed for triboelectric pressure sensors. Stretchable and flexible electrodes are highly required for flexible, wearable mechanical sensors and e-skins. Nonetheless, the sophisticated fabrication procedures, high cost, and stiff properties of most conductive materials obstruct their applications for flexible electronics. Recently, a stretchable self-powered tactile sensor fabricated with the polymer–nanoparticle composite electrode was reported by Yang et al. [60]. This stretchable, self-powered tactile interface is based on the triboelectric nanogenerator, which includes a stretchable electrode, patterned elastic polymer friction, and encapsulation layer. The self-powered tactile sensor shows excellent performance in detecting the external pressures. The prepare procedure of the pressure sensor is shown in Figure 5.14. Before the fabrication of the tactile sensor devices is the preparation of the polymer nanocomposite composed of liquid metal (LM)-NP materials. The LM-NPs material composed of gallium, indium, and tin is synthesized as shown in Figure 5.14a. The LM-NPs are based on the core–shell structure, where nanoparticles consists of gallium, indium, and stannum work as the core and gallium oxide surrounded by a carbon coating works as the shell, respectively. The electrical stability and the excellent mechanical of the LM-NPs were protected by the oxide shell. The optical image of the LM-NPs in solution and the SEM image LM-NPs are shown in Figure 5.14b,c, respectively, showing uniform structure of the LM-NPs. First,

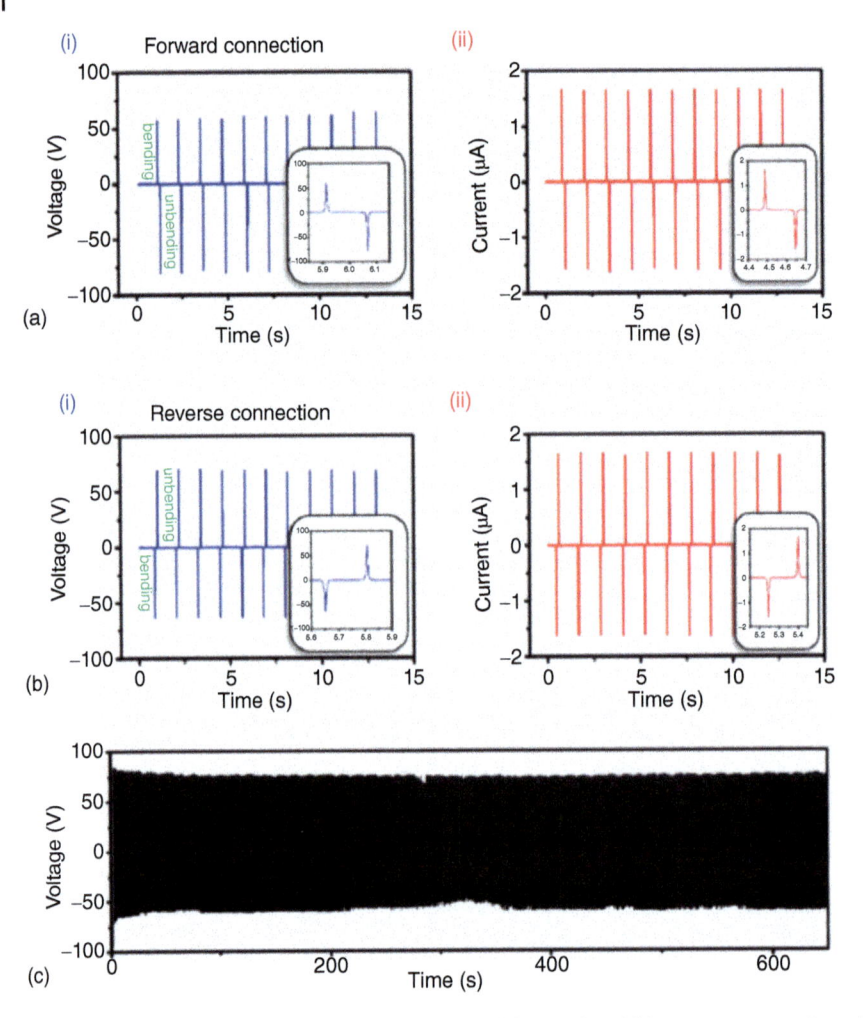

Figure 5.13 The output of the PNG device with (a) forward and (b) reverse connections. (c) The durability test of the PNG device. Source: Lee et al. [41]. Copyright 2018, Royal Society of Chemistry.

the suspension was drop casted onto the PDMS substrate followed by dried in ambient condition. Then, the pre-polymer of PDMS was poured on the galinstan films and cured to form the polymer nanocomposite film. Sufficient pressure is applied to the composite film to change the LM-NPs to the continuous liquid phase by a mechanical sintering method, achieving the highly stretchable and conductive electrodes. Another piece of PDMS layer was spin-coated onto the surface of the LM film, and the surface of the PDMS layer was patterned with microstructures by using different templates during the thermal annealing process to improve the charge density during triboelectrification. The LM-NPs-based TENG can work as the energy harvesting device, where the output current and voltage response to the mechanical stimuli. Therefore, the TENG device can be employed as the stretchable tactile sensor to detect the external forces. A shown in Figure 5.15, the output

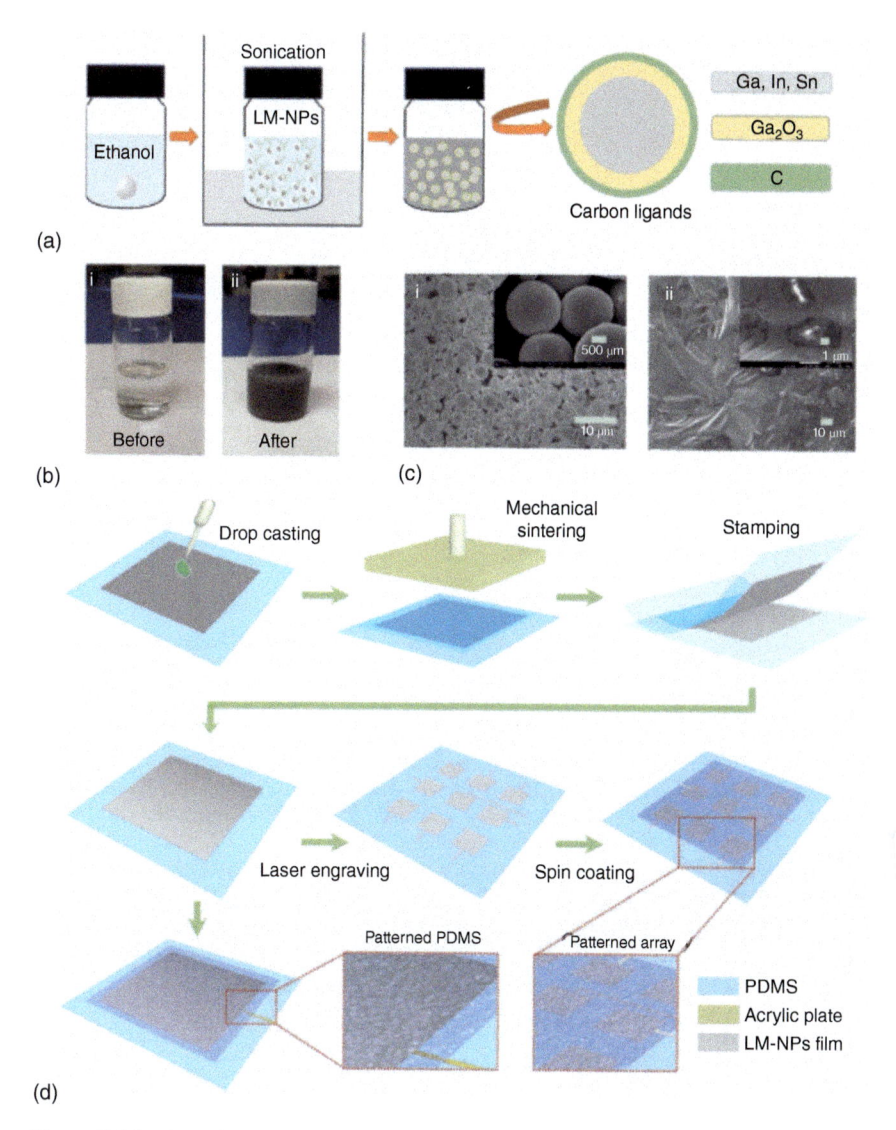

Figure 5.14 (a–c) Preparation procedure and photograph of LM-NPs. (d) Schematic illustration of the fabrication process of self-powered sensor and sensor arrays. Source: (a, c, d) Yang et al. [60], (b) Reproduced with permission Yang et al. [60]. Copyright 2020, Wiley-VCH.

voltage increases with the increased external forces. The output voltage increases from 40 to 256 V when the pressure increases from 3 to 200 kPa, which is attributed to the increased effective contact area under larger external forces and result in the enhanced output performance. The output performance of the tactile sensors with and without the patterns is compared as shown in Figure 5.15b,c, implying that the patterns play critical roles in improving the device performance. As shown in Figure 5.15d,e, the tactile sensor based on the patterned electrode shows 2.54 and 0.43 V kPa^{-1}, respectively, which are higher than those of the tactile sensor without

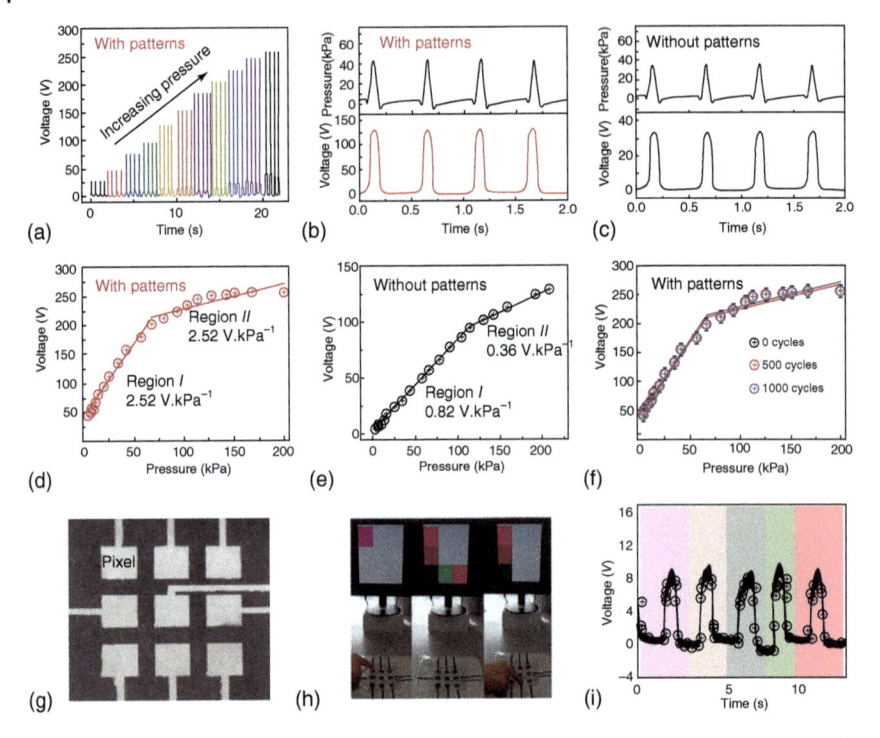

Figure 5.15 Characterization of the self-powered tactile interface and sensor arrays. (a) Output voltage with applied pressure. Output voltage of the sensor (b) with and (c) without patterns. (d) and (e) Pressure sensitivity of the tactile sensor with and without patterns, respectively. (f) Stability test of the sensor under external pressures. (g) Photograph of the tactile sensor array. (h) The tactile sensor arrays working as the tapping board. (i) Collected electrical signal in real time. Source: (a–f, h, i) Yang et al. [60], (g) Reproduced with permission Yang et al. [60]. Copyright 2020, Wiley-VCH.

patterns. 3 × 3 tactile sensor array is fabricated to demonstrate its application for tactile interactive interface. As shown in Figure 5.15h, the applied touch patterns are clearly displayed with measured sensing signals in real time.

5.4 Applications of the Polymer Nanocomposite Based Pressure Sensors

5.4.1 Human Wrist Pulse Detection

In nowadays, the number of people with cardiovascular diseases is increasing gradually. The early stage of atherosclerotic cardiovascular disease is associated with increased mortality and is generally characterized by the thickening of blood vessel walls. Therefore, it is important to detect the cardiovascular disease at the early stage to prevent the serious diseases. Human wrist pulses can lead to a subtle epidermal vibration that could be detected by pressure sensors, and the collected data supplies useful information about the health of the cardiovascular system for

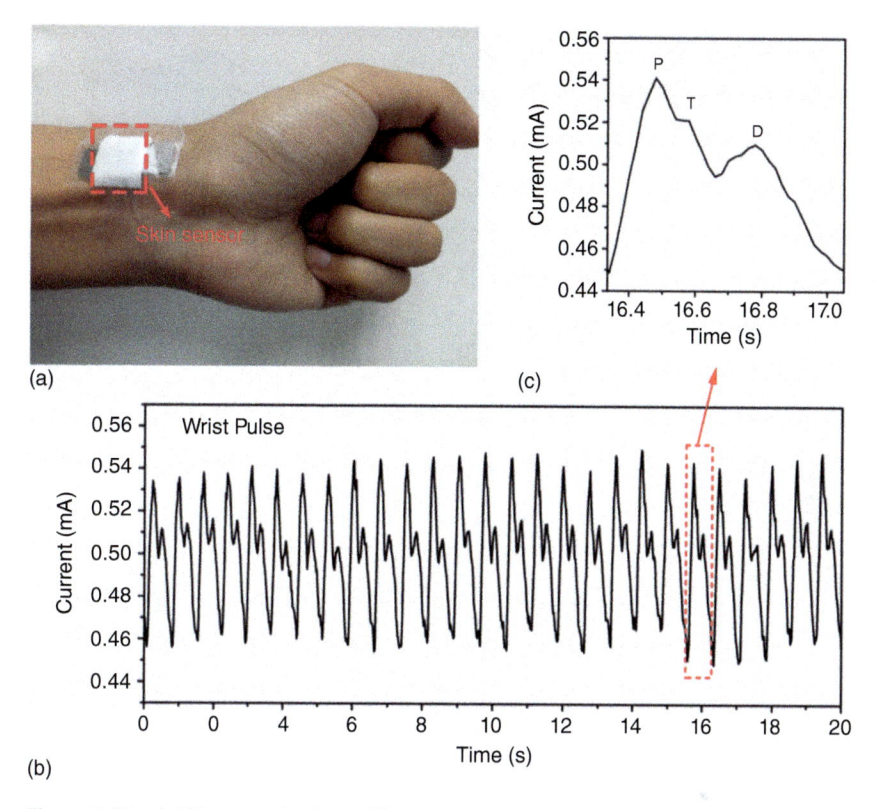

Figure 5.16 (a) Photograph of the skin sensor on human wrist. (b) Wrist pulse collection. (c) A pulse waveform showing three peaks. Source: (a) Reproduced with permission Sun et al. [58]. Copyright 2019, Wiley-VCH, (b, c) Sun et al. [58].

future diagnosis and treatment. We have presented a piezoresistive skin sensor based on the G/PDMS polymer nanocomposite. The pressure sensor was affixed onto the human wrist as shown in Figure 5.16a. The pressure vibration caused by the wrist pulse can change the resistance of the piezoresistive pressure sensor. Owing to the high sensitivity and fast response speed of the skin sensor, the wrist pulse waveforms were collected with high signal resolution by our skin sensor. The current to time curves collected by our skin sensor is shown in Figure 5.16b. It is observed that the measured pulse signals are under normal condition, indicating an estimated beat frequency of 79 bpm, which is in the normal range of adult from 60 to 100 bpm. A pulse waveform is shown in Figure 5.16c, and three characteristic peaks including percussion, tidal, and diastolic wave are exhibited clearly, demonstrating the application of the pressure sensor for disease diagnosis and healthcare monitoring in real time.

5.4.2 Subtle Human Motion Detection

The pressure sensors can be affixed onto skin to monitor the motions of human body because of their excellent flexibility of the polymer nanocomposite based pressure

150 | *5 Polymer Nanocomposites for Pressure Sensors*

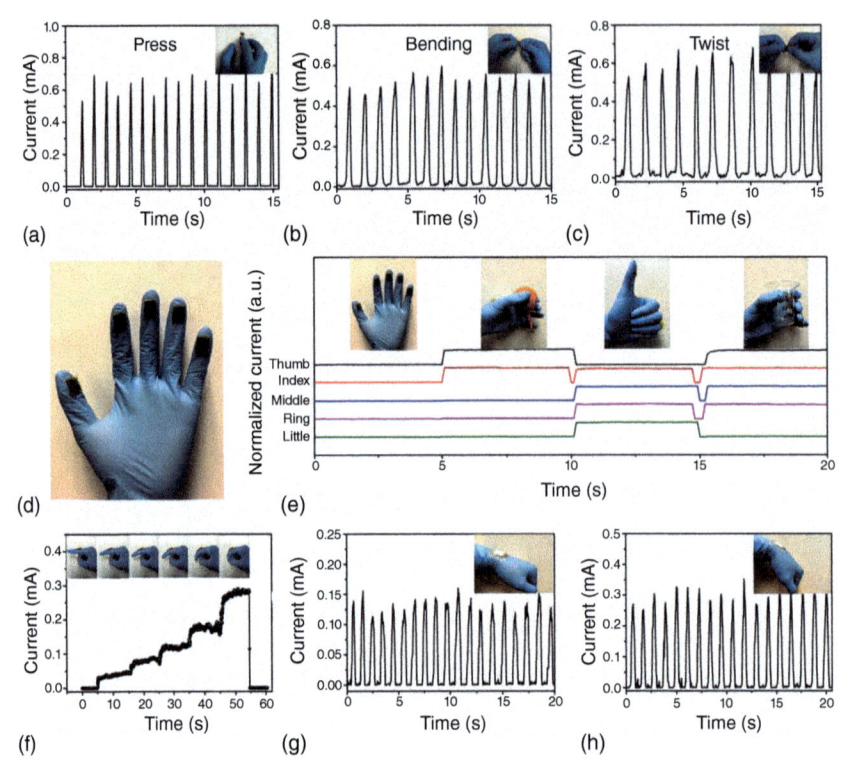

Figure 5.17 (a–c) Detection of press, bending, and twist forces. (d) Photograph of a rubber glove with five pressure sensors. (e) Detection of hand gestures. (f) Detection of finger bending angles. (g, h) Wrist bending under bending angles of 30 and 60. Source: (a) Reproduced with permission Sun et al. [58]. Copyright 2019, Wiley-VCH, (b, c, e–h) Sun et al. [58].

sensors. For our previously reported pressure sensor, not only can it detect different forces including press, bending, and twist forces, but also its versatile potential in subtle human body motion detection has been proved. As depicted in Figure 5.17a, when the piezoresistive pressure sensor is compressed, the contact area between the ITO electrode and the polymer nanocomposite film increases, and the pores in the polymer composite are simultaneously compressed to generate more conductive pathways, resulting in the current increase. The resistance of pressure sensor returned to its initial value immediately after removing the press force, implying fast response time and low-energy consumption in the standby state. Additionally, as depicted in Figure 5.17b,c, the skin sensor was demonstrated with the capability in detecting bending and twist forces, respectively. To demonstrate their application in robots, five pressure sensor devices were affixed onto a rubber glove (Figure 5.17d) to mimicking the grasping and holding objects. Our pressure sensor can detect the bending states and motions of every finger, and distinguished current patterns were obtained to assess the hand gestures (Figure 5.17e). Furthermore, the pressure sensors were used to detect the bending states of index finger and wrist by attaching the pressure on the related parts. As depicted in Figure 5.17f–h, the contacts between the electrode and composite film were enlarged once the index finger or wrist was bent, and the current was improved correspondingly. The current altitude of the pressure sensor can be used to detect the bending angles of human wrist.

5.4.3 Texture Roughness Detection

It is notable that the biological fingertip skin can detect the surface texture of the object. Mimicking the biological skin, our pressure sensor based on the porous G/PDMS polymer nanocomposite can measure the tiny friction forces generated in the process of sliding the pressure sensor on the surface of the objects to reveal the roughness. Schematic device configuration and work principle of the sensor are shown in Figure 5.18a–c, respectively. As shown in Figure 5.18b, there are almost no contacts between the polymer nanocomposite and ITO electrodes when no pressure is applied. As shown in Figure 5.18c, the contact area and conductive pathways between the polymer nanocomposite films increased when the pressure sensor slides across the surface of the test objects, and the resistance of the device decreased simultaneously. Sandpapers and printing paper are used as the test materials for the measurement. The surface morphology of the papers is characterized by SEM and shown in Figure 5.18d–f. The texture roughness measurement was carried out by affixing the skin sensor on the index finger and horizontally moving across the material surface. During the test, a tiny press pressure was added to the

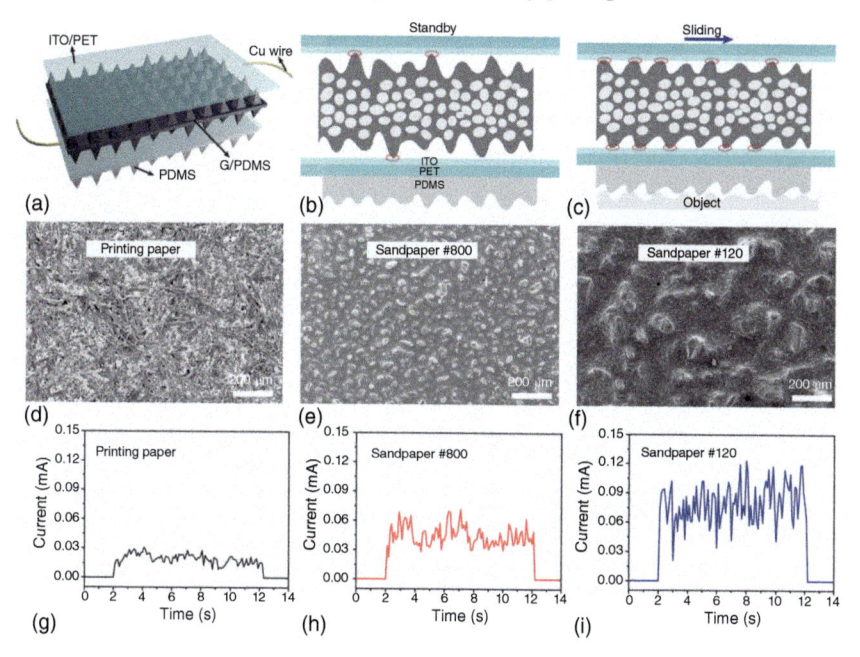

Figure 5.18 (a–c) Schematic diagram of texture roughness recognition of the skin sensor fabricated with G/PDMS polymer nanocomposite, in which, (a) schematic device structure, (b and c) sensing mechanism in roughness detection. (d–f) SEM images of printing paper and sandpapers. (g–i) Current signals measured from different textures. Source: (a–c, g–i) Sun et al. [58], (d) Reproduced with permission Sun et al. [58]. Copyright 2019, Wiley-VCH.

sensor to make the contact between the sensor and the test materials. As shown in Figure 5.18g–i, when the pressure sensor slides across the surface of the papers with different roughness, different friction forces between the sensor and the object were generated. Meanwhile, distinct current signal patterns were obtained. From Figure 5.18g–i, we observe that when the pressure sensor slides across the objects, the device shows the smallest current fluctuation of 52% for the smooth printing paper, an increased fluctuation of 240% for the #800 sandpaper. The highest current fluctuation of 500% was observed for the #120 sandpaper, demonstrating that our device can be used to detect the roughness of texture.

5.4.4 E-skin Application

Because of the stretchable and flexible property of polymer nanocomposite based pressure sensor, they can be used as the artificial skin for robots. Recently, various works have demonstrated the pressure sensor arrays fabricated with the polymer nanocomposites for electronic skin applications. For example, a PVDF/reduced graphene oxide (rGO) polymer composite nanofiber film was fabricated by Lou et al. for flexible piezoresistive pressure sensor [2]. The rGO wrapped PVDF NF film was prepared by a feasible solution process. A 5×5 pressure sensor array was prepared based on the rGO/PVDF nanofibers and the optical image of the sensor arrays wrapped around human wrist is shown in Figure 5.19a, indicating the excellent flexibility. Figure 5.19b shows the equivalent circuit of the pressure sensor array. When three different metal letters C, A, and S are put on the pressure sensor array, the pressure distribution of these letters applied on the pressure sensor array is revealed by the relative current change as shown in Figure 5.19c–h. The pressure distribution is well revealed by the pressure sensor array, implying its potential application as e-skin for robots.

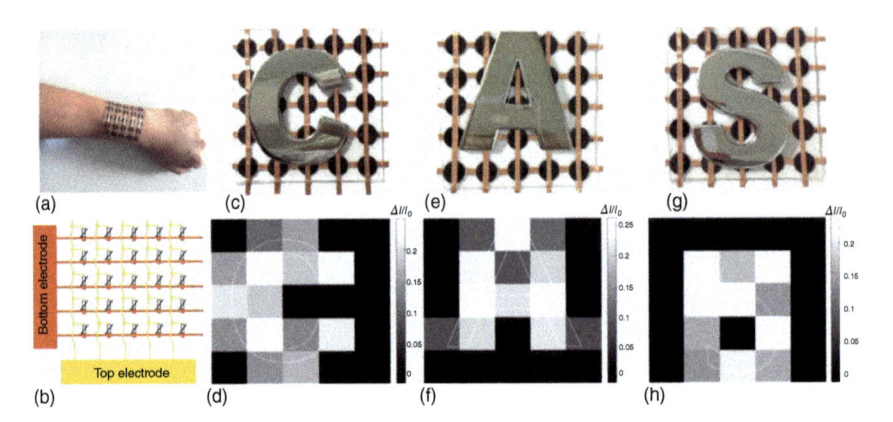

Figure 5.19 Pressure sensor array based on PVDF/rGO polymer nanocomposite for spatial pressure detection. (a) Pressure sensor arrays wrapped on human wrist. (b) Logic circuits of the pressure sensor arrays. (c, e and g) Photographs of metal letters "C" "A" "S" loaded on the pressure sensor arrays. (d, f, and h) Corresponding spatial pressure distribution detected by the pressure sensor arrays. Source: (a) Reproduced with permission Lou et al. [2]. Copyright 2016, Elsevier, (b–h) Lou et al. [2].

5.5 Performance of Pressure Sensors with the Polymer Nanocomposites Reported Over the Past Decade | 153

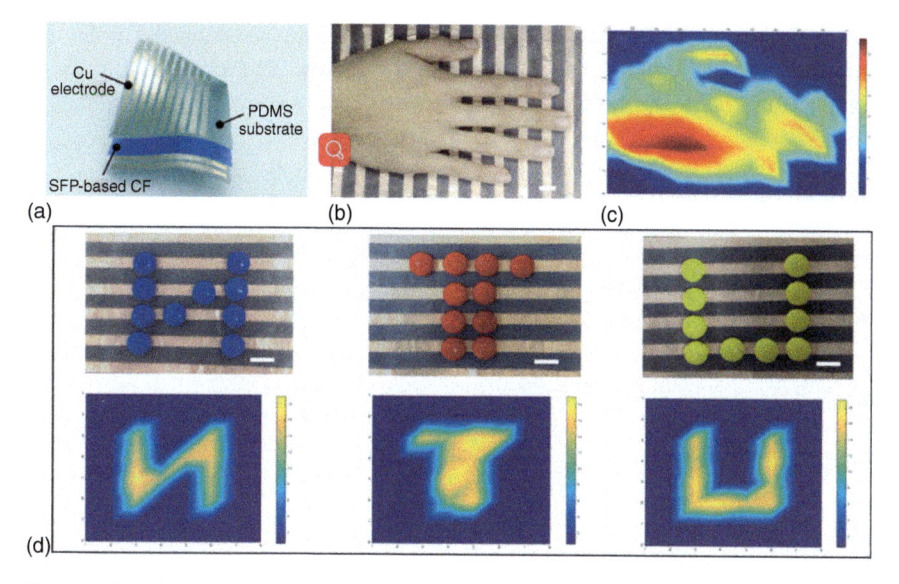

Figure 5.20 Pressure sensor array fabricated with natural microcapsule actuator based polymer nanocomposite for spatial pressure detection. (a) Schematic diagram of the pressure sensor arrays. (b) Photograph of a human hand loaded on the pressure sensor arrays. (c) Spatial pressure distribution revealed by the current variation. (d) Photograph of Letters "N" "T" "U" and related current mapping. Source: (a) Reproduced with permission Wang et al. [61]. Copyright 2017, Elsevier, (b–d) Wang et al. [61].

In another example, a flexible pressure sensor based on natural microcapsule actuator was developed by Wang et al. [61]. The sensor exhibits an LOD of 1.6 Pa and can discriminate both static and dynamic tactile stimuli. Additionally, a 12 × 8 pixel array with a rubber composite film was fabricated on a microstructured PDMS substrate as depicted in Figure 5.20a. When the hand is put on the pressure sensor arrays (Figure 5.20b), the spatial pressure distribution on the pressure sensor array is revealed by the current variation mapping as shown in Figure 5.20c. Additionally, three letters of "N," "T," and "U" organized by individual objects were put on top of the pressure sensor arrays (Figure 5.20d). Distinguished relative current change at each location clearly indicates that the sensor array can reveal spatial pressure successfully with resolved images, demonstrating it promising application for flexible electronic skin in future.

5.5 Performance of Pressure Sensors with the Polymer Nanocomposites Reported Over the Past Decade

The nanomaterial fillers employed in the polymer nanocomposites are mainly metal-based nanomaterials (e.g. Au NWs, Ag NWs, and Ag NPs), carbon-based nanomaterials [e.g. graphene, GO, rGO, carbon nanowires (CNWs), and carbon nanotubes (CNTs)], and polymers. The recently developed pressure sensors based on the polymer nanocomposites are listed in Table 5.1. The sensitivity, LOD, and response time of the pressure sensors are important in selecting the polymer nanocomposite materials for pressure sensor devices. Therefore, the related parameters of the pressure sensor devices are compared as shown in Table 5.1.

5.6 Conclusion

This chapter presents a brief introduction and overview of the recent research on the flexible pressure sensors based on the polymer nanocomposite materials. Various polymer nanocomposites have been developed into the pressure sensors in forms of capacitive, piezoresistive, piezoelectric, and triboelectric sensors. The sensing mechanisms of the pressure sensors based on different working principles are introduced and discussed in this chapter. Additionally, the representative examples for different types of pressure sensors are listed and discussed as well.

The wearable and flexible pressure sensors for healthcare monitoring, subtle human motion detection, and spatial pressure distribution mapping have been demonstrated in many previously reported works. The potential applications of the flexible pressure sensors have led to increasing interest in the replacement of traditional MEMS-based pressure sensors. For the traditional MEMS sensors, the devices are generally constructed using silicon (and its derivatives) as both sensing and substrate materials. For flexible and wearable pressure sensors, various nanomaterials such as graphene, GO, rGO, and CNWs have been imparted to the polymer matrix as sensing layer for pressure sensor construction. As introduced in this chapter, many polymer nanocomposite based pressure sensors with excellent performance have been demonstrated for various applications. Regarding sensor output signals, considerable achievements have been realized in the development of highly sensitive, rapidly responsive pressure sensor devices, and many sensors have already met the pressure magnitude and frequency response requirements for real-time wearable applications.

Despite the achievements in the development of polymer nanocomposite based pressure sensors, much work remains to be done to ensure the stability and durability of polymer nanocomposite based pressure sensor devices. Additionally, the development of pressure sensors that are large-scale production compatible is far from required. Therefore, extensive research activities have been devoted to developing flexible and wearable sensors by using cost-effective and large-scale compatible approaches. Moreover, a lot of work is required to integrate the pressure sensor devices with other sensor devices and the power supply together. We firmly believe that pressure sensors based on the polymer nanocomposites will expand to diverse of wearable applications in the near future, including wearable pressure pulse-sensing devices, e-skin, and phonation rehabilitation.

References

1 Ho, D.H., Sun, Q., Kim, S.Y. et al. (2016). *Adv. Mater.* 28: 2601.
2 Lou, Z., Chen, S., Wang, L. et al. (2016). *Nano Energy* 23: 7.

3 Mu, C., Song, Y., Huang, W. et al. (2018). *Adv. Funct. Mater.* 28: 1707503.
4 Nela, L., Tang, J., Cao, Q. et al. (2018). *Nano Lett.* 18: 2054.
5 Ramuz, M., Tee, B.C., Tok, J.B., and Bao, Z. (2012). *Adv. Mater.* 24: 3223.
6 Someya, T., Sekitani, T., Iba, S. et al. (2004, 101). *Proc. Natl. Acad. Sci. USA*: 9966.
7 Takei, K., Takahashi, T., Ho, J.C. et al. (2010). *Nat. Mater.* 9: 821.
8 Tee, B.C.K., Wang, C., Allen, R., and Bao, Z. (2012). *Nat. Nanotechnol.* 7: 825.
9 Wang, Q., Jian, M., Wang, C., and Zhang, Y. (2017). *Adv. Funct. Mater.* 27: 1605657.
10 Boutry, C.M., Nguyen, A., Lawal, Q.O. et al. (2015). *Adv. Mater.* 27: 6954.
11 Choong, C.L., Shim, M.B., Lee, B.S. et al. (2014). *Adv. Mater.* 26: 3451.
12 Gao, Y., Ota, H., Schaler, E.W. et al. (2017). *Adv. Mater.* 29: 1701985.
13 Gong, S., Schwalb, W., Wang, Y. et al. (2014). *Nat. Commun.* 5: 3132.
14 Han, X., Chen, X., Tang, X. et al. (2016). *Adv. Funct. Mater.* 26: 3640.
15 Shin, K.-Y., Lee, J.S., and Jang, J. (2016). *Nano Energy* 22: 95.
16 Song, Y., Chen, H., Su, Z. et al. (2017). *Small* 13: 1702091.
17 Trung, T.Q. and Lee, N.E. (2016). *Adv. Mater.* 28: 4338.
18 Wang, X., Gu, Y., Xiong, Z. et al. (2014). *Adv. Mater.* 26: 1336.
19 Wang, Y., Wang, L., Yang, T. et al. (2014). *Adv. Funct. Mater.* 24: 4666.
20 Yang, T., Jiang, X., Zhong, Y. et al. (2017). *ACS Sensors* 2: 967.
21 Zang, Y., Zhang, F., Di, C.-a., and Zhu, D. (2015). *Mater. Horiz.* 2: 140.
22 Zang, Y., Zhang, F., Huang, D. et al. (2015). *Nat. Commun.* 6: 6269.
23 Gerratt, A.P., Michaud, H.O., and Lacour, S.P. (2015). *Adv. Funct. Mater.* 25: 2287.
24 Kim, J., Lee, M., Shim, H.J. et al. (2014). *Nat. Commun.* 5: 5747.
25 Lei, Z., Wang, Q., Sun, S. et al. (2017). *Adv. Mater.* 29: 1700321.
26 Parida, K., Bhavanasi, V., Kumar, V. et al. (2017). *Nano Res.* 10: 3557.
27 Chun, S., Hong, A., Choi, Y. et al. (2016). *Nanoscale* 8: 9185.
28 Kang, M., Kim, J., Jang, B. et al. (2017). *ACS Nano* 11: 7950.
29 Luo, Y., Shao, J., Chen, S. et al. (2019). *ACS Appl. Mater. Interfaces* 11: 17796.
30 Shuai, X., Zhu, P., Zeng, W. et al. (2017). *ACS Appl. Mater. Interfaces* 9: 26314.
31 Wang, J., Jiu, J., Nogi, M. et al. (2015). *Nanoscale* 7: 2926.
32 Hu, N., Karube, Y., Arai, M. et al. (2010). *Carbon* 48: 680.
33 Kim, K., Jung, M., Kim, B. et al. (2017). *Nano Energy* 41: 301.
34 Kim, K.-H., Hong, S.K., Jang, N.-S. et al. (2017). *ACS Appl. Mater. Interfaces* 9: 17499.
35 Liu, Y., Zhang, D., Wang, K. et al. (2016). *Composites Part A* 80: 95.
36 Luo, Y., Wu, D., Zhao, Y. et al. (2019). *Org. Electron.* 67: 10.
37 Tang, Y., Zhao, Z., Hu, H. et al. (2015). *ACS Appl. Mater. Interfaces* 7: 27432.
38 Wang, L. (2016). *Sens. Actuators A: Phys.* 252: 89.
39 Zhang, R., Deng, H., Valenca, R. et al. (2013). *Compos. Sci. Technol.* 74: 1.
40 Bhaskar, D., Hyun, K.D., Krishna, B.L., and Su, Y.J. (2018). *Appl. Energy* 230: 865–874.
41 Eun Jung Lee, T.Y.K., Kim, S.-W., Jeong, S. et al. (2018). *Energy Environ. Sci.* 11: 1425.
42 Huan, Y., Zhang, X., Song, J. et al. (2018). *Nano Energy* 50: 62.

43 Chun, J., Kang, N.-R., Kim, J.-Y. et al. (2015). *Nano Energy* 11: 1.
44 Li, J., Chen, S., Liu, W. et al. (2019). *J. Phys. Chem. C* 123: 11378.
45 Shin, D.-J., Ji, J.-H., Kim, J. et al. (2019). *J. Alloys Compd.* 802: 562.
46 Cheon, S., Kang, H., Kim, H. et al. (2018). *Adv. Funct. Mater.* 28: 1703778.
47 Garcia, C., Trendafilova, I., and Sanchez del Rio, J. (2019). *Nano Energy* 56: 443.
48 Shi, K., Huang, X., Sun, B. et al. (2019). *Nano Energy* 57: 450.
49 Wang, S., Liu, B., Duan, Z. et al. (2019). *Nano Energy* 58: 312.
50 Luo, N., Huang, Y., Liu, J. et al. (2017). *Adv. Mater.* 29: 1702675.
51 Qiu, A., Li, P., Yang, Z. et al. (2019). *Adv. Funct. Mater.* 29: 1806306.
52 Ding, Y., Xu, T., Onyilagha, O. et al. (2019). *ACS Appl. Mater. Interfaces* 11: 6685.
53 Li, Y., Han, D., Jiang, C. et al. (2019). *Adv. Mater. Technol.* 4: 1800504.
54 Zhu, B., Niu, Z., Wang, H. et al. (2014). *Small* 10: 3625.
55 Nie, P., Wang, R., Xu, X. et al. (2017). *ACS Appl. Mater. Interfaces* 9: 14911.
56 Oh, J.H., Hong, S.Y., Park, H. et al. (2018). *ACS Appl. Mater. Interfaces* 10: 7263.
57 Lee, D., Lee, H., Jeong, Y. et al. (2016). *Adv. Mater.* 28: 9364.
58 Sun, Q.J., Zhao, X.H., Zhou, Y. et al. (2019). *Adv. Funct. Mater.* 29: 1808829.
59 Wang, Z.L. and Song, J. (2006). *Science* 312: 242.
60 Yang, Y., Han, J., Huang, J. et al. (2020). *Adv. Funct. Mater.*: 1909652.
61 Wang, L., Jackman, J.A., Tan, E.-L. et al. (2017). *Nano Energy* 36: 38.
62 Tien, N.T., Jeon, S., Kim, D.I. et al. (2014). *Adv. Mater.* 26: 796–804.
63 Graz, I., Krause, M., Bauer-Gogonea, S. et al. (2009). *J. Appl. Phys.* 106: 034503.
64 Persano, L., Dagdeviren, C., Su, Y. et al. (2013). *Nat. Commun.* 4: 1633.
65 Bin Yao, H., Ge, J., Wang, C.F. et al. (2013). *Adv. Mater.* 25: 6692–6698.
66 Pan, L., Chortos, A., Yu, G. et al. (2014). *Nat. Commun.* 5: 3002.
67 Hou, C., Wang, H., Zhang, Q. et al. (2014). *Adv. Mater.* 26: 5018–5024.
68 Jung, S., Kim, J.H., Kim, J. et al. (2014). *Adv. Mater.* 26: 4825–4830.
69 Park, J., Lee, Y., Hong, J. et al. (2014). *ACS Nano* 8: 4689–4697.
70 Park, H., Jeong, Y.R., Yun, J. et al. (2015). *ACS Nano* 9: 9974.
71 Bae, G.Y., Pak, S.W., Kim, D. et al. (2016). *Adv. Mater.* 28: 5300.
72 Xia, K., Wang, C., Jian, M. et al. (2018). *Nano Res.* 11: 1124.
73 Jian, M., Xia, K., Wang, Q. et al. (2017). *Adv. Funct. Mater.* 27: 1606066.
74 Liu, M., Pu, X., Jiang, C. et al. (2017). *Adv. Mater.* 29: 1703700.
75 Huang, M., Pascal, T.A., Kim, H. et al. (2011). *Nano Lett.* 11: 1241–1246.
76 Wan, S., Bi, H., Zhou, Y. et al. (2017). *Carbon* 114: 209–216.
77 Viry, L., Levi, A., Totaro, M. et al. (2014). *Adv. Mater.* 26: 2659–2664.
78 Lee, J., Kwon, H., Seo, J. et al. (2015). *Adv. Mater.* 27: 2433.
79 Yao, S. and Zhu, Y. (2014). *Nanoscale* 6: 2345.
80 Kwon, D., Lee, T.-I., Shim, J. et al. (2016). *ACS Appl. Mater. Interfaces* 8: 16922.
81 Park, S., Kim, H., Vosgueritchian, M. et al. (2014). *Adv. Mater.* 26: 7324–7332.
82 Pang, C., Koo, J.H., Nguyen, A. et al. (2015). *Adv. Mater.* 27: 634–640.
83 Mannsfeld, S.C.B., Tee, B.C.K., Stoltenberg, R.M. et al. (2010). *Nat. Mater.* 9: 859–864.
84 Schwartz, G., Tee, B.C.-K., Mei, J. et al. (2013). *Nat. Commun.* 4: 1859.
85 Li, T., Luo, H., Qin, L. et al. (2016). *Small* 12: 5042–5048.

6

The Application of Polymer Nanocomposites in Energy Storage Devices

Ningyuan Nie[1,2,3], Mengmeng Hu[1,2], Jie Liu[1,2], Jiangqi Wang[1,2], Panpan Wang[1,2], Hua Wang[1,2], Zhenyuan Ji[1,2], Zhe Chen[1,2], and Yan Huang[1,2,3]

[1] Harbin Institute of Technology (Shenzhen), Department of Materials Science and Engineering, State Key Laboratory of Advanced Welding and Joining, Pingshan 1st Road, Nanshan District, Shenzhen, Guangdong 518055, China
[2] Harbin Institute of Technology (Shenzhen), Department of Materials Science and Engineering, Flexible Printed Electronics Technology Centre, Pingshan 1st Road, Nanshan District, Shenzhen, Guangdong 518055, China
[3] Harbin Institute of Technology (Shenzhen), School of Materials Science and Engineering, Department of Materials Science and Engineering, Pingshan 1st Road, Nanshan District, Shenzhen, Guangdong 518055, China

6.1 Introduction

Due to the increasing energy crisis of traditional fossil fuel consumption, the development of renewable energy storage resources such as wind energy and solar energy is urgently needed [1–4]. Because of the intermittent nature of these renewable energy sources, reliable energy storage systems are in urgent demand to store and output energy in a stable way [3]. Thus, scientists have made a lot of efforts to the research of materials for batteries and supercapacitors which are the main energy storage systems, including the exploration of new materials for electrodes, electrolytes and separators, which can improve the electrochemical performance and physical property of these energy storage devices.

Polymer nanocomposites (PNCs), defined as a kind of multiphase material made of two or more kinds of materials, such as metal, ceramic, or other inorganic non-metallic materials besides polymers, exhibit excellent physicochemical properties that cannot be obtained by using any individual components alone. To be specific, the PNCs always provide higher specific capacity and better cycling stabilities for energy storage devices when they are applied in electrodes. What's more, the PNCs electrodes sometimes contribute other functions for energy storage devices such as flexibility and self-healing capability [5, 6]. As for electrolytes consisting PNCs, most of them can deliver higher ion conductivity than those that are made of individual components for both batteries and supercapacitors [7–11]. Moreover, batteries and supercapacitors are often equipped with separators made of PNCs with better cycling performance and mechanical properties.

Polymer Nanocomposite Materials: Applications in Integrated Electronic Devices, First Edition.
Edited by Ye Zhou and Guanglong Ding.
© 2021 WILEY-VCH GmbH. Published 2021 by WILEY-VCH GmbH.

Here, we discuss the topic of PNCs applied in energy storage devices by summarizing the recent development of relevant research. We first give an introduction on the definition and development of PNCs and energy storage device. According to the description of energy storage devices, they can be classified to battery and supercapacitor, which both have three main sections: electrode, electrolyte, and separator. Electrodes will be first discussed according to their use and composition. After that, discussion about electrolytes and separators will be introduced in detail.

6.2 Electrodes

Electrode materials are the most critical component of energy storage devices, which play a decisive role in the electrochemical performance of the whole devices. We will discuss electrode materials for batteries and supercapacitors separately.

6.2.1 For Battery

Batteries, with their high capacity, high energy density and good rate performance, have been widely investigated and applied in the growing energy consumption of electric vehicles and hybrid electric vehicles, including lithium-ion batteries, lithium–sulfur batteries, sodium ion batteries, and so on [12–22]. Thus, thousands of kinds of electrodes have been designed by researchers for the sake of higher capacity, longer cycling life, more environmental benignancy, and lower cost [16, 23–26]. Though inorganic materials such as lithium transition-metal oxides or phosphates for lithium ion batteries, lithium polysulfide intermediates for lithium–sulfur batteries [27], P_2-$Na_{2/3}$[$Fe_{1/2}Mn_{1/2}$]O_2 for sodium ion batteries [17], and organic materials such as poly(anthraquinonyl sulfide) (PAQS), disodium terephthalate hold good electrochemical performance, they are limited by their single component and function. However, PNCs, as a multiphase material, are always used to combine the advantages of two or more kinds of materials to achieve the higher performance and multifunctionality. These PNCs are discussed by their composition.

6.2.1.1 Polymer–Graphene/Carbon Nanotube

Rechargeable lithium batteries are considered as one of the most promising energy storage technologies, especially as power sources for emerging applications, such as plug-in hybrid electric vehicles and electric vehicles. In these applications, besides high energy density, high power density is also essential [28, 29]. In traditional lithium-ion batteries, cathode materials are usually lithium transition metal oxides or phosphates (such as $LiCOO_2$ or $LiFePO_4$), which can be reversibly de-/re-embedded with Li^+ ions. However, the diffusion kinetics of lithium ion in inorganic materials is slow, which will impair charge and discharge performance of lithium battery [30, 31]. In addition to the traditional inorganic materials, organic cathode materials, including small molecules and polymers, have become the research hotspot of the new generation of green lithium battery electrodes in recent years due to their sustainability and environmental friendliness [32, 33]. Nevertheless, the poor conductivity of organic materials limits the rate performance of electrodes.

To obtain better materials for electrodes and optimize the electrochemical performance of lithium batteries, researchers have made many attempts and PNCs are commonly used. Graphene and graphene sheets (GNS) are two-dimensional thin sheets with one or more atomic thicknesses. They are composed of sp^2 carbon atoms arranged in honeycomb structure. Because of their special electronic conductivity, they are considered as excellent conductive additives in nanocomposites [34, 35]. What's more, graphene and GNS hold high specific surface area (theoretical value 2630 $m^2 g^{-1}$), which improves the interface contact between active materials and the substrate. Guo et al. intended to enhance the electrochemical performance of nitroxide free radical polymer cathode by using graphene in dispersion–deposition process. However, the loading rate of active materials in the composite electrode is only 10%, while that of graphene is as high as 60% [36]. Taking PAQS and polyimide (PI) as examples, polymer–graphene nanocomposites with high dispersion graphene were introduced by Song et al. via one-pot synthesis (Figure 6.1a). Functionalized graphene sheets (FGSs) were prepared by graphene thermal expansion method because of their higher electrical conductivity than chemically reduced graphene oxide (RGO). FGS were first dispersed in 1-methyl-2-pyrrolidinone (NMP) by sonication. Then, 1,5-dichloroanthraquinone (DCAQ) or 1,4,5,8-naphthalene triformic anhydride (NCTDA) were added as precursors of PAQS or PI, and completely dissolved in NMP. Polymer chains are formed by condensation polymerization at high temperatures after the addition of a condensate (anhydrous sodium sulfide for PAQS or ethylenediamine for PI). In the final product, FGSs are wrapped and uniformly embedded in the polymer matrix. In order to find the morphology of the polymer–graphene nanocomposites, the samples were characterized by scanning electron microscope (SEM). The SEM image in Figure 6.1b indicates that each FGS is several microns in size, with a uniform and thick PAQS coating on both sides of the FGS. For PI-FGS nanocomposites, the SEM image in Figure 6.1c showed that the FGS was uniformly coated with flower-like PI polymer particles, which is smaller than those found in pure PI. The electrochemical performance of PI samples was improved. Although the theoretical capacity of PI based on the four-electron redox process is 367 mA h g^{-1}, only two electrons can normally be reversibly transferred in the actual charging and discharging process, reaching about half of the theoretical capacity. As shown in Figure 6.1d, the discharge capacity of pure PI at 0.1 C is 156 mA h g^{-1}, and the utilization rate is only 42% compared with the theoretical capacity of 367 mA h g^{-1}. With the addition of 6% graphene, the capacity of PI-FGS increased to 172 mA h g^{-1} with a corresponding polymer utilization rate of 49%. In other words, graphene can increase the number of electrons per unit of PI from 1.7 to 2. Similar performance improvement was also observed for the PAQS-FGS samples [37].

Besides better electrochemical performance, PNCs are also explored for the multifunctionality of battery. Flexible and wearable electronic products have

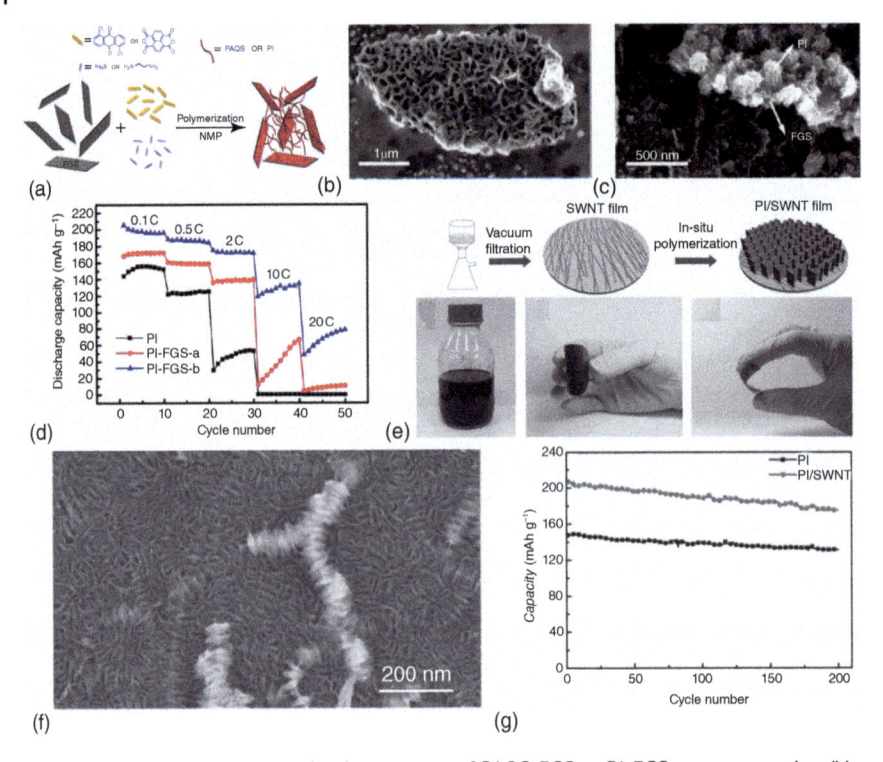

Figure 6.1 (a) In situ polymerization process of PAQS-FGS or PI-FGS nanocomposite. (b) SEM image of PAQS-FGS-a. (c) SEM image of PI-FGS-b. (d) Cycling performance of PI and its composites at different C-rates. Source: (a, d) Song et al. [37], (b) Reproduced with permission Song et al. [37]. Copyright 2012, American Chemical Society Publication. (e) Schematic of the preparation process of PI/SWNT film; photographs of SWNT aqueous dispersion, SWNT film, and PI/SWNT film. (f) SEM images of PI/SWNT film after reaction for six hours. (g) Cycling performance of PI and PI/SWNT at a rate of 0.5 C. Source: (e) Reproduced with permission Wu et al. [38]. Copyright 2014, Wiley-VCH, (g) Wu et al. [38].

attracted extensive attention from academia and industry due to their broad application prospects in medical monitoring, portable military equipment, intelligent textiles, etc. Thus, flexible energy storage devices are in demand to power those electronic devices. Lithium batteries are potential flexible power sources with high volumetric energy density and long cycle life [12, 13, 39–43]. However, owing to their inherent fragility, inorganic electrode materials cannot provide enough flexibility even they were loaded on curvilinear surfaces such as carbon nanotubes (CNTs), carbon nanofiber, graphene paper, and graphene foam [44–53]. Thus, polymers were considered because of their inherent flexibility, including conducting polymers such as polyacetylene, radical polymers such as nitroxides, and so on [54, 55]. Among all the organic materials considered, PI is esteemed to be a potential material for the electrodes because of its high theoretical capacity [56]. Wu et al. successfully built an electrode with single-wall carbon nanotube (SWCNT) film as current collector and PI as active materials [38]. As shown in Figure 6.1e, the SWNT film was fabricated through vacuum filtration of SWNT aqueous dispersion. SWNT film was fixed in a polytetrafluoroethylene (PTFE) container and then PI was polymerized in situ on one side of SWNT film. The film in Figure 6.1e held good flexibility. And the SEM image in Figure 6.1f indicated that SWNT bundles were covered by PI nanoflakes fully and evenly. What's more, the PI/SWNT also obtained better electrochemical performance besides better flexibility. Using Swagelok cells, researchers testified that PI/SWNT exhibited higher capacity (206 mA h g^{-1}) compared with pure PI. And it held better cycling stability which retained 85% (175 mA h g^{-1}) after 200 cycles (Figure 6.1g).

6.2.1.2 Polymer Inorganic

At present, lithium ion batteries are considered to be promising energy storage devices, which can solve the problems related to the use of renewable energy rather than fossil fuel. However, the specific capacity and energy density of lithium ion batteries with graphite as anode and lithium cobaltic oxide as cathode are usually limited. Thus, with their high theoretical capacity (1672 mA h g^{-1}), which is 6 times more than that of current lithium ion batteries, environmentally friendliness, and low operation costs, lithium–sulfur batteries are one of the most promising candidate batteries to solve these problems [14–16, 57–59]. Nevertheless, the widespread development of lithium–sulfur batteries is impeded severely by several considerable challenges, including insulating nature of sulfur, volume expansion, and dissolution of lithium polysulfide (active materials) [23, 27, 57, 60–62]. Several approaches have been invented to mitigate these issues, such as the use of conductive polymers, metal–organic frameworks, and carbon-based materials to increase the conductivity of the cathode and impede the loss of active sulfur [16, 23, 24, 63–81]. Among all of approaches stated, the copolymerization of sulfur with polymerizable monomers or polymers with reactive groups through radical polymerization has attracted researchers' attention owing to their good retention ability of active material and excellent conductivity, which is also a kind of PNC made of polymer and inorganic material [82–91].

In order to impede the dissolution of sulfur, Oschmann et al. introduce an S-P3HT copolymer that is copolymerized by allyl-terminated poly(3-hexylthiophene-2,5-diyl) (P3HT) with an excess of molten sulfur radicals. Figure 6.2a shows the synthetic approach for the copolymerization of allyl-terminated P3HT and sulfur and proposed microstructure of the sample. The SEM image with high magnification of S-P3HT sulfur composites (8 : 2) is shown in Figure 6.2b, indicating that sulfur was copolymerized on the surface of P3HT smoothly and homogeneously. To further investigate the improvement of electrochemical performance of lithium–sulfur battery, which uses S-P3HT as active material, the authors designed three different electrodes as cathode material: sulfur and carbon black (S/CB), sulfur, P3HT, and CB (S/P3HT/CB), and sulfur, S-P3HT, and CB (S/S-P3HT/CB). We can indicate from Figure 6.2c that although they held similar initial specific capacities, the system S/S-P3HT/CB exhibited a prime cycling performance compared with S/CB and S/P3HT/CB [90]. It is attributed to the strong interaction between the polysulfides and polythiophene.

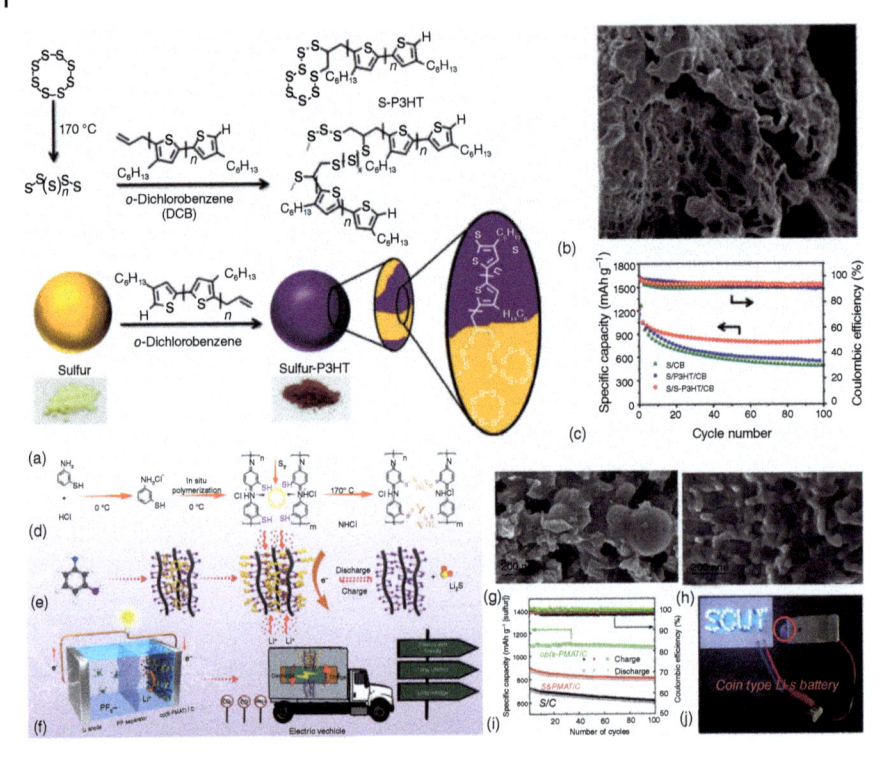

Figure 6.2 (a) Synthetic approach for the copolymerization of allyl-terminated P3HT and sulfur and proposed microstructure of the sample. (b) SEM image of S-P3HT sulfur composites (8 : 2) at high magnification. Scale bar: 2 μm. (c) Cycling performance. Source: (a, c) Oschmann et al. [90]. (d) The synthesis of the cp(S-PMAT) copolymer. (e) The structural evolution in the synthesis of cp(S-PMAT) and discharge–charge process. (f) The device configuration of a Li–S battery with cp(S-PMAT) as the cathode material. (g) SEM images of S&PMAT. (h) SEM images of cp(S-PMAT). (i) Cycling performances and coulombic efficiencies of different cathodes at 0.2 C. (j) LEDs lit by a coin-type Li–S battery with a cp(S-PMAT)/C cathode. Source: (d–f, i, j) Zeng et al. [89], (g) Zeng et al. [89]. Copyright 2017, Wiley-VCH.

Zeng et al. synthesize an organosulfur electrode – cp(S-PMAT) (PMAT, poly(*m*-aminothiophenol)) by the direct copolymerization of sulfur with prepared PMAT through inverse vulcanization, which is shown in Figure 6.2d. Figure 6.2e shows the reaction of the cp(S-PMAT) during the charging and discharging. This kind of copolymer not only provides a continuous way for electron transport, but also inhibits the diffusion of polysulfide due to the chemical bond between mercaptan group and sulfur of conducting polymer. Therefore, it can be used as an electroactive material with carbon (C) as substrate to improve the rate performance and cycle stability of lithium sulfur battery (Figure 6.2f). The SEM images (Figure 6.2g,h) show that after the copolymerization of sulfur and PMAT, the surface of cp(S-PMAT) is much smoother that of S&PMAT sample, which held homogeneous interconnected nanoparticles of approximately 50 nm diameter on its surface. The Li–S battery with cp(S-PMAT)/C as cathode at 0.2 C holds much higher specific capacity (1085 mA h g^{-1}) than that with S&PMAT/C (892 mA h g^{-1}) or S/C (702 mA h g^{-1}) as cathode. What's more, it is worth noting that the specific capacity (1074 mA h g^{-1}) of cp(S-PMAT)/C retained 99.0% after 100 discharge/charge cycles, which corresponds to a decrease of only 0.013% per cycle, an order of magnitude lower than S&PMAT/C (0.103%) and S/C (0.155%) (Figure 6.2i). A coin-type prototype of cp(S-PMAT) as a cathode material for Li–S batteries was fabricated to demonstrate the performance of the electrode, which is shown in Figure 6.2j [89]. The conductive and stable network in the PNCs shows effective ion transport pathways and impedes the dissolution of sulfur, which in turn enhances the electrochemical performance of Li–S battery.

6.2.1.3 Polymer–Organic Salt Graphene

With the continuous development of lithium ion batteries, lithium resources are less and less, and the cost of lithium resources is more and more expansive [92]. Thus, researchers have paid a lot of effort to develop other kinds of batteries. At present, sodium ion battery is regarded as a substitute of lithium ion battery because of its abundant raw materials and low standard electrode potential [17–22]. But several challenges have been brought by the large radius of the Na ion, such as the difficulty in preventing volume change and structure pulverization during charging/discharging and slow Na ion diffusion kinetics [93].

To solve the problems and challenges, researchers have taken organic materials account but then abandon it due to its poor conductivity, cycle stability, and rate performances [94, 95]. Nevertheless, the combination of the organic materials with carbon-based materials, such as RGO and CNT, has attracted researchers' attention because it could not only enhance the conductivity but also promote the decentralization of active electrode materials [37]. Li et al. introduced a novel and advanced cathode for sodium ion battery: porous sodium salt of poly(2,5-dihydroxy-*p*-benzoquinonyl sulfide)/RGO (Na$_2$PDHBQS/RGO). The feature picture Na$_2$PDHBQS/RGO is shown in Figure 6.3a. To further investigate the morphology of Na$_2$PDHBQS/RGO, transmission electron microscope (TEM) and field emission scanning electron microscope (FESEM) were used to detect the structure of its surface (Figure 6.3b,c). The Na$_2$PDHBQS flakes grew uniformly on the RGO without significant aggregation. The presence of holes in the nanocomposite material (in the range of 100–300 nm) can effectively avoid the agglomeration of the Na$_2$PDHBQS sheet and increase the contact area between the cathode and the electrolyte, thereby improving the electrochemical performance of the sodium ion battery. The initial reversible capacity of Na$_2$PDHBQS/RGO is 179 mA h g^{-1}, reaching 183 mA h g^{-1} after 150 cycles, higher than that of single Na$_2$PDHBQS (123 mA h g^{-1} in the first cycle, 121 mA h g^{-1} after 150 cycles) (Figure 6.3d,e). At the same time, the specific capacity of Na$_2$PDHBQS/RGO and Na$_2$PDHBQS is much higher than disodium chloranilate (NaCL) (the initial specific capacity is 43 and only 14 mA h g^{-1} is left after 150 cycles), which proves that the polymerization process is an effective method to improve the electrochemical performance [93].

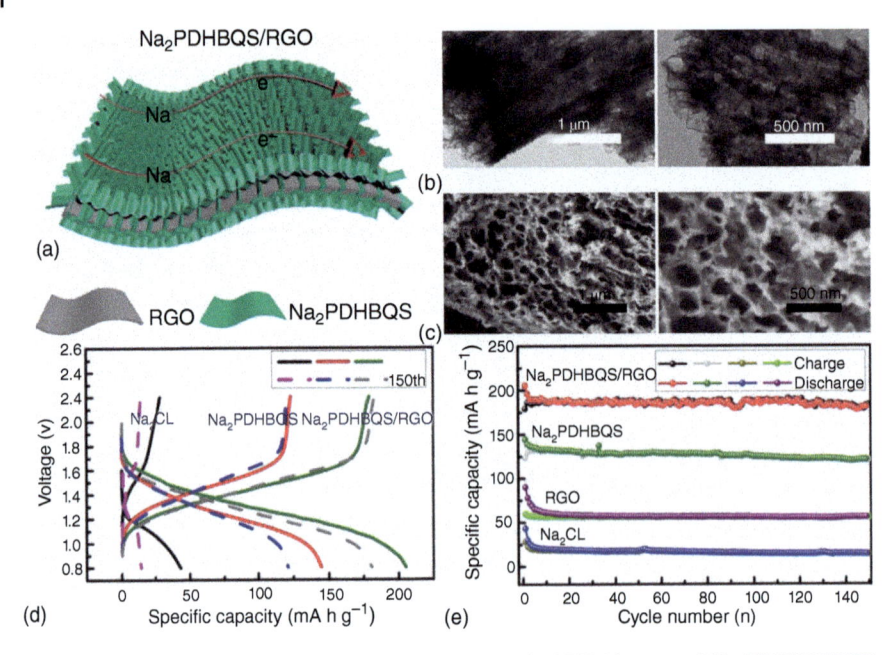

Figure 6.3 (a) Schematic of Na$_2$PDHBQS/RGO. (b) Typical TEM images of Na$_2$PDHBQS/RGO. (c) The FESEM image of Na$_2$PDHBQS/RGO. (d) The charge/discharge curves of sample. (e) The cycle performance at 100 mA g^{-1}. Source: Li et al. [93]. Copyright 2017, Elsevier.

6.2.2 For Supercapacitor

With great attention paid to energy and its usability, it is urgent to develop an efficient energy storage device that can be charged in a short time. Although the capacitor can be charged and discharged in a short pulse and the battery can store a huge amount of energy, they cannot reach the requirements of storing enough energy and have the ability to charge and discharge in a short time. Thus, supercapacitor that holds the ability to fast charge–discharge, high power density, and long cycle life has been invented to meet the requirement. There are three main types of supercapacitor: electrical double-layer capacitors (EDLCs), pseudo-capacitors, and hybrid capacitors, which combine both EDLC and pseudo-capacitor [96–101]. Since these three types of supercapacitor have different operating mechanism, different kinds of materials for electrodes are needed. In EDLCs, the ions move from an electrolyte onto limited area of electrodes during the charge. Thus, different carbon-based materials such as carbon nanofibers and nanotubes and graphene are considered due to their high surface area and high conductivity [102, 103]. The operating mechanism of pseudo-capacitors is storing the energy by rapid reversible redox reaction taking place on the interface between the electrode and the electrolyte, and metal oxides and conductive polymers are regarded as the ideal material owing to their ability to have rapid reversible redox reactions [104–112]. Hybrid capacitors, in which both the operating mechanism of EDLC and pseudo-capacitor works, always use the composites composed of all three types of materials above [113–119]. Herein, we will discuss three main types of PNCs for the electrodes of supercapacitors.

6.2.2.1 Polymer–Metal Oxide

Metal oxides are considered as the promising materials for pseudo-capacitor since their multiple stable valance states lead to excellent energy density [39, 120, 121]. Nevertheless, they are sometimes limited by their nature of low electronic conductivity and poor mechanical stability [122–124]. To improve the conductivity and mechanical stability of electrode material, researchers combine the metal oxide with conductive polymers, which possess advantageous properties such as high electrical conductivity and high flexibility.

Maheswari and Muralidharan prepared CeO_2/polyaniline (PANI) nanocomposites by simple solution mixing method (Figure 6.4a). The attachment of PANI to CeO_2 particles was revealed by the SEM image in Figure 6.4b. The authors used symmetric supercapacitor made of CeO_2/PANI electrodes with polymer gel electrolyte to evaluate the electrochemical performance of the material. Figure 6.4c displays the excellent cyclic stability PANI/CeO_2 composite, which exhibits no noticeable fading even after 1500 cycles at a current density of $20\,A\,g^{-1}$. Figure 6.4d,e show the cyclic voltammetry (CV) and galvanostatic charge–discharge (GCD) profiles of the supercapacitor under different scan rates and current densities. The highest capacitance of the supercapacitor is $170\,F\,g^{-1}$ at $2\,mV\,s^{-1}$ and $126\,F\,g^{-1}$ at a current density of $2\,A\,g^{-1}$, respectively. Obviously, the combination of metal oxide and polymer brings a type of promising material for supercapacitors.

With the development of multiple flexible electronic devices such as future smart clothes and epidermal electronics, people need to create flexible supercapacitors to meet the increasing requirement of energy supply for such devices [127–133]. Taking polypyrrole's (PPy) advantage of great tolerance to strain and perfect matching with the conductive mesh, Huang et al. designed MnO_2 nanosheets penetrated by the PPy (Figure 6.4f). The specific capacitances as functions of scan rates of PPy-penetrated MnO_2 electrode compared with PPy electrode and MnO_2 electrode under 0% and 20% strain are shown in Figure 6.4g to demonstrate the advantage of PPy penetration. The PPy-penetrated MnO_2 held the highest capacitances among all three types of electrodes. And the normalized capacitances (obtained by dividing the capacitance under the 20% strain by the one under the 0% strain at the same scan rate) under different scan rates were also studied to show the enhancement in capacitance retention of MnO_2 when penetrated by PPy (Figure 6.4h). While the PPy film decreases from over 1.3 to 0.96 and the number of MnO_2 nanosheets is always lower than 0.8, the PPy-penetrated MnO_2 electrode retains a stable number between 0.93 and 0.95, showing an effectively enhancement in tolerance of stretch. The SEM image shown in Figure 6.4i indicates the cover and the penetration of PPy on MnO_2, leading to more materials matching between the substrate and electrochemically active materials when the supercapacitors are under stretch [126]. It is clearly seen from the aforementioned examples that the combination of polymer and metal oxide leads to the better electrochemical performance of relevant supercapacitors.

6.2.2.2 Polymer–Graphene/Carbon Nanotube

Graphene and CNTs hold excellent characteristic such as high conductivity and large surface area, which lead to its application in EDLCs [39, 102, 103, 134].

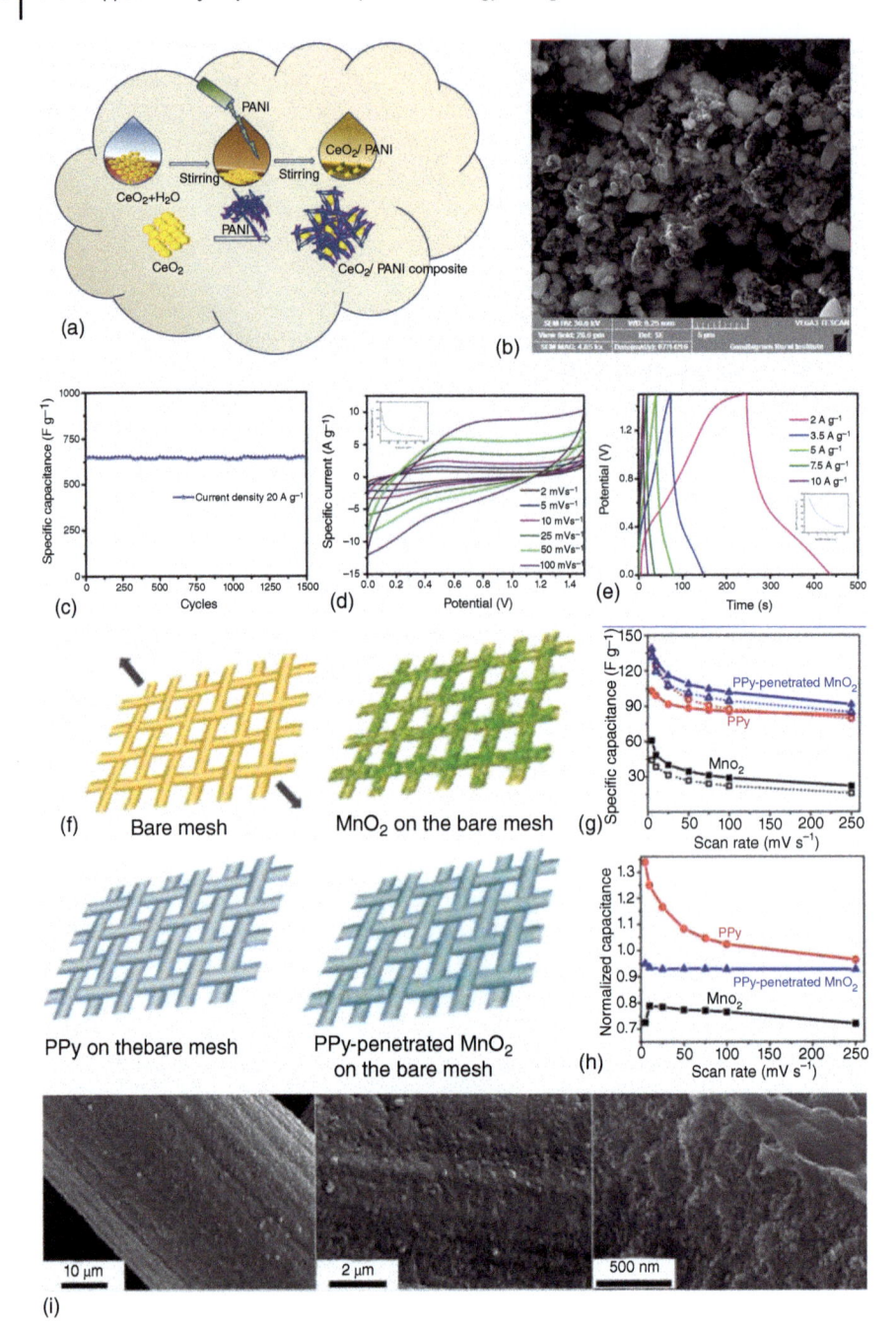

166 | *6 The Application of Polymer Nanocomposites in Energy Storage Devices*

Figure 6.4 (a) Schematic illustration of the formation of CeO$_2$/PANI nanocomposites. (b) SEM images of CeO$_2$/PANI. (c) Cyclic stability of CeO$_2$/PANI electrode calculated at 20 A g^{-1} (vs. Ag/AgCl). (d) CV plots of symmetric supercapacitor at different scan rates. The inset shows variation of specific capacitance with scan rates. (e) Charge–discharge curves at different current densities. The inset shows variation of specific capacitance with current densities. Source: Maheswari et al. [125]. Copyright 2018, Elsevier. (f) Schematic illustrations of fabricated electrodes: tailored stretchable mesh (stress is along the direction of arrows), MnO$_2$ nanosheet electrodeposited mesh, PPy film electrodeposited mesh, PPy-penetrated MnO$_2$ electrodeposited mesh. (g) Specific capacitances as functions of scan rates under 0% (solid lines) and 20% strain (dotted lines). (h) Normalized specific capacitances as functions of scan rates. (i) Top views of the electrode surface and side view of the PPy-penetrated MnO$_2$ composite electrode structure. Source: Huang et al. [126]. Copyright 2015, American Chemical Society.

However, they are always limited by the electrostatic nature of charge and no redox properties, thus showing relative low specific capacitance [135]. Conductive polymers have high flexibility and are also good materials of pseudo-capacitors owing to their ability for fast and reversible redox reaction process, but the huge capacity loss during successive charge/discharge cycles restricts their application [136, 137]. Therefore, there is a good solution, which is combining the graphene/CNTs with conductive polymers, where graphene/CNTs provide large surface area and fix polymers and polymers provide great capacitance.

PPy/GNS/MWCNT composites were successfully prepared by the facile method of in-situ polymerization of pyrrole by Hoe-Seung Kim and his colleagues. The synthesis method of the PPy/GNS/MWCNT nanocomposites is depicted in Figure 6.5a. FESEM images of GNS, pure PPy, and PPy/GNS/MWCNT nanocomposites are shown in Figure 6.5b, from which we can see that PPy was coated on the surface of the graphene and CNTs. To testify the electrochemical performance of PPy/GNS/MWCNT composites, CV test of the pure PPy, PPy/GNS, and PPy/GNS/MWCNT composites electrode was conducted (Figure 6.5c), the result of which showed that the composite sample had a higher capacitance than pure PPy or PPy/GNS composite. And the same result is also shown in Figure 6.5d, which includes the specific capacitance of PPy, PPy/GNS, and PPy/GNS/MWCNT as a function of the various current densities. The GNS and CNTs provide large surface area for PPy and CNTs also improve the conductivity of the composites, which act as the bridge between GNS.

Polymer–graphene/CNTs are often used to satisfy the flexibility requirement of devices. Liu et al. prepared a highly flexible, bendable, and conductive films, which consisted of poly(3,4-ethylenedioxythiophene) (PEDOT), poly(styrenesulfonate) (PSS), and RGO. The composite was prepared by the addition of PEDOT/PSS into the RGO dispersions. As shown in Figure 6.5e, RGO is covered by PEDOT/PSS, and there are some re-stackings of several layers of RGO sheets that are separated by PEDOT/PSS, leading to the increase of the porosity and surface area. A device was made by a poly(vinyl alcohol) (PVA)/H$_3$PO$_4$ solid-state electrolyte between two symmetric RGO-PEDOT/PSS films (Figure 6.5f). The electrochemical performance of the supercapacitor is shown in Figure 6.5g–i. The CV curve of this device depicts

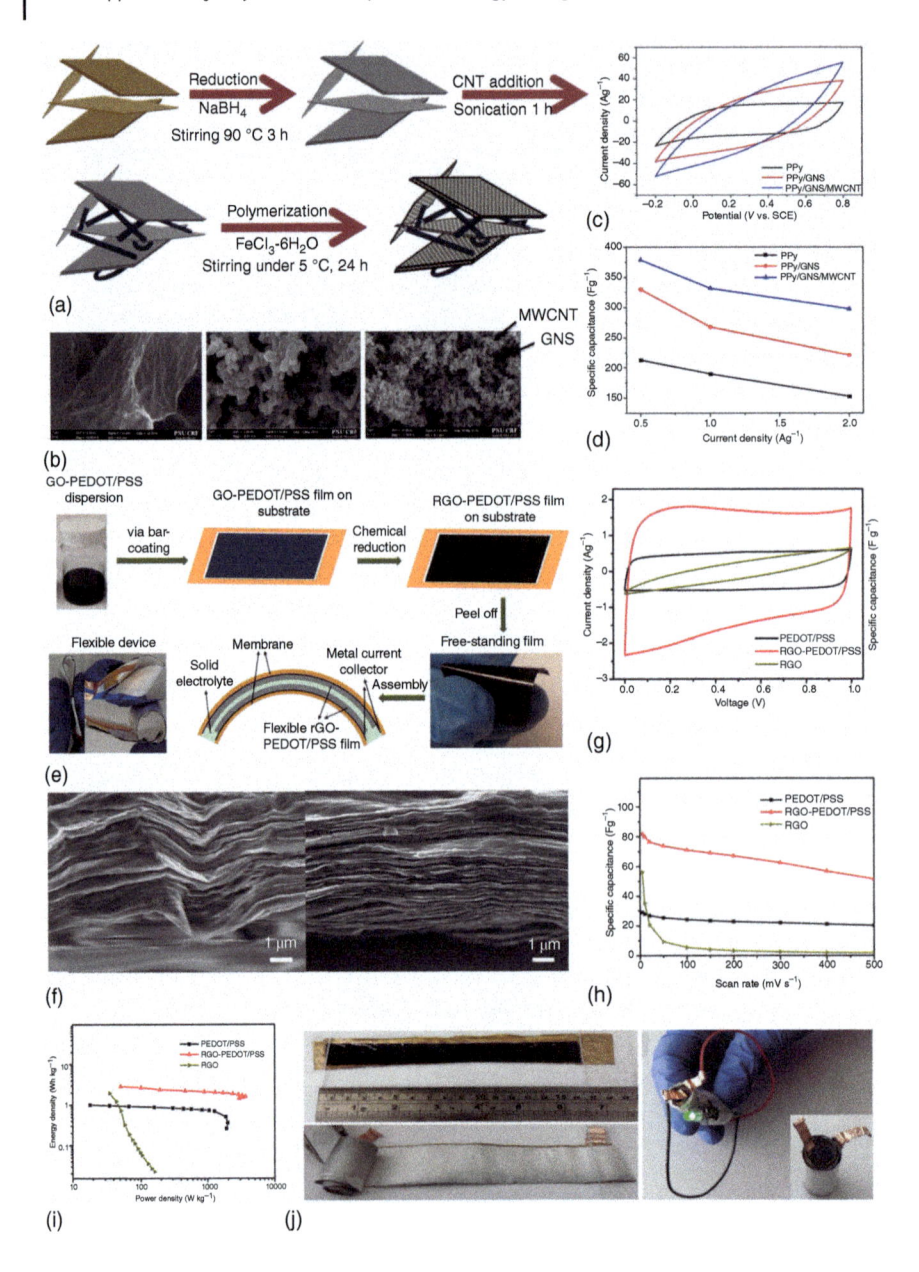

168 | *6 The Application of Polymer Nanocomposites in Energy Storage Devices*

Figure 6.5 (a) Schematic diagram of the synthesis route of the PPy/GNS/MWCNT composites. PPy: polypyrrole; GNS: graphene nanosheets; MWCNT: multiwalled carbon nanotubes. (b) FESEM images of GNS, pure PPy, and PPy/GNS/MWCNT. (c) CV curves of the samples at a scan rate of 100 mV s^{-1} in 1 M NaNO$_3$ electrolyte. (d) Specific capacitance of PPy, PPy/GNS, and PPy/GNS/MWCNT as a function of the various current densities. Source: (a, c, d) Kim et al. [138]. © 2017, Korean Carbon Society, (b) Kim et al. [138]. Copyright 2017, Korean Carbon Society. (e) SEM images of RGO and RGO-PEDOT/PSS. (f) Schematic illustration of the preparation process of RGO-PEDOT/PSS films and the structure of assembled supercapacitor devices. (g) CVs of pure RGO, PEDOT/PSS, and RGO-PEDOT/PSS devices at a scan rate of 50 mV s^{-1}. (h) Specific capacitance of RGO, PEDOT/PSS, and RGO-PEDOT/PSS electrodes calculated from CV. (i) Ragone plot of RGO, PEDOT/PSS, and RGO-PEDOT/PSS electrodes. (j) A demo device used to power a green light-emitting diode (LED). Source: (e, g–i) Liu et al. [139]. Copyright 2015, Nature Publishing Group.

more rectangular shape and larger area than that of the PEDOT/PSS and RGO devices (Figure 6.5g). And the specific capacitance of supercapacitor is higher than either of the pure PEDOT/PSS or RGO at all scan rates of 5–500 mV s^{-1} (Figure 6.5h). What's more, the Ragone plot in Figure 6.5i indicates the improvement in the power and energy densities of RGO-PEDOT/PSS composite electrodes. A supercapacitor was also fabricated to power a light emitting diode (LED) to demonstrate the ability of the supercapacitor [139].

Two examples earlier clearly show the progress that the polymer–graphene/CNTs make compared with separate polymer or graphene/CNTs.

6.2.2.3 Polymer–Metal Oxide–Graphene/Carbon Nanotubes

The discussion previous indicates that the electrochemical performance of supercapacitors made by the combination of two or more kinds of materials universally improves. By synthetizing ternary composites with these three kinds of materials, they have advantages [39, 120–122, 136, 137, 140]. Many types of composites have been invented, in which graphene and CNTs with high conductivity and large surface area always serve as support materials, and polymer and metal oxide work as the functional materials.

Lv et al. prepared a ternary composite with 3D porous structure by electrochemical deposition which could serve as the electrode material. Aligned carbon nanotubes (ACNTs) were grown on carbon fabric (CF) by chemical vapor deposition (CVD) in a reactor corundum tube furnace. And then, MnO$_2$ and PEDOT were electrochemically deposited on CF-ACNTs for 10 minutes and 1 minute in order. Finally, the CF-ACNT/MnO$_2$(10)/PEDOT(1) nanocomposite was prepared (Figure 6.6a). To demonstrate the enhancement in electrochemical performance of CF-ACNT/MnO$_2$(10)/PEDOT(1) electrode, the CV and galvanostatic charge–discharge curves were tested for CF-ACNT, CF-ACNT/MnO$_2$(10), CF-ACNT/PEDOT(1), and CF-ACNT/MnO$_2$(10)/PEDOT(1) electrodes (Figure 6.6b,c). The CV curve of CF-ACNT/MnO$_2$(10)/PEDOT(1) is more like a rectangular in shape. Calculated based on the CV curves, CF-ACNT, CF-ACNT/MnO$_2$(10), CF-ACNT/PEDOT(1), and CF-ACNT/MnO$_2$(10)/PEDOT(1)

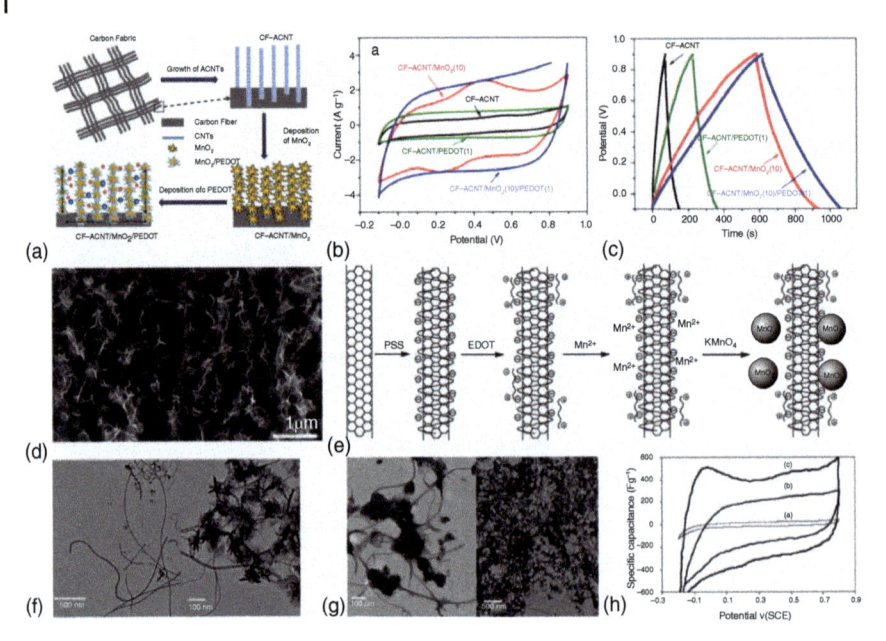

Figure 6.6 (a) Schematic illustration of the preparation of 3D porous ternary composites. (b) CV curves of CF-ACNT, CF-ACNT/PEDOT(1), CF-ACNT/MnO$_2$(10), and CF-ACNT/MnO$_2$(10)/PEDOT(1). (c) Galvanostatic charge–discharge of CF-ACNT, CF-ACNT/PEDOT(1), CF-ACNT/MnO$_2$(10), and CF-ACNT/MnO$_2$(10)/PEDOT(1). (d) SEM image of CF-ACNT/MnO$_2$(10)/PEDOT(1). Source: (a–c) Lv et al. [141], (d) Lv et al. [141]. Copyright 2012, Elsevier. (e) Schematic illustration of synthesizing MnO$_2$/PEDOT nanostructures on MWCNT-PSS. (f) HRTEM micrographs of MWCNT-PSS and MWCNT-PSS/MnO$_2$. (g) HRTEM micrographs MWCNT-PSS/PEDOT/MnO$_2$ and the low magnification picture of MWCNTPSS/PEDOT/MnO$_2$. (h) Cyclic voltammograms of the electrodes recorded in 0.5 M Na$_2$SO$_4$ at 5 mV s^{-1}: (a) MWCNT-PSS, (b) MWCNT-PSS/MnO$_2$, and (c) MWCNT-PSS/PEDOT/MnO$_2$ nanocomposite. Source: (e–h) Sharma et al. [142], (f) Sharma et al. [142]. Copyright 2009, Elsevier.

hold capacitance of 71, 151, 404, and 481 F g^{-1}, respectively. The GCD curves show similar result. SEM image of CF-ACNT/MnO$_2$(10)/PEDOT(1) is exhibited in Figure 6.6d, indicating that the surface of MnO$_2$ nanosheets and CNTs are covered by PEDOT and the interval between MnO$_2$ and CNTs are filled with PEDOT too [141].

A novel multi-walled carbon nanotube (MWCNT)-PSS supported PEDOT/MnO$_2$ nanocomposite for electrochemical electrode is prepared by Raj Kishore Sharma and his colleagues (Figure 6.6e). PSS was wrapped around MWCNT firstly via sonication. And then MWCNT-PSS was stirred for 12 hours in manganese acetate solution and then oxidized by KMnO$_4$ solution to synthesize MWCNT-PSS/PEDOT/MnO$_2$. High resolution transmission electron microscope (HRTEM) was conducted to figure out the specific morphology of the nanocomposites (Figure 6.6f,g). The de-bundled MWCNT/PSS were first dispersed and interconnected, and then MnO$_2$ was grown in the form of nanorods on PSS layers (Figure 6.6f). After the deposition of PEDOT, the MnO$_2$ nanorods, MnO$_2$-PSS, and MWCNT were interconnected by PEDOT (Figure 6.6g). Figure 6.6h shows the improvement in electrochemical performance of nanocomposites compared with former composites. The CV curve of MWCNT-PSS/PEDOT/MnO$_2$ nanocomposite more resembles a rectangle and has larger area, indicating higher capacitance compared with MWCNT-PSS and MWCNT-PSS/PEDOT [142].

6.3 Electrolytes

As a key constituent of energy storage device, electrolyte plays an important role in electrochemical characteristics and mechanical properties. We will discuss electrolyte by the type of the energy storage devices.

6.3.1 For Battery

For batteries, the volatilization and flammability of traditional liquid electrolytes and the danger of battery explosion are the main problems that hinder their large-scale application [143–154]. Thus, polymer electrolytes have attracted a lot of attention owing to their ability to overcome battery leakage, corrosion, and portability issues such as poly(ethylene oxide) (PEO)-based electrolyte [155, 156]. However, PEO-based electrolyte sometimes shows high crystallinity and low ionic conductivity [157, 158]. Many methods have been probed to solve these problems, such as using conductive polymers, polymer blending, and ionic liquids (IL) as fillers.

Li et al. designed a type of solid electrolyte consisting 1-vinyl-3-methylimidazolium bis(trifluoromethylsulfonyl)imide [[VMIM]TFSI] monomer (IL), PEO (polymer), lithium bis(trifluoromethanesulfonyl)imide (LiTFSI), azodiisobutyronitrile (AIBN), and poly(ethyleneglycol) diacrylate (PEGDA) (cross-linkers) and the way of production is shown in Figure 6.7a. The PIL-PEO semi-interpenetrating polymer networks (IPN) solid electrolytes (PIL-PEO SE) were divided into three types (1, 2, 3) by different weight ratios of IL monomers, LiTFSI, and PEO. The FESEM image shown in Figure 6.7b indicates the thickness of the electrolyte. What's more, the filling of additives after copolymerization in the former PEO matrix was observed from the cross-sectional (magnification) SEM image of PIL-PEO SE (Figure 6.7c). The linear curves of log(σ) as function of 1000/T manifest that ionic conductivities of PIL-PEO SEs are higher than PEO SE (Figure 6.7d) and they also increase as temperature rises (0.42×10^{-4} and 6.12×10^{-4} S cm^{-1} at 25 and 55 °C). What's more, the Li/PIL-PEO SE-2/LiFePO$_4$ and Li/PEO SE/LiFePO$_4$ cells were tested to testify the electrochemical performance of PIL-PEO SE (Figure 6.7e). The Li/PIL-PEO SE-2/LiFePO$_4$ cell exhibits 147 mA h g^{-1} at 0.2 C, higher than that for the PEO SE. After 50 cycles, the delivered capacity of the battery still retains at 114 mA h g^{-1} and the average coulombic efficiency value keeps above 99%, indicating the good cycling reversibility of the cell. Therefore, the combination of ion liquid and polymer provides a type of electrolytes with wide operation temperature window, high conductivity, and high cycle ability, contributed by the high conduction of Li$^+$ ions and the stability of poly ionic liquid [11].

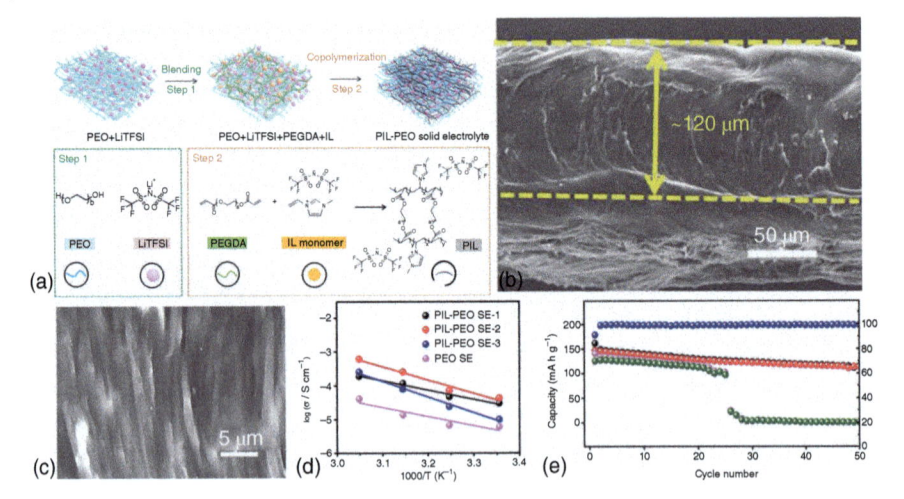

Figure 6.7 (a) The schematic illustration for copolymerization route of PIL-PEO SE. (b) The cross-sectional SEM image of PIL-PEO SE-2. (c) The magnification SEM image cross-sectional image of PIL-PEO SE-2. (d) The temperature-dependent ionic conductivity of IL-PEO SEs with different IL concentrations. (e) Cycle performance and coulombic efficiency of Li/PIL-PEO SE-2/LiFePO$_4$ and Li/PEO SE/LiFePO$_4$ cell at a current rate of 0.2 C at 55 °C. Source: (a, d, e) Li et al. [11], (b) Li et al. [11]. Copyright 2019, Elsevier.

6.3.2 For Supercapacitor

Supercapacitors that holds the ability to fast charge–discharge, high power density, and long cycle life have attracted researchers' attention [159–162]. Novel gel polymer electrolytes are in need to satisfy the increasing requirement of higher ionic conductivity and higher power density, or to satisfy the multiple function besides electrochemical performance such as flexibility, self-healing ability, etc. [163–165]. PNCs have been investigated because of their combination of multiple materials, which could help to achieve these requirements.

Yu et al. doped redox additive KI into PVA–KOH polymer electrolyte (Figure 6.8a). From the electrochemical test, we can figure out that the doping of KI brings the supercapacitor with higher capacitance (Figure 6.8b,c). The appearance of the redox peaks in CV curves manifests that the doping of KI in the electrolyte causes a redox process during the operation of the supercapacitor, leading to the increase of the capacitance [7].

To get with the fast development of flexible electronics such as electronic skin and smart energy storage clothes, supercapacitors that are flexible and have self-healing ability as well as high capacitance are needed to power those devices [39, 130, 131, 167]. At the same time, the aforementioned requirements also become the demanding on electrolytes. Huang et al. fabricated a multifunctional electrolyte by polyacrylic acid (PAA) dual crosslinked by hydrogen bonding and vinyl hybrid silica nanoparticles (VSNPs). The procedure of the fabrication is shown in Figure 6.8d. A supercapacitor was formed with VSNPs-PAA film as electrolyte and CNT paper-based PPy as symmetric electrodes. The picture in Figure 6.8e shows

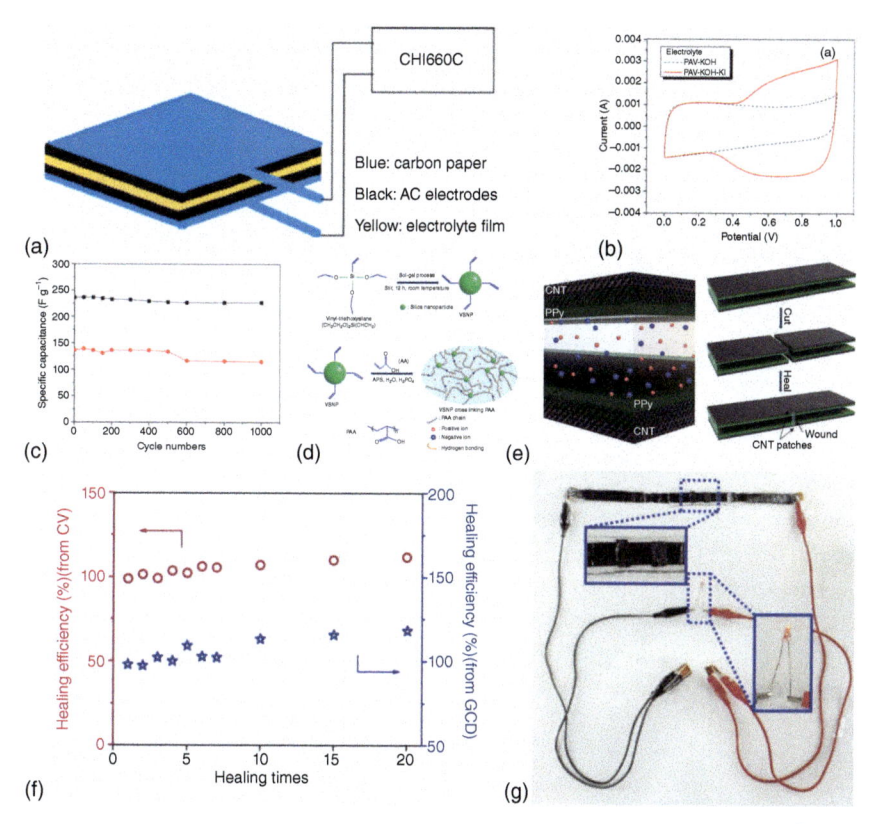

Figure 6.8 (a) Schematic representation of the supercapacitor. (b) CVs of supercapacitor with different electrolytes at a scan rate of 5 mV s^{-1}. (c) The specific capacitances of the supercapacitors with PVA–KOH (red) and PVA–KOH–KI (black) electrolyte during long-term cycling. Source: Yu et al. [7]. (d) Preparation of VSNPs from vinyltriethoxysilane and preparation of the VSNPs-PAA electrolyte from VSNPs (crosslinker), acrylic acid (AA, main monomer), ammonium persulfate (APS, initiator), and phosphoric acid (pH and water content regulator). (e) Schematic of the supercapacitor comprising the VSNPs-PAA polyelectrolyte and CNT paper-based PPy electrodes. (f) Healing efficiency calculated from CV and GCD curves. (g) A photo of three supercapacitors connected in series, one of which has self-healed twice to power an LED bulb after self-healing (Insets show enlarged profiles of the twice self-healed supercapacitor and the lit LED bulb.). Source: (d–f) Huang et al. [166]. Licensed under CC-BY-4.0, (g) Huang et al. [166]. Copyright 2015, Nature publishing group.

that the cut wound on the supercapacitor recovers to restore the conductivity, which means that the supercapacitor is self-healable. To further investigate the specific self-healability of the supercapacitor, the electrochemical performance was tested before and after self-healing. The healing efficiency got from CV curves and GCD curves indicates the same trend that the healing efficiency is ∼100% during all breaking/healing cycles (Figure 6.8f). A demo of the supercapacitor is also shown to demonstrate its ability to light up an LED after self-healing (Figure 6.8g) [166].

Above all, PNCs that add excellent and multifunctional additive components always exhibit better ability and become multifunctional.

6.4 Separator

The primary function of separators, located between two electrodes of energy storage devices, is to separate electrodes. However, separators also have the duty to transport ions during operations of energy devices. There are some similar requirements for different types of energy storage devices, such as small resistance for ion transfer while strong electronic insulating capability, high electrochemical and thermal stability and good mechanical strength to provide device durability [60, 168]. Nevertheless, we will discuss them by the type of the energy devices since there are still some distinction between different types of the energy devices.

6.4.1 For Battery

A nonwoven fabric is a very promising material to meet the requirement, because it has high porosity and can easily composite with other materials including ceramic powders [169, 170]. With its own natural characteristic, ceramic exhibits high dielectric constant, high ionic conductivity, and good thermal stability. However, nonwoven fabrics have some drawbacks such as low tensile strength and thermal stability, while ceramic exhibits unstable pore size [171]. In order to remove the shortcomings

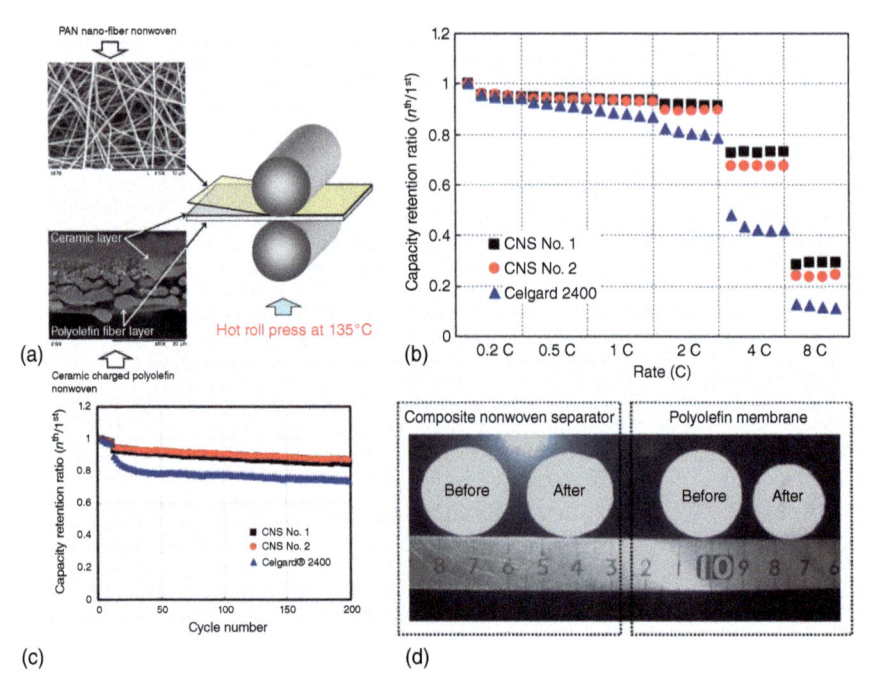

Figure 6.9 (a) Schematic depiction of combining process. (b) Results of rate capability tests for the batteries with the Celgard® 2400, the CNS No. 1 and No. 2. (c) Discharge capacities vs. cycle numbers of the test batteries. (d) Photographs of the CNS No. 1 and a conventional microporous membrane before and after hot oven test at 150 °C for 10 minutes. Source: Cho et al. [171]. Copyright 2010, Elsevier.

and combine the advantages, Tae-Hyung Cho et al. fabricated a new type nonwoven composite separator by nanofiber and ceramic (Figure 6.9a). The combining process is shown in separator technology. To investigate the electrochemical performance of the separators, electrochemical tests were conducted on batteries with those three types of separators. Figure 6.9b shows the rate capabilities of battery with the separators. The rate capabilities of the batteries with the CNSs are much better, which can be ascribed to their high porosities and low Gurley values than the separator of the Celgard. What's more, the capacity retention ability of the batteries with the CNSs is also greater than that of Celgard since the pore sizes of CNSs are smaller and much more stable (Figure 6.9c). A hot oven test of different types of separators was conducted to demonstrate the thermal stability of CNS. The smaller thermal shrinkage of CNS after the roast indicates that CNS holds better thermal stability (Figure 6.9d) [171].

6.4.2 For Supercapacitors

Since the optimizations of electrodes and electrolytes of supercapacitors have attracted a lot of attention and got plenty of achievement, the enhancement of separator is a good opportunity to further increase the performance and efficiency of supercapacitors [172, 173]. For supercapacitors, separators that exhibit good electrochemical performance and high flexibility are in demand because of the quick development of flexible electronics and other devices that need flexible energy storage devices.

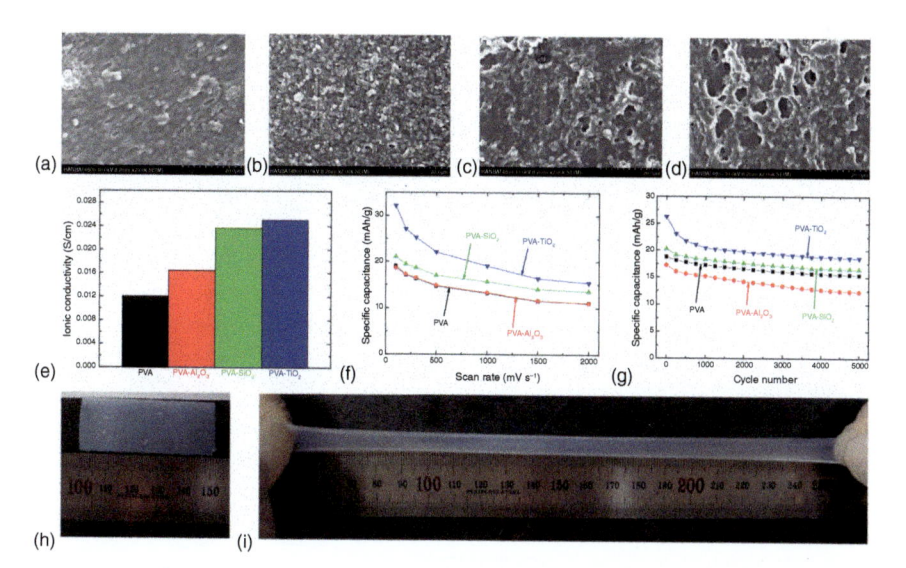

Figure 6.10 SEM images at a magnification of 2000× of (a) bare PVA, (b) PVA–Al_2O_3 composite, (c) PVA–SiO_2 composite, and (d) PVA–TiO_2 composite separators. (e) Ionic conductivity of electrolytes through separators. (f) Specific capacitance as a function of scan rate at a potential window of 0–1.5 V. (g) Cycle life test of the various PCC separators at a scan rate of 200 mV s^{-1} and a potential window of 0–1.5 V for 1000 cycles. (h, i) Stretching test of PVA–TiO_2 composite separators. Source: Bon et al. [174]. Copyright 2018, Elsevier.

To obtain the flexibility and great performance in energy storage at the same time, Bon et al. reported a PVA-based gel-type separator. To increase the ion conductivity of PVA, authors added ceramic particles to increase porosity by disrupting the crystallinity of the polymer matrix. With the excellent flexibility of PVA, PVA-ceramic composites (PVA-CC) are promising materials for separators of supercapacitors. The SEM images show that the mix of ceramic particles [alumina (Al_2O_3), silica (SiO_2), and titania (TiO_2)] increase the number of pores in the PVA gels (Figure 6.10a–d), leading to the enhancement of the ion conductivity in PVA gels (Figure 6.10e). To further explore the electrochemical performance of the PVA-CC separator, electrochemical tests were conducted on the assembled supercapacitor. Figure 6.10f,g indicate that PVA-CCs are better on rate performance as well as cycling performance. The excellent flexibility of the PVA-CC separators is shown in Figure 6.10h,i. The PVA–TiO_2 composite separator can be stretched to 5 times of their original length [174].

6.5 Conclusion

In this chapter, we focus on the recent reports about the application of PNCs in energy storage devices with respect to the electrode, electrolyte, and separator for batteries and supercapacitors. PNCs are widely accepted and utilized owing to their ability to remove the drawbacks of separate materials and combine the advantages of those materials. However, there are still some challenges to conquer and big progress to make.

First, the commercial value of the PNCs is often limited by their complicated synthetic methods, which lead to high cost and low preparation efficiency. In addition, the cost of several components such as CNTs and graphene is very expensive for the use as the raw materials of energy devices. What's more, most composites discussed earlier are still in laboratory research, which are far away from industry productions.

Second, the electrochemical performance of PNCs needs to be further enhanced. To satisfy the ever-increasing demand of energy consumption in electronics whose functions are growing, batteries and supercapacitors need higher energy density and higher power density, which means higher requirements for electrodes, electrolytes, and separators.

Third, safety and stability of energy storage devices have become extremely important since the appearance of explosion caused by lithium ion batteries. The fractures of the devices are still dangerous even that PNCs are always in solid state. Thus, PNCs with higher flexibility are needed to adapt to the stochastic deformations of electronics which utilize those energy storage devices.

In conclusion, the research on PNCs is a meaningful field that connects polymers with other kinds of materials. The combination of excellent properties of polymers and other kinds of material improve the utilization ratio of polymers as well as other materials serving as the reinforcer in devices. Even though many extraordinary progresses have been made in this area, there is still a long way to make PNCs used in real life rather than laboratories as the increasing demand from various electronic products with higher performance requirements.

References

1 Simon, P. and Gogotsi, Y. (2008). Materials for electrochemical capacitors. *Nat. Mater.* 7 (11): 845–854.
2 Wang, G.P., Zhang, L., and Zhang, J.J. (2012). A review of electrode materials for electrochemical supercapacitors. *Chem. Soc. Rev.* 41 (2): 797–828.
3 Diaz-Gonzalez, F., Sumper, A., Gomis-Bellmunt, O., and Villafafila-Robles, R. (2012). A review of energy storage technologies for wind power applications. *Renewable Sustainable Energy Rev.* 16 (4): 2154–2171.
4 Cook, T.R., Dogutan, D.K., Reece, S.Y. et al. (2010). Solar energy supply and storage for the legacy and non legacy worlds. *Chem. Rev.* 110 (11): 6474–6502.
5 Hanemann, T. and Szabo, D.V. (2010). Polymer-nanoparticle composites: from synthesis to modern applications. *Materials* 3 (6): 3468–3517.
6 Meer, S., Kausar, A., and Iqbal, T. (2016). Trends in conducting polymer and hybrids of conducting polymer/carbon nanotube: a review. *Polym. Plast. Technol. Eng.* 55 (13): 1416–1440.
7 Yu, H.Y., Wu, J.H., Fan, L.Q. et al. (2011). Improvement of the performance for quasi-solid-state supercapacitor by using PVA-KOH-KI polymer gel electrolyte. *Electrochim. Acta* 56 (20): 6881–6886.
8 Chodankar, N.R., Dubal, D.P., Lokhande, A.C., and Lokhande, C.D. (2015). Ionically conducting PVA-LiClO$_4$ gel electrolyte for high performance flexible solid state supercapacitors. *J. Colloid Interface Sci.* 460: 370–376.
9 Wang, K., Zhang, X., Li, C. et al. (2015). Chemically crosslinked hydrogel film leads to integrated flexible supercapacitors with superior performance. *Adv. Mater.* 27 (45): 7451–7457.
10 Lu, F., Gao, X.P., Wu, A.L. et al. (2017). Lithium-containing zwitterionic poly(ionic liquid)s as polymer electrolytes for lithium-ion batteries. *J. Phys. Chem. C* 121 (33): 17756–17763.
11 Li, Y.H., Sun, Z.J., Shi, L. et al. (2019). Poly(ionic liquid)-polyethylene oxide semi-interpenetrating polymer network solid electrolyte for safe lithium metal batteries. *Chem. Eng. J.* 375: 121925.
12 Koo, M., Park, K.I., Lee, S.H. et al. (2012). Bendable inorganic thin-film battery for fully flexible electronic systems. *Nano Lett.* 12 (9): 4810–4816.
13 Dunn, B., Kamath, H., and Tarascon, J.M. (2011). Electrical energy storage for the grid: a battery of choices. *Science* 334 (6058): 928–935.
14 Borchardt, L., Oschatz, M., and Kaskel, S. (2016). Carbon materials for lithium sulfur batteries-ten critical questions. *Chem. Eur. J.* 22 (22): 7324–7351.
15 Pope, M.A. and Aksay, I.A. (2015). Structural design of cathodes for Li-S batteries. *Adv. Energy Mater.* 5 (16): 1500124.

16 Cheng, H. and Wang, S.P. (2014). Recent progress in polymer/sulphur composites as cathodes for rechargeable lithium-sulphur batteries. *J. Mater. Chem. A* 2 (34): 13783–13794.

17 Yabuuchi, N., Kajiyama, M., Iwatate, J. et al. (2012). P2-type Na$_x$[Fe$_{1/2}$Mn$_{1/2}$]O$_2$ made from earth-abundant elements for rechargeable Na batteries. *Nat. Mater.* 11 (6): 512–517.

18 Wang, Y.S., Yu, X.Q., Xu, S.Y. et al. (2013). A zero-strain layered metal oxide as the negative electrode for long-life sodium-ion batteries. *Nat. Commun.* 4: 2365.

19 Xu, Y.L., Swaans, E., Basak, S. et al. (2016). Reversible Na-ion uptake in Si nanoparticles. *Adv. Energy Mater.* 6 (2): 1501436.

20 Wang, S.W., Wang, L.J., Zhu, Z.Q. et al. (2014). All organic sodium-ion batteries with Na$_4$C$_8$H$_2$O$_6$. *Angew. Chem. Int. Ed.* 53 (23): 5892–5896.

21 Park, Y., Shin, D.S., Woo, S.H. et al. (2012). Sodium terephthalate as an organic anode material for sodium ion batteries. *Adv. Mater.* 24 (26): 3562–3567.

22 Wang, C.L., Xu, Y., Fang, Y.G. et al. (2015). Extended pi-conjugated system for fast-charge and -discharge sodium-ion batteries. *J. Am. Chem. Soc.* 137 (8): 3124–3130.

23 Wang, L., Wang, D., Zhang, F.X., and Jin, J. (2013). Interface chemistry guided long-cycle-life Li-S battery. *Nano Lett.* 13 (9): 4206–4211.

24 Wang, J.L., Yang, J., Wan, C.R. et al. (2003). Sulfur composite cathode materials for rechargeable lithium batteries. *Adv. Funct. Mater.* 13 (6): 487–492.

25 Reddy, M.V., Rao, G.V.S., and Chowdari, B.V.R. (2013). Metal oxides and oxysalts as anode materials for Li ion batteries. *Chem. Rev.* 113 (7): 5364–5457.

26 Li, A.H., Xu, L.Q., Li, S.L. et al. (2015). One-dimensional manganese borate hydroxide nanorods and the corresponding manganese oxyborate nanorods as promising anodes for lithium ion batteries. *Nano Res.* 8 (2): 554–565.

27 Etacheri, V., Marom, R., Elazari, R. et al. (2011). Challenges in the development of advanced Li-ion batteries: a review. *Energy Environ. Sci.* 4 (9): 3243–3262.

28 Tarascon, J.M. (2008). Towards sustainable and renewable systems for electrochemical energy storage. *ChemSusChem* 1 (8–9): 777–779.

29 Larcher, D. and Tarascon, J.M. (2015). Towards greener and more sustainable batteries for electrical energy storage. *Nat. Chem.* 7 (1): 19–29.

30 Okubo, M., Hosono, E., Kim, J. et al. (2007). Nanosize effect on high-rate Li-ion intercalation in LiCoO$_2$ electrode. *J. Am. Chem. Soc.* 129 (23): 7444–7452.

31 Hosono, E., Kudo, T., Honma, I. et al. (2009). Synthesis of single crystalline spinel LiMn$_2$O$_4$ nanowires for a lithium ion battery with high power density. *Nano Lett.* 9 (3): 1045–1051.

32 Chen, H., Armand, M., Demailly, G. et al. (2008). From biomass to a renewable Li$_x$C$_6$O$_6$ organic electrode for sustainable Li-ion batteries. *ChemSusChem* 1 (4): 348–355.

33 Koshika, K., Sano, N., Oyaizu, K., and Nishide, H. (2009). An ultrafast chargeable polymer electrode based on the combination of nitroxide radical and aqueous electrolyte. *Chem. Commun.* (7): 836–838.

34 Allen, M.J., Tung, V.C., and Kaner, R.B. (2010). Honeycomb carbon: a review of graphene. *Chem. Rev.* 110 (1): 132–145.

35 Verdejo, R., Bernal, M.M., Romasanta, L.J., and Lopez-Manchado, M.A. (2011). Graphene filled polymer nanocomposites. *J. Mater. Chem.* 21 (10): 3301–3310.

36 Guo, W., Yin, Y.X., Xin, S. et al. (2012). Superior radical polymer cathode material with a two-electron process redox reaction promoted by graphene. *Energy Environ. Sci.* 5 (1): 5221–5225.

37 Song, Z.P., Xu, T., Gordin, M.L. et al. (2012). Polymer-graphene nanocomposites as ultrafast-charge and -discharge cathodes for rechargeable lithium batteries. *Nano Lett.* 12 (5): 2205–2211.

38 Wu, H.P., Shevlin, S.A., Meng, Q.H. et al. (2014). Flexible and binder-free organic cathode for high-performance lithium-ion batteries. *Adv. Mater.* 26 (20): 3338–3343.

39 Hu, L.B., Pasta, M., La Mantia, F. et al. (2010). Stretchable, porous, and conductive energy textiles. *Nano Lett.* 10 (2): 708–714.

40 Jost, K., Perez, C.R., McDonough, J.K. et al. (2011). Carbon coated textiles for flexible energy storage. *Energy Environ. Sci.* 4 (12): 5060–5067.

41 Jost, K., Dion, G., and Gogotsi, Y. (2014). Textile energy storage in perspective. *J. Mater. Chem. A* 2 (28): 10776–10787.

42 Huang, Q.Y., Wang, D.R., and Zheng, Z.J. (2016). Textile-based electrochemical energy storage devices. *Adv. Energy Mater.* 6 (22): 1600783.

43 Choi, C., Sim, H.J., Spinks, G.M. et al. (2016). Elastomeric and dynamic MnO_2/CNT core-shell structure coiled yarn supercapacitor. *Adv. Energy Mater.* 6 (5): 1502119.

44 Pushparaj, V.L., Shaijumon, M.M., Kumar, A. et al. (2007). Flexible energy storage devices based on nanocomposite paper. *Proc. Natl. Acad. Sci. U.S.A.* 104 (34): 13574–13577.

45 Hu, L.B., Wu, H., La Mantia, F. et al. (2010). Thin, flexible secondary Li-Ion paper batteries. *ACS Nano* 4 (10): 5843–5848.

46 Jia, X.L., Yan, C.Z., Chen, Z. et al. (2011). Direct growth of flexible $LiMn_2O_4$/CNT lithium-ion cathodes. *Chem. Commun.* 47 (34): 9669–9671.

47 Luo, S., Wang, K., Wang, J.P. et al. (2012). Binder-free $LiCoO_2$/carbon nanotube cathodes for high-performance lithium ion batteries. *Adv. Mater.* 24 (17): 2294–2298.

48 Wang, K., Luo, S., Wu, Y. et al. (2013). Super-aligned carbon nanotube films as current collectors for lightweight and flexible lithium ion batteries. *Adv. Funct. Mater.* 23 (7): 846–853.

49 Wu, M.S., Lee, J.T., Chiang, P.C.J., and Lin, J.C. (2007). Carbon-nanofiber composite electrodes for thin and flexible lithium-ion batteries. *J. Mater. Sci.* 42 (1): 259–265.

50 Wei, D., Andrew, P., Yang, H.F. et al. (2011). Flexible solid state lithium batteries based on graphene inks. *J. Mater. Chem.* 21 (26): 9762–9767.

51 Gwon, H., Kim, H.S., Lee, K.U. et al. (2011). Flexible energy storage devices based on graphene paper. *Energy Environ. Sci.* 4 (4): 1277–1283.

52 Liu, F., Song, S.Y., Xue, D.F., and Zhang, H.J. (2012). Folded structured graphene paper for high performance electrode materials. *Adv. Mater.* 24 (8): 1089–1094.

53 Li, N., Chen, Z.P., Ren, W.C. et al. (2012). Flexible graphene-based lithium ion batteries with ultrafast charge and discharge rates. *Proc. Natl. Acad. Sci. U.S.A.* 109 (43): 17360–17365.

54 Macinnes, D., Druy, M.A., Nigrey, P.J. et al. (1981). Organic batteries - reversible n-type and p-type electrochemical doping of polyacetylene, (Ch)X. *J. Chem. Soc., Chem. Commun.* (7): 317–319.

55 Suga, T., Konishi, H., and Nishide, H. (2007). Photocrosslinked nitroxide polymer cathode-active materials for application in an organic-based paper battery. *Chem. Commun.* (17): 1730–1732.

56 Song, Z.P., Zhan, H., and Zhou, Y.H. (2010). Polyimides: promising energy-storage materials. *Angew. Chem. Int. Ed.* 49 (45): 8444–8448.

57 Bresser, D., Passerini, S., and Scrosati, B. (2013). Recent progress and remaining challenges in sulfur-based lithium secondary batteries - a review. *Chem. Commun.* 49 (90): 10545–10562.

58 Evers, S. and Nazar, L.F. (2013). New approaches for high energy density lithium-sulfur battery cathodes. *Acc. Chem. Res.* 46 (5): 1135–1143.

59 Bruce, P.G., Freunberger, S.A., Hardwick, L.J., and Tarascon, J.M. (2012). Li-O_2 and Li-S batteries with high energy storage (vol 11, pg 19, 2012). *Nat. Mater.* 11 (1): 19–29.

60 Xiang, Y.Y., Li, J.S., Lei, J.H. et al. (2016). Advanced separators for lithium-Ion and lithium-sulfur batteries: a review of recent progress. *ChemSusChem* 9 (21): 3023–3039.

61 Mikhaylik, Y.V. and Akridge, J.R. (2004). Polysulfide shuttle study in the Li/S battery system. *J. Electrochem. Soc.* 151 (11): A1969–A1976.

62 Manthiram, A., Fu, Y.Z., and Su, Y.S. (2013). Challenges and prospects of lithium-sulfur batteries. *Acc. Chem. Res.* 46 (5): 1125–1134.

63 Zhou, J.W., Yu, X.S., Fan, X.X. et al. (2015). The impact of the particle size of a metal-organic framework for sulfur storage in Li-S batteries. *J. Mater. Chem. A* 3 (16): 8272–8275.

64 Yang, H., Bradley, S.J., Chan, A. et al. (2016). Catalytically active bimetallic nanoparticles supported on porous carbon capsules derived from metal-organic framework composites. *J. Am. Chem. Soc.* 138 (36): 11872–11881.

65 Li, G., Sun, J.H., Hou, W.P. et al. (2016). Three-dimensional porous carbon composites containing high sulfur nanoparticle content for high-performance lithium-sulfur batteries. *Nat. Commun.* 7: 10601.

66 Song, J.X., Xu, T., Gordin, M.L. et al. (2014). Nitrogen-doped mesoporous carbon promoted chemical adsorption of sulfur and fabrication of high- areal-capacity sulfur cathode with exceptional cycling stability for lithium- sulfur batteries. *Adv. Funct. Mater.* 24 (9): 1243–1250.

67 Chang, C.H., Chung, S.H., and Manthiram, A. (2016). Effective stabilization of a high-loading sulfur cathode and a lithium-metal anode in Li-S batteries utilizing SWCNT-modulated separators. *Small* 12 (2): 174–179.

68 Chung, S.H., Han, P., Singhal, R. et al. (2015). Electrochemically stable rechargeable lithium-sulfur batteries with a microporous carbon nanofiber filter for polysulfide. *Adv. Energy Mater.* 5 (18): 1500738.

69 Singhal, R., Chung, S.H., Manthiram, A., and Kalra, V. (2015). A free-standing carbon nanofiber interlayer for high-performance lithium-sulfur batteries. *J. Mater. Chem. A* 3 (8): 4530–4538.

70 Elazari, R., Salitra, G., Garsuch, A. et al. (2011). Sulfur-impregnated activated carbon fiber cloth as a binder-free cathode for rechargeable Li-S batteries. *Adv. Mater.* 23 (47): 5641–5644.

71 Behzadirad, M., Lavrova, O., and Busani, T. (2016). Demonstration of 99% capacity retention in Li/S batteries with a porous hollow carbon cap nanofiber-graphene structure through a semi-empirical capacity fading model. *J. Mater. Chem. A* 4 (20): 7830–7840.

72 Tang, C., Li, B.Q., Zhang, Q. et al. (2016). CaO-templated growth of hierarchical porous graphene for high-power lithium-sulfur battery applications. *Adv. Funct. Mater.* 26 (4): 577–585.

73 Yang, Y., Yu, G.H., Cha, J.J. et al. (2011). Improving the performance of pithium-sulfur batteries by conductive polymer coating. *ACS Nano* 5 (11): 9187–9193.

74 Wang, J.L., Yang, J., Xie, J.Y., and Xu, N.X. (2002). A novel conductive polymer-sulfur composite cathode material for rechargeable lithium batteries. *Adv. Mater.* 14 (13–14): 963–965.

75 Wu, F., Chen, J.Z., Chen, R.J. et al. (2011). Sulfur/polythiophene with a core/shell structure: synthesis and electrochemical properties of the cathode for rechargeable lithium batteries. *J. Phys. Chem. C* 115 (13): 6057–6063.

76 He, G., Ji, X.L., and Nazar, L. (2011). High "C" rate Li-S cathodes: sulfur imbibed bimodal porous carbons. *Energy Environ. Sci.* 4 (8): 2878–2883.

77 Ji, X.L., Lee, K.T., and Nazar, L.F. (2009). A highly ordered nanostructured carbon-sulphur cathode for lithium-sulphur batteries. *Nat. Mater.* 8 (6): 500–506.

78 Zheng, G.Y., Yang, Y., Cha, J.J. et al. (2011). Hollow carbon nanofiber-encapsulated sulfur cathodes for high specific capacity rechargeable lithium batteries. *Nano Lett.* 11 (10): 4462–4467.

79 Cao, Y.L., Li, X.L., Aksay, I.A. et al. (2011). Sandwich-type functionalized graphene sheet-sulfur nanocomposite for rechargeable lithium batteries. *Phys. Chem. Chem. Phys.* 13 (17): 7660–7665.

80 Evers, S. and Nazar, L.F. (2012). Graphene-enveloped sulfur in a one pot reaction: a cathode with good coulombic efficiency and high practical sulfur content. *Chem. Commun.* 48 (9): 1233–1235.

81 Zhang, D., Tu, J.P., Xiang, J.Y. et al. (2011). Influence of particle size on electrochemical performances of pyrite FeS_2 for Li-ion batteries. *Electrochim. Acta* 56 (27): 9980–9985.

82 Talapaneni, S.N., Hwang, T.H., Je, S.H. et al. (2016). Elemental-sulfur-mediated facile synthesis of a covalent triazine framework for high-performance lithium-sulfur batteries. *Angew. Chem. Int. Ed.* 55 (9): 3106–3111.

83 Chung, W.J., Griebel, J.J., Kim, E.T. et al. (2013). The use of elemental sulfur as an alternative feedstock for polymeric materials. *Nat. Chem.* 5 (6): 518–524.

84 Diez, S., Hoefling, A., Theato, P., and Pauer, W. (2017). Mechanical and electrical properties of sulfur-containing polymeric materials prepared via inverse vulcanization. *Polymers* 9 (2): 59.

85 Tsuda, T. and Takeda, A. (1996). Palladium-catalysed cycloaddition copolymerisation of diynes with elemental sulfur to poly(thiophene)s. *Chem. Commun.* (11): 1317–1318.

86 Ding, Y. and Hay, A.S. (1997). Copolymerization of elemental sulfur with cyclic(arylene disulfide) oligomers. *J. Polym. Sci., Part A: Polym. Chem.* 35 (14): 2961–2968.

87 Kim, H., Lee, J., Ahn, H. et al. (2015). Synthesis of three-dimensionally interconnected sulfur-rich polymers for cathode materials of high-rate lithium-sulfur batteries. *Nat. Commun.* 6: 7278.

88 Fu, Y.Z. and Manthiram, A. (2012). Enhanced cyclability of lithium-sulfur batteries by a polymer acid-doped polypyrrole mixed ionic-electronic conductor. *Chem. Mater.* 24 (15): 3081–3087.

89 Zeng, S.B., Li, L.G., Xie, L.H. et al. (2017). Conducting polymers crosslinked with sulfur as cathode materials for high-rate, ultralong-life lithium-sulfur batteries. *ChemSusChem* 10 (17): 3378–3386.

90 Oschmann, B., Park, J., Kim, C. et al. (2015). Copolymerization of polythiophene and sulfur to improve the electrochemical performance in lithium-sulfur batteries. *Chem. Mater.* 27 (20): 7011–7017.

91 Dirlam, P.T., Simmonds, A.G., Kleine, T.S. et al. (2015). Inverse vulcanization of elemental sulfur with 1,4-diphenylbutadiyne for cathode materials in Li-S batteries. *RSC Adv.* 5 (31): 24718–24722.

92 Kim, J.K., Lim, Y.J., Kim, H. et al. (2015). A hybrid solid electrolyte for flexible solid-state sodium batteries. *Energy Environ. Sci.* 8 (12): 3589–3596.

93 Li, A.H., Feng, Z.Y., Sun, Y. et al. (2017). Porous organic polymer/RGO composite as high performance cathode for half and full sodium ion batteries. *J. Power Sources* 343: 424–430.

94 Song, Z.P., Qian, Y.M., Liu, X.Z. et al. (2014). A quinone-based oligomeric lithium salt for superior Li-organic batteries. *Energy Environ. Sci.* 7 (12): 4077–4086.

95 Ma, T., Zhao, Q., Wang, J.B. et al. (2016). A sulfur heterocyclic quinone cathode and a multifunctional binder for a high-performance rechargeable lithium-ion battery. *Angew. Chem. Int. Ed.* 55 (22): 6428–6432.

96 Shao, Y.L., El-Kady, M.F., Wang, L.J. et al. (2015). Graphene-based materials for flexible supercapacitors. *Chem. Soc. Rev.* 44 (11): 3639–3665.

97 Kim, Y. and Kim, S. (2015). Direct growth of cobalt aluminum double hydroxides on graphene nanosheets and the capacitive properties of the resulting composites. *Electrochim. Acta* 163: 252–259.

98 Oh, M. and Kim, S. (2012). Synthesis and electrochemical analysis of polyaniline/TiO_2 composites prepared with various molar ratios between aniline monomer and para-toluenesulfonic acid. *Electrochim. Acta* 78: 279–285.

99 Frackowiak, E. and Beguin, F. (2001). Carbon materials for the electrochemical storage of energy in capacitors. *Carbon* 39 (6): 937–950.

100 Sahoo, S., Zhang, S.J., and Shim, J.J. (2016). Porous ternary high performance supercapacitor electrode based on eeduced graphene oxide, $NiMn_2O_4$, and polyaniline. *Electrochim. Acta* 216: 386–396.

101 Kowsari, E., Ehsani, A., Najafi, M.D., and Bigdeloo, M. (2018). Enhancement of pseudocapacitance performance of p-type conductive polymer in the presence of newly synthesized graphene oxide-hexamethylene tributylammonium iodide nanosheets. *J. Colloid Interface Sci.* 512: 346–352.

102 Kim, T., Jung, G., Yoo, S. et al. (2013). Activated graphene-based carbons as supercapacitor electrodes with macro- and mesopores. *ACS Nano* 7 (8): 6899–6905.

103 Huang, Y., Liang, J.J., and Chen, Y.S. (2012). An overview of the applications of graphene-based materials in supercapacitors. *Small* 8 (12): 1805–1834.

104 Aghazadeh, M., Maragheh, M.G., Ganjali, M.R., and Norouzi, P. (2016). One-step electrochemical preparation and characterization of nanostructured hydrohausmannite as electrode material for supercapacitors. *RSC Adv.* 6 (13): 10442–10449.

105 Aghazadeh, M. and Ganjali, M.R. (2018). One-step electro-synthesis of Ni^{2+} doped magnetite nanoparticles and study of their supercapacitive and superparamagnetic behaviors. *J. Mater. Sci. - Mater. Electron.* 29 (6): 4981–4991.

106 Chen, D., Wang, Q.F., Wang, R.M., and Shen, G.Z. (2015). Ternary oxide nanostructured materials for supercapacitors: a review. *J. Mater. Chem. A* 3 (19): 10158–10173.

107 Salehifar, N., Shayeh, J.S., Siadat, S.O.R. et al. (2015). Electrochemical study of supercapacitor performance of polypyrrole ternary nanocomposite electrode by fast Fourier transform continuous cyclic voltammetry. *RSC Adv.* 5 (116): 96130–96137.

108 Chen, H., Hu, L.F., Chen, M. et al. (2014). Nickel-cobalt layered double hydroxide nanosheets for high-performance supercapacitor electrode materials. *Adv. Funct. Mater.* 24 (7): 934–942.

109 Kim, Y. and Kim, S. (2015). Microstructural modification of NiAl layered double hydroxide electrodes by adding graphene nanosheets and their capacitative property. *Bull. Korean Chem. Soc.* 36 (2): 665–671.

110 Mondal, S., Rana, U., and Malik, S. (2015). Graphene quantum dot-doped polyaniline nanofiber as high performance supercapacitor electrode materials. *Chem. Commun.* 51 (62): 12365–12368.

111 Zhang, H.R., Wang, J.X., Shan, Q.J. et al. (2013). Tunable electrode morphology used for high performance supercapacitor: polypyrrole nanomaterials as model materials. *Electrochim. Acta* 90: 535–541.

112 Kim, J. and Kim, S. (2014). Preparation and electrochemical analysis of graphene/polyaniline composites prepared by aniline polymerization. *Res. Chem. Intermed.* 40 (7): 2519–2525.

113 Wang, W.J., Lei, W., Yao, T.Y. et al. (2013). One-pot synthesis of graphene/SnO$_2$/PEDOT ternary electrode material for supercapacitors. *Electrochim. Acta* 108: 118–126.

114 Moyseowicz, A., Sliwak, A., Miniach, E., and Gryglewicz, G. (2017). Polypyrrole/iron oxide/reduced graphene oxide ternary composite as a binderless electrode material with high cyclic stability for supercapacitors. *Composites Part B* 109: 23–29.

115 Wu, Q.H., Chen, M., Wang, S.S. et al. (2016). Preparation of sandwich-like ternary hierarchical nanosheets manganese dioxide/polyaniline/reduced graphene oxide as electrode material for supercapacitor. *Chem. Eng. J.* 304: 29–38.

116 Shiri, H.M. and Ehsani, A. (2016). A novel and facile route for the electrosynthesis of Ho$_2$O$_3$ nanoparticles and its nanocomposite with p-type conductive polymer: characterisation and electrochemical performance. *Bull. Chem. Soc. Jpn.* 89 (10): 1201–1206.

117 Shiri, H.M. and Ehsani, A. (2016). Pulse electrosynthesis of novel wormlike gadolinium oxide nanostructure and its nanocomposite with conjugated electroactive polymer as a hybrid and high efficient electrode material for energy storage device. *J. Colloid Interface Sci.* 484: 70–76.

118 Nguyen, V.H. and Shim, J.J. (2015). Ultrasmall SnO$_2$ nanoparticle-intercalated graphene@polyaniline composites as an active electrode material for supercapacitors in different electrolytes. *Synth. Met.* 207: 110–115.

119 Jiang, L.L., Lu, X., Xie, C.M. et al. (2015). Flexible, free-standing TiO$_2$-graphene-polypyrrole composite films as electrodes for supercapacitors. *J. Phys. Chem. C* 119 (8): 3903–3910.

120 Wu, Z.S., Ren, W.C., Wang, D.W. et al. (2010). High-energy MnO$_2$ nanowire/graphene and graphene asymmetric electrochemical capacitors. *ACS Nano* 4 (10): 5835–5842.

121 Chen, S., Zhu, J.W., Wu, X.D. et al. (2010). Graphene oxide-MnO$_2$ nanocomposites for supercapacitors. *ACS Nano* 4 (5): 2822–2830.

122 Tang, X.H., Liu, Z.H., Zhang, C.X. et al. (2009). Synthesis and capacitive property of hierarchical hollow manganese oxide nanospheres with large specific surface area. *J. Power Sources* 193 (2): 939–943.

123 Ma, S.B., Nam, K.W., Yoon, W.S. et al. (2008). Electrochemical properties of manganese oxide coated onto carbon nanotubes for energy-storage applications. *J. Power Sources* 178 (1): 483–489.

124 Kek-Merl, D., Lappalainen, J., and Tuller, H.L. (2006). Electrical properties of nanocrystalline CeO$_2$ thin films deposited by in situ pulsed laser deposition. *J. Electrochem. Soc.* 153 (3): J15–J20.

125 Maheswari, N. and Muralidharan, G. (2018). Fabrication of CeO$_2$/PANI composites for high energy density supercapacitors. *Mater. Res. Bull.* 106: 357–364.

126 Huang, Y., Huang, Y., Meng, W.J. et al. (2015). Enhanced tolerance to stretch-induced performance degradation of stretchable MnO$_2$-based supercapacitors. *ACS Appl. Mater. Interfaces* 7 (4): 2569–2574.

127 Li, X., Gu, T.L., and Wei, B.Q. (2012). Dynamic and galvanic stability of stretchable supercapacitors. *Nano Lett.* 12 (12): 6366–6371.

128 Filiatrault, H.L., Porteous, G.C., Carmichael, R.S. et al. (2012). Stretchable light-emitting electrochemical cells using an elastomeric emissive material. *Adv. Mater.* 24 (20): 2673–2678.

129 Lipomi, D.J., Vosgueritchian, M., Tee, B.C.K. et al. (2011). Skin-like pressure and strain sensors based on transparent elastic films of carbon nanotubes. *Nat. Nanotechnol.* 6 (12): 788–792.

130 Kim, D.H., Lu, N.S., Ma, R. et al. (2011). Epidermal electronics. *Science* 333 (6044): 838–843.

131 Park, S. and Jayaraman, S. (2003). Smart textiles: wearable electronic systems. *MRS Bull.* 28 (8): 585–591.

132 Lacour, S.P., Wagner, S., Huang, Z.Y., and Suo, Z. (2003). Stretchable gold conductors on elastomeric substrates. *Appl. Phys. Lett.* 82 (15): 2404–2406.

133 Bowden, N., Brittain, S., Evans, A.G. et al. (1998). Spontaneous formation of ordered structures in thin films of metals supported on an elastomeric polymer. *Nature* 393 (6681): 146–149.

134 Emmenegger, C., Mauron, P., Sudan, P. et al. (2003). Investigation of electrochemical double-layer (ECDL) capacitors electrodes based on carbon nanotubes and activated carbon materials. *J. Power Sources* 124 (1): 321–329.

135 Largeot, C., Portet, C., Chmiola, J. et al. (2008). Relation between the ion size and pore size for an electric double-layer capacitor. *J. Am. Chem. Soc.* 130 (9): 2730–2731.

136 Wang, D.W., Li, F., Chen, Z.G. et al. (2008). Synthesis and electrochemical property of boron-doped mesoporous carbon in supercapacitor. *Chem. Mater.* 20 (22): 7195–7200.

137 Wang, D.W., Li, F., Liu, M. et al. (2008). 3D aperiodic hierarchical porous graphitic carbon material for high-rate electrochemical capacitive energy storage. *Angew. Chem. Int. Ed.* 47 (2): 373–376.

138 Kim, H.S., Jung, Y., and Kim, S. (2017). Capacitance behaviors of conducting polymer-coated graphene nanosheets composite electrodes containing multi-walled carbon nanotubes as additives. *Carbon Lett.* 23 (1): 63–68.

139 Liu, Y.Q., Weng, B., Razal, J.M. et al. (2015). High-performance flexible all-solid-state supercapacitor from large free-standing graphene-PEDOT/PSS films. *Sci. Rep.* 5: 17045.

140 Zheng, H.J., Kang, W., Fengming, Z. et al. (2010). Studies on mechanism of carbon nanotube and manganese oxide nanosheet self-sustained thin film for electrochemical capacitor. *Solid State Ionics* 181 (37–38): 1690–1696.

141 Lv, P., Feng, Y.Y., Li, Y., and Feng, W. (2012). Carbon fabric-aligned carbon nanotube/MnO$_2$/conducting polymers ternary composite electrodes with high utilization and mass loading of MnO$_2$ for super-capacitors. *J. Power Sources* 220: 160–168.

142 Sharma, R.K. and Zhai, L. (2009). Multiwall carbon nanotube supported poly(3,4-ethylenedioxythiophene)/manganese oxide nano-composite electrode for super-capacitors. *Electrochim. Acta* 54 (27): 7148–7155.

143 Song, Z.Y., Duan, H., Li, L.C. et al. (2019). High-energy flexible solid-state supercapacitors based on O, N, S-tridoped carbon electrodes and a 3.5 V gel-type electrolyte. *Chem. Eng. J.* 372: 1216–1225.

144 Hu, Z.L., Zhang, S., Dong, S.M. et al. (2018). Self-stabilized solid electrolyte interface on a host-free Li-metal anode toward high areal capacity and rate utilization. *Chem. Mater.* 30 (12): 4039–4047.

145 Du, H.P., Li, S.Z., Qu, H.T. et al. (2018). Stable cycling of lithium-sulfur battery enabled by a reliable gel polymer electrolyte rich in ester groups. *J. Membr. Sci.* 550: 399–406.

146 Ma, Q., Zhang, H., Zhou, C.W. et al. (2016). Single lithium-ion conducting polymer electrolytes based on a super-delocalized polyanion. *Angew. Chem. Int. Ed.* 55 (7): 2521–2525.

147 Li, X.W., Zhang, Z.X., Li, S.J. et al. (2016). Polymeric ionic liquid-plastic crystal composite electrolytes for lithium ion batteries. *J. Power Sources* 307: 678–683.

148 Zhou, R., Liu, W.S., Leong, Y.W. et al. (2015). Sulfonic acid- and lithium sulfonate-grafted poly(vinylidene fluoride) electrospun mats as ionic liquid host for electrochromic device and lithium-ion battery. *ACS Appl. Mater. Interfaces* 7 (30): 16548–16557.

149 Zhao, H., Jia, Z., Yuan, W. et al. (2015). Fumed silica-based single-ion nanocomposite electrolyte for lithium batteries. *ACS Appl. Mater. Interfaces* 7 (34): 19335–19341.

150 Zhang, P.F., Li, M.T., Yang, B.L. et al. (2015). Polymerized ionic networks with high charge eensity: quasi-solid electrolytes in lithium-metal batteries. *Adv. Mater.* 27 (48): 8088–8094.

151 Kammoun, M., Berg, S., and Ardebili, H. (2015). Flexible thin-film battery based on graphene-oxide embedded in solid polymer electrolyte. *Nanoscale* 7 (41): 17516–17522.

152 Kalhoff, J., Eshetu, G.G., Bresser, D., and Passerini, S. (2015). Safer electrolytes for lithium-ion batteries: state of the art and perspectives. *ChemSusChem* 8 (17): 2765–2765.

153 Zhang, Y.F., Rohan, R., Cai, W.W. et al. (2014). Influence of chemical microstructure of single-ion polymeric electrolyte membranes on performance of lithium-ion batteries. *ACS Appl. Mater. Interfaces* 6 (20): 17534–17542.

154 Kuo, P.L., Wu, C.A., Lu, C.Y. et al. (2014). High performance of transferring lithium ion for polyacrylonitrile-interpenetrating cross linked polyoxyethylene network as gel polymer electrolyte. *ACS Appl. Mater. Interfaces* 6 (5): 3156–3162.

155 Kimura, K. and Tominaga, Y. (2017). Ionic liquid-containing composite poly(ethylene oxide) electrolyte reinforced by electrospun silica nanofiber. *J. Electrochem. Soc.* 164 (13): A3357–A3361.

156 Shin, J.H., Henderson, W.A., and Passerini, S. (2003). Ionic liquids to the rescue? Overcoming the ionic conductivity limitations of polymer electrolytes. *Electrochem. Commun.* 5 (12): 1016–1020.

157 Tang, W.J., Tang, S., Zhang, C.J. et al. (2018). Simultaneously enhancing the thermal stability, mechanical modulus, and electrochemical performance of

solid polymer electrolytes by incorporating 2D sheets. *Adv. Energy Mater.* 8 (24): 1800866.
158 He, W.S., Cui, Z.L., Liu, X.C. et al. (2017). Carbonate-linked poly(ethylene oxide) polymer electrolytes towards high performance solid state lithium batteries. *Electrochim. Acta* 225: 151–159.
159 Miller, J.R., Outlaw, R.A., and Holloway, B.C. (2010). Graphene double-layer capacitor with ac line-filtering performance. *Science* 329 (5999): 1637–1639.
160 Zhang, L.L. and Zhao, X.S. (2009). Carbon-based materials as supercapacitor electrodes. *Chem. Soc. Rev.* 38 (9): 2520–2531.
161 Lota, G. and Frackowiak, E. (2009). Striking capacitance of carbon/iodide interface. *Electrochem. Commun.* 11 (1): 87–90.
162 Stoller, M.D., Park, S.J., Zhu, Y.W. et al. (2008). Graphene-based ultracapacitors. *Nano Lett.* 8 (10): 3498-3502.
163 Zhang, Z.T., Deng, J., Li, X.Y. et al. (2015). Superelastic supercapacitors with high performances during stretching. *Adv. Mater.* 27 (2): 356–362.
164 Wang, H., Zhu, B.W., Jiang, W.C. et al. (2014). A mechanically and electrically self-healing supercapacitor. *Adv. Mater.* 26 (22): 3638–3643.
165 Sun, H., You, X., Jiang, Y.S. et al. (2014). Self-healable electrically conducting wires for wearable microelectronics. *Angew. Chem. Int. Ed.* 53 (36): 9526–9531.
166 Huang, Y., Zhong, M., Huang, Y. et al. (2015). A self-healable and highly stretchable supercapacitor based on a dual crosslinked polyelectrolyte. *Nat. Commun.* 6: 10310.
167 Sun, J.Y., Keplinger, C., Whitesides, G.M., and Suo, Z.G. (2014). Ionic skin. *Adv. Mater.* 26 (45): 7608–7614.
168 Zhong, C., Deng, Y.D., Hu, W.B. et al. (2015). A review of electrolyte materials and compositions for electrochemical supercapacitors. *Chem. Soc. Rev.* 44 (21): 7484–7539.
169 Cho, T.H., Tanaka, M., Onishi, H. et al. (2008). Silica-composite nonwoven separators for lithium-ion battery: development and characterization. *J. Electrochem. Soc.* 155 (9): A699–A703.
170 Kritzer, P. and Cook, J.A. (2007). Nonwovens as separators for alkaline batteries - an overview. *J. Electrochem. Soc.* 154 (5): A481–A494.
171 Cho, T.H., Tanaka, M., Ohnishi, H. et al. (2010). Composite nonwoven separator for lithium-ion battery: development and characterization. *J. Power Sources* 195 (13): 4272–4277.
172 Szubzda, B., Szmaja, A., Ozimek, M., and Mazurkiewicz, S. (2014). Polymer membranes as separators for supercapacitors. *Appl. Phys. A* 117 (4): 1801–1809.
173 Tonurist, K., Thomberg, T., Janes, A. et al. (2013). Influence of separator properties on electrochemical performance of electrical double-layer capacitors. *J. Electroanal. Chem.* 689: 8–20.
174 Bon, C.Y., Mohammed, L., Kim, S. et al. (2018). Flexible poly(vinyl alcohol)-ceramic composite separators for supercapacitor applications. *J. Ind. Eng. Chem.* 68: 173–179.

7

Functional Polymer Nanocomposite for Triboelectric Nanogenerators

Xingyi Dai, Jiancheng Han, Qiuqun Zheng, Cheng-Han Zhao, and Long-Biao Huang

Shenzhen University, College of Physics and Optoelectronic Engineering, Shenzhen 518060, China

7.1 Introduction

With the rapid development of human civilization, the huge demands for portable and wearable electronic devices have given rise to a sharp rise in energy consumption [1]. The traditional fossil fuel are unrenewable, and have brought about environmental contamination. Therefore, it's of great significance to harvest green and environmentally friendly energy. Since 2012, triboelectric nanogenerator (TENG), as an emerging technology, invented by Zhong Lin Wang and coworkers [2] has shown a prospect to mitigate energy crisis and environmental pollution problems. Relying on the coupling effect of contact electrification and electrostatic induction, TENGs can effectively convert mechanical energy into electrical energy and signal, which have been widely used in energy harvesters and self-powdered sensors [3, 4]. Compared with other energy harvesting technology and power source, such as piezoelectric nanogenerators, batteries, and capacitors, TENGs present the advantages of simple manufacturing process, flexible designability, diverse material availability, and low cost [5].

Polymer as one of key material in TENGs has the merits of flexibility, abundant raw materials, rich variety, and cost-effectiveness. Polymer plays an important role in the device structure of TENGs. Moreover, the performance of TENGs can be enhanced by optimizing materials. On the other side, during practical operation, TENGs are prone to suffering from mechanical damage and harsh conditions, resulting in the damage of materials, the reduction of performance, and even the complete failure of device, which may be discharged as environmental waste. As a result, it is urgent to improve the reliability, durability, and environmental friendliness of TENGs. Functional polymer-based nanocomposites, including self-healing [6], shape memory [7], and biodegradable materials [8], might be the ideal candidates to satisfy the aforementioned requirements. Polymer nanocomposite materials exhibit unique and excellent functions by taking advantages of each component. Moreover, through the utilization of the functional materials,

Polymer Nanocomposite Materials: Applications in Integrated Electronic Devices, First Edition.
Edited by Ye Zhou and Guanglong Ding.
© 2021 WILEY-VCH GmbH. Published 2021 by WILEY-VCH GmbH.

the TENGs can be imparted with outstanding performances and the application of devices will be largely broadened.

In this chapter, we mainly focus on the fabrication of TENGs by functional polymer nanocomposite including self-healing, shape memory, or biodegradable materials. The operating mechanisms of TENGs and the approaches to enhance the electrical output performances are firstly described. Subsequently, the methods of preparing polymer nanocomposite and improving dispersion of nanomaterial in polymer matrix are introduced. Moreover, brief introductions of self-healing, shape memory, and biodegradable polymer nanocomposite are provided, respectively, and their application for TENGs is discussed. This chapter provides a guideline for the prospective development of TENGs based on functional polymer nanocomposite.

7.2 Triboelectric Nanogenerators

Triboelectric effect used to be considered negative for it may cause some accidents, such as radio equipment interference, electrostatic spark, fire, and explosion. However, through the combination of triboelectric effect and electrostatic induction, Wang's group invented TENGs in 2012, which can effectively convert mechanical energy into electricity [2]. The first TENG was fabricated by polyester (PET) and Kapton films serving as triboelectric layers, on the back of which Au alloy films were coated as electrodes. When two materials come into contact, positive and negative triboelectric charges will be generated on the contact surfaces due to electrification. During cyclic contact separation, driven by electrostatic induction, the potential difference changes and the induced charges can flow in external circuit. Consequently, the conversion of mechanical energy to electrical energy is accomplished. The commonly used triboelectric materials for TENGs include polytetrafluoroethylene (PTFE), polyimide (Kapton), polyurethane (PU), and polydimethylsiloxane (PDMS), and electrode materials include Ag, Au, Cu, and Al. In our daily life, mechanical energy is ubiquitous in surrounding environments, such as human motion, vibration, wind, raindrop, and ocean wave, which can be harvested by TENGs [9, 10]. Moreover, TENGs serving as self-powered sensors can be utilized for human–machine interface, vibration and biomedical monitoring, chemical and environmental monitoring, and tracking moving objects [11, 12].

Currently, there are four main working modes of TENG, including vertical contact-separation (CS), lateral-sliding (LS), single-electrode (SE), and freestanding triboelectric-layer (FT) mode. The operation mechanisms of all four working modes originate from the couple of contact electrification and electrostatic induction. In vertical CS mode, the relative motion is perpendicular to the surface of material. A typical example is schematically illustrated in Figure 7.1a [13]. Under pressure, poly(methyl methacrylate) (PMMA) and Kapton come into contact. Due to triboelectric effect, the equivalent amount of charges with opposite polarities is generated on the contact surfaces. The surface electron affinity of Kapton is higher than that of PET, leading to Kapton being negatively charged. At this point, there is no electric potential difference between two surfaces because of the distance

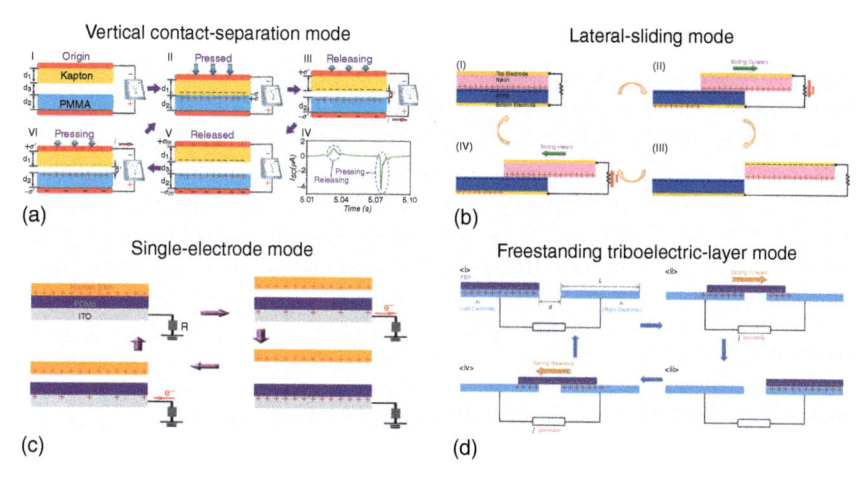

Figure 7.1 Four working modes of TENG. (a) Vertical contact-separation mode. Source: Reproduced with permission Zhu et al. [13]. Copyright 2012, American Chemical Society. (b) Lateral-sliding mode. Source: Reproduced with permission Wang et al. [14]. Copyright 2013, American Chemical Society. (c) Single-electrode mode. Source: Reproduced with permission Yang et al. [15]. Copyright 2013, American Chemical Society. (d) Freestanding triboelectric-layer mode. Source: Reproduced with permission Wang et al. [16]. Copyright 2014, Wiley-VCH.

between two surfaces is close to zero. Due to the insulating property of polymer, the triboelectric charges will remain on the surfaces. After releasing pressure, the gap between two charged surfaces increased, resulting in the formation of electric potential difference between the two back electrodes. Meanwhile, the induced charges are generated in the electrodes to compensate static charges on the surfaces of triboelectric layers, which drive electrons to flow, generating an instantaneous electrical current in the external circuit. When the films completely reverts to the original position, the positive and negative charges are balanced, resulting in no current. Once the PET is pressed to contact with Kapton immediately, the electrons will flow in the opposite direction. By the periodical contact-separation between the two triboelectric layers, an alternative current will be continuously generated.

In LS mode, the relative sliding is parallel to the contact interface with the change of relative displacement. As shown in Figure 7.1b, when the two triboelectric layers are fully and tightly overlapped, the surfaces of polyamide 6,6(nylon) and PTFE are positively and negatively charged, respectively [14]. Upon a layer sliding outward, the contact surface area gradually decreases, leading to the in-plane charge separation and the formation of potential difference. The electrons flow from the bottom electrode to the top electrode, thereby generating current. The amount of induced charges continues to increase until the two layers are completely separated. When the layer is sliding backward, the flow of electrons is driven in the opposite direction. Grating structure can be effectively applied in the LS-based TENG [17]. Moreover, a plane rotation-induced sliding mode is further derived based on LS mode [18].

The SE mode is developed on the CS mode. In an SE-based device, there is only one electrode, and ground is usually taken as reference electrode. This mode can be conveniently utilized in wearable and portable devices, since human beings are good nature conductors. And it can effectively harvest energy from human motion, such hand typing, walking, and running. As shown in Figure 7.1c, when hand taps on the surface of PDMS, PDMS is negatively charged [15]. Upon the separation, the potential difference is built and the electrons flow from indium tin oxide (ITO) to ground. When the induced charges on ITO and electrostatic charges on PDMS are balanced, there is no electric signal. Once the skin approaching back to PDMS, the electrons reversely flow, completing a cycle. Similarly, a sliding SE mode can also be implemented [19].

The FT mode is improved on the basis of aforementioned modes. In FT mode, the sliding triboelectric layers don't need to be attached to the electrodes, providing convenience and feasibility in numerous application scenarios. As depicted in Figure 7.1d, two stationary Al electrodes are connected, and a fluorinated ethylene propylene (FEP) film as the freestanding triboelectric layer is on top of Al films [16]. When FEP is in full contact with the left Al electrode, Al is positively charged and FEP is negatively charged. Subsequently, as the FEP slides forward, electrons flow from right to left electrode. When the FEP is completely overlapped with right Al electrode, the charges are balanced. Upon FEP sliding backward, the electrons flow back. Grating and rotating structure have been broadly operated in FT mode-based TENGs [20, 21].

Based on the working mechanisms of TENGs, a wide range of applications have been developed and shown attractive performance. Not limited to a single operating mode, a TENG device can simultaneously work in multiple modes, with increased charges and energy delivered [22]. Although TENGs can convert environmental mechanical energy into electrical energy, TENGs cannot directly power some portable and wearable devices because of its low/unstable power density/output. Therefore, improving the electrical output performances of TENGs, such as output voltage, current, and power density, is of great significance for the practical applications of TENGs. From the working principle, it can be found that the process of triboelectrification will influence the output performances of TENGs. Therefore, the capability of material to gain and lose electrons deeply influences the performance of TENG and has a strong relationship to its own polarity. In order to attain ideal output, it is better to choose tribomaterials with large difference in polarity, resulting in the increment of the amount of transferred charges. The reported triboelectric series [23] provide a guide to select suitable friction materials. Recently, numerous endeavors have been fulfilled to develop new materials and increase the output performances in various aspects, such as optimization in material composition and improvement in device structure.

The strategies of surface functionalization and bulk composite can be utilized to optimize material composition. As triboelectric charges are generated and distributed on the surface of friction layer, triboelectric effect can be largely enhanced through chemically modifying the surface of triboelectric material with functional groups, without the change of main materials. For example, treating the surface by inductive-coupled plasma etching using mixture of carbon tetrafluoride (CF_4) and oxygen (O_2) gases [24] and grafting $1H,1H,2H,2H$-perfluorooctyltrichlorosilane

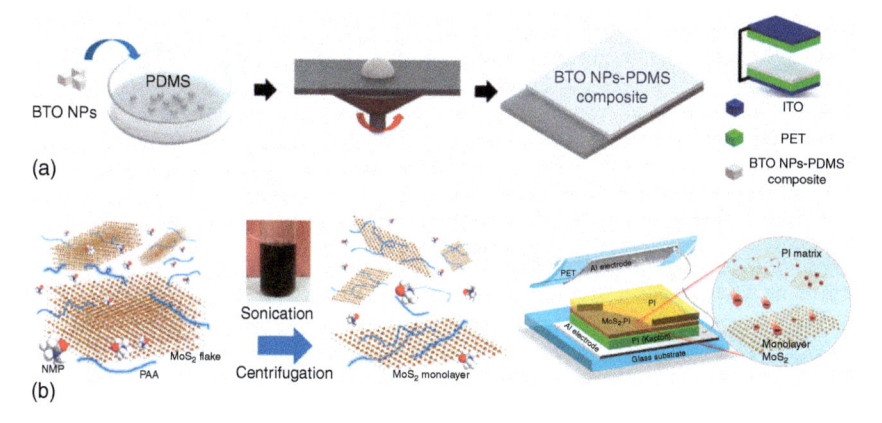

Figure 7.2 Bulk composite. (a) BaTiO$_3$ nanoparticles (BTO NPs) embedded into PDMS. Source: Reproduced with permission Kwon et al. [26]. Copyright 2016, Elsevier. (b) Mixture of monolayer MoS$_2$ and polyamic acid (PAA) used to fabricate negative triboelectric layer. Source: Reproduced with permission Wu et al. [28]. Copyright 2017, American Chemical Society.

onto the friction surface [25] have been demonstrated to significantly enhance triboelectric charge density due to the strong electron affinity of fluorine groups. In addition, polymer composite, through incorporating functional nanomaterials into polymer matrix, can be used as good tribomaterials for TENG. The introduction of high dielectric filler such as BaTiO$_3$ [26] and SrTiO$_3$ [27] can enhance both triboelectric effect and permittivity of materials, which are key parameters to the output performances of TENGs. Kwon et al. embedded BaTiO$_3$ nanoparticles (BTO NPs) into PDMS to form BTO NPs–PDMS composite thin films (Figure 7.2a). The output power densities of BTO NPs–PDMS composite based TENG were 5–6 times larger than that of the TENG without BTO NPs [26]. Besides, introducing carbon nanotubes (CNTs) [29], reduced graphene oxide (rGO) [30], and molybdenum disulfide (MoS$_2$) [28] into the triboelectric layer can improve output performances of TENGs, by taking advantage of the abilities of nanoparticles to capture triboelectric electrons. Wu et al. employed the mixture of monolayer MoS$_2$ and polyamic acid (PAA) to fabricate the negative triboelectric layer (Figure 7.2b). The power density of TENGs with monolayer MoS$_2$ improved up to 120 times higher, by comparison with the TENG without monolayer MoS$_2$ [28].

Optimizing device structure has also been proved to be an effective approach to improve the output performances of TENGs by enlarging the effective contact area, which can be accomplished by surface morphology modification, liquid–solid interface, and structure design. Enlargement of the surface roughness can increase the amount of triboelectric charges. By chemical etching and pattern-transfer printing process, the surface of friction layer can be modified with nanostructure such as nanowires, nanofibers, nanorods, and nanoparticles [31–34], as well as micro or nano patterning such as pyramid-, cube-, and hemisphere-based arrays [35, 36]. Comparing with solid–solid interface, liquid–solid interface can enhance a larger contact area [37]. Normally, the contact area between two solid layers cannot reach to 100% due to the surface roughness and matching degree. Therefore, the complete and intimate contact between liquid and solid interface can be achieved due to the wettability and shape adaptability of liquid and result in the increment of charge density.

In addition, the ingenious structure design can enable the device to improve the output performances. Recently, several new structures have been designed for TENGs. For example, grating and rotating structures [38, 39] are widely used in LS and FT mode based TENGs, which can enable the device continue to generate electric signals during motion. Besides, unique structure designs such as wave- [40], honeycomb- [41], and multilayered-structured TENGs can effectively increase the contact area [42], which have been developed for highly effective energy harvesting and self-powered sensing.

7.3 Functional Polymer Nanocomposite

Taking advantage of components, polymer nanocomposites can be endowed with unique and excellent features through incorporating nanomaterials into polymer matrix. Enormous functional polymer and polymer nanocomposite have been exploited and widely utilized in triboelectric layers and electrodes. Functional polymer nanocomposites can be effectively used in triboelectric layer to improve the output performance of TENGs [43]. Especially, functional polymer nanocomposite plays a significant role in electrode materials. Compared with conventional metal conductors such as gold, silver, copper, and aluminum foils, polymer nanocomposite-based electrodes have shown immense potential in future self-powered devices with flexibility, robustness, lightweight, transmittance, portability, and tunability [44].

There are tremendous advances in the preparation of functional polymer nanocomposites for TENG. The fabrication strategies of polymer nanocomposite can be mainly classified into three categories: solution blending, melt blending, and in situ polymerization. Mixing nanomaterials with polymer matrix is a very simple and extensive method to prepare polymer nanocomposites. For solution blending, nanomaterial is dispersed in a good solvent, followed by adding the polymer into the solution. After the nanofillers and polymer being well mixed, nanocomposite films could be formed in the mold or on the surface of substrate through the process of solution casting, spin coating, or electrospinning. However, the major disadvantages of solution blending are the toxicity of organic solvent and environmental pollution due to the volatilization of solvent. Solution casting is the easiest method without external instruments. Fan et al. dispersed the CNTs in ethyl alcohol and then mixed them with PDMS matrix in a mold. After evaporating alcohol and curing polymer, a CNT–elastomer hybrid nanocomposite for harvesting mechanical energy was prepared [45]. Spin-coating is extensively used to prepare uniform and thin nanocomposite films. Cao et al. doped ZrO_2 nanoparticles into the solution of poly(vinylidene fluoride)–trifluoroethylene (PVDF–TrFE). Finally, the transparent ferroelectric polymer composite film as friction layer was formed

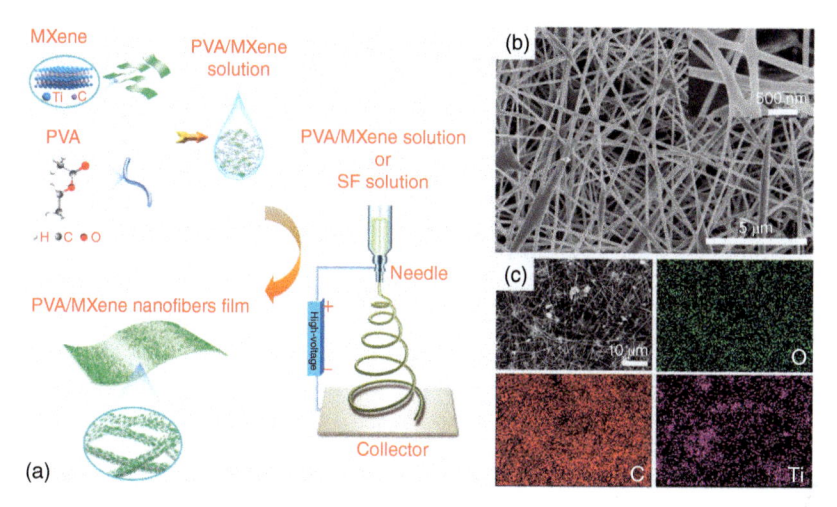

Figure 7.3 (a) Schematic illustration of the fabrication process of PVA/MXene nanofibers film by electrospinning. (b) SEM images of PVA/MXene nanofibers with an inset of the morphology at a higher magnification. (c) Elemental mapping images of PVA/MXene nanofibers for O, C, and Ti. Source: Reproduced with permission Jiang et al. [47]. Copyright 2019, Elsevier.

on a substrate by spinning coating at 2000 rpm for 30 seconds [46]. Electrospinning is an efficient technique to manufacture nanofiber at large scale. Jiang et al. homogeneously mixed MXene with poly(vinyl alcohol) (PVA) aqueous solution. Subsequently, electrospinning was conducted at voltage of 18 kV to obtain a thin and flexible PVA/MXene nanofiber film as negative friction layer (Figure 7.3) [47].

In melt blending, nanomaterials are mixed with the polymer melt in viscous flow state. Then the mixture can be processed through compression molding, extrusion, or injection molding. The main advantage of melt blending is that nanocomposites can be prepared in large quantities without solvent pollution. Compared with solution blending, the viscosity of melt polymer is much higher, which makes the uniform dispersion of nanofillers more difficult. In addition, the processing temperature must be strictly controlled to enable polymer in a good flowing state and prevent polymer from decomposing under high temperature. At present, melt blending is rarely applied in the fabrication of TENGs due to the restriction of nanostructure. While, combination of the melt blending processing and nanofabrication technology has been demonstrated to hold great potential for developing cost-effective and large-scale TENGs. Zheng et al. indicated nanoporous PVDF bulks for piezoelectric and triboelectric hybrid nanogenerators (PTNGs). The nanoporous PVDF bulks were quickly and easily fabricated by hot compressing the mixture of PVDF and SiO_2 in a die, followed by the removal of SiO_2 nanospheres (Figure 7.4) [48]. In the extrusion process, the polymer matrix is thermoplastic. Yan et al. prepared the poly(ethylene-*co*-poly(vinyl alcohol)) nanofibers via extrusion method using a twin-screw extruder with the shear rate of 50 rpm at 200–220 °C. Then the triboelectric composite pair was achieved by post-fabrication modifications on the nanofiber membranes [49].

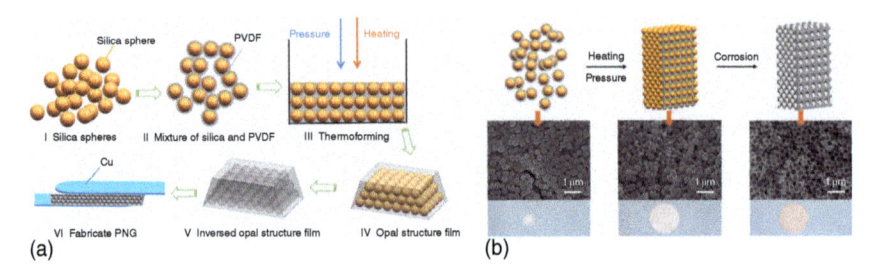

Figure 7.4 (a) Fabrication process of porous PTNG using thermoforming-corrosion method. (b) Schematic diagram of materials and cross-section SEM images of powder mixture, blocks after thermoforming, and porous blocks. The insets show their structure and morphology. Source: Reproduced with permission Zheng et al. [48]. Copyright 2019, Elsevier.

In in situ polymerization, nanomaterials are well mixed with monomers, and then polymerization is initiated. This method is much suitable for the insoluble cross-linking and thermally unstable polymeric systems. Additionally, the viscosity of monomer is much lower comparing with polymer, leading to the better dispersion of nanofillers. During the polymerization process, the nanofillers and the polymer could form an interpenetrating network, which can improve the compatibility between the nanoparticles and the polymer matrix. Wang et al. reported a Ce-doped ZnO–polyaniline (PANI) nanocomposite film through in-situ polymerization to fabricate the respiration-driven TENG. The Ce-doped ZnO nanoparticles were dispersed in aniline solution for polymerizing. After drying, the nanocomposite film was formed [50].

Homogeneously dispersing nanomaterials into polymer matrix is a critical parameter to obtain high-performance polymer nanocomposites. However, it's a challenge to avoid the aggregation of nanofillers. Due to the high specific surface area of nanomaterials, the atoms on the surface of nanomaterials are in a highly activated state, resulting in high surface energy, which makes these atoms easily combine with each other. Therefore, it's of great significance to improve the interfacial interactions between polymer matrix and nanofillers, allowing for the enhancement of their compatibility. Various approaches including physical or chemical methods have been adopted to overcome the above problems. Ultrasonic dispersion is an easy and convenient technique for the disaggregation of nanomaterials. Under the influence of ultrasound, nanofillers tend to uniformly dispersed in the medium with smaller aggregates due to the strong impact and shear forces. Moreover, surface modification of nanomaterials can effectively facilitate well dispersion. Driven by non-covalent interaction, dispersant agent such as sodium dodecyl benzene sulfonate, ammonium citrate, and polyvinylpyrrolidone can be absorbed onto the particle surface, leading to strong repulsive force between particles based on the effect of electrostatic interaction and steric hindrance. On the other hand, grafting functional groups onto the surface of nanomaterials through covalent interaction can create strong interaction with polymer matrix and promote the uniform and stable dispersion of nanofillers in matrix.

To further avoid the uneven dispersion of nanomaterials in the polymer matrix and obtain high-performance nanocomposite, three-dimensional (3D) polymer

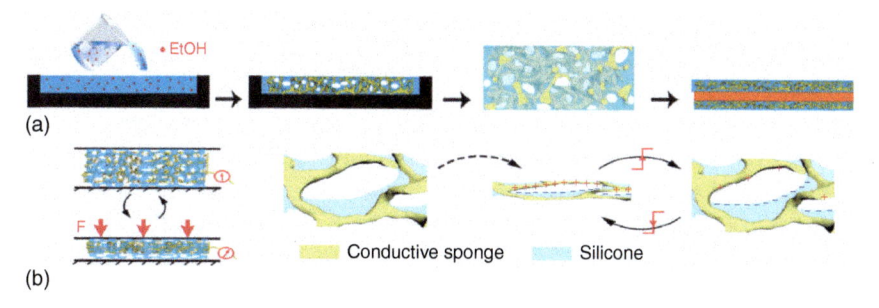

Figure 7.5 (a) Fabrication process and (b) working principle of the conductive sponge/porous silicone-based TENG. Source: Reproduced with permission Chen et al. [52]. Copyright 2019, Wiley-VCH.

nanocomposites have attracted considerable interest due to their unique structures and properties. The 3D networks are mainly in the form of aerogel, sponge, and foam through in situ self-assembly, chemical vapor deposition (CVD), freeze–drying, and template methods [51]. Then, 3D polymer nanocomposites can be achieved by pouring polymer into the network of porous nanomaterial [52], in situ polymerization on the surface of assembled nanomaterial [53], and coating nanomaterials on the surface of porous polymer-based template [54]. Chen et al. reported a conductive sponge/porous silicone-based TENG as a tactile sensor. The conductive sponge was immersed into the mixture of liquid silicone and ethanol. After curing under heating, the ethanol evaporated and expanded to form porous structure (Figure 7.5) [52]. The merits of 3D polymer nanocomposites include light-weight, large surface area, high specific strength, high flexibility and deformability, and good electrical conductivity. Especially for TENGs, the use of 3D porous polymer nanocomposite can effectively enhance the electrical output performance due to the increase in contact area between the porous structures, leading to additional charges on the porous surface. Kim et al. fabricated a TENG composed of porous conductive polymer (PCP–TENG). Compared with flat-structured contact separation TENG (F-TENG), PCP–TENG exhibited much higher short-circuit current and could more effectively harvest mechanical energy source with varying amplitude of vibration [55].

7.4 Self-healing Triboelectric Nanogenerators

During the operation of TENGs, frequent friction, such as pressing, bending, twisting, and sliding, will give rise to the fracture of materials and further destroy the devices. Furthermore, it's difficult to repair or replace the broken components of devices. Since most of TENGs are based on polymer materials, self-healing polymers could solve the aforementioned problems. After self-healing materials being damaged, the generated invisible micro-cracks can be repaired, and the performances of self-healing materials can partially or entirely revert to their original state. Following the self-healing actions of polymer materials, the device can extend lifetime, save resources, and enable safety and reliability. Self-healing materials have been utilized in chemical sensors, electronic skins, supercapacitor, batteries, coatings, and other fields [56]. The TENGs with self-healing capability have shown promising application prospects. Therefore, it's important to further design and construct self-healing TENGs for the practical applications.

Self-healing materials can be mainly divided into two categories: extrinsic and intrinsic type. In extrinsic systems, the repairing agents are embedded into the polymer matrix in the form of microcapsules, microvessels, or fibers [57]. When the material is damaged, the repairing agent is released to the crack area to complete self-healing through curing reactions between repair agent and matrix. Compared with extrinsic systems, the intrinsically self-healing materials are more preferred for electric devices, owing to their flexibility, deformability, and repeated healing features. In intrinsic systems, self-healing materials contain reversible covalent bonds or non-covalent bonds, which can reversibly break and reform under certain conditions. When polymer material is damaged, the spontaneous healing process can be achieved through bond reformation and metathesis reaction. Reversible covalent bonds include imine bonds [58], acylhydrazone bonds [59], disulfide bonds [60], and Diels–Alder reactions [61]. Compared with non-covalent systems, reversible covalent bonds with stronger chemical interactions enable self-healing materials to have higher mechanical strength and dimensional stability. Non-covalent systems are physically reversible, including hydrogen bonding [62], metal–ligand coordination [63], and host–guest interactions [64]. The formed physical crosslinking points are liable to be affected by pH, temperature, and mechanical forces. The material containing non-covalent bonds can achieve relatively fast self-healing through rapid disassociation–association process with the change of external conditions.

The self-healing polymer-based nanocomposites play an important role in the electrode of TENGs, which can recover not only mechanical property, but also electrical performance. Driven by the self-healing of polymer matrix, the conductive networks can be reconstructed and achieve the recovery of electrical performance. The conductive nanomaterials are dispersed into healable polymer matrix or deposited on surface of healable polymer to construct self-healing electrode. The good compatibility (dispersibility, interfacial bonding, etc.) between the conductive nanomaterials and polymer matrix is crucial for superior performances of polymer composite. Wu et al. prepared ultrafast self-repairing nanocomposites and applied them in flexible electronics [61]. The functionalized graphene nanosheets were covalently linked with the self-healing polymer matrix based on the Diels–Alder reaction. The mechanical healing efficiency of the nanocomposites was more than 96% after one minutes near-infrared (NIR) irradiation, and the conductivity could be completely recovered. Yan et al. reported highly stretchable self-healing conductive bilayer composite films, which were composed of hydrogen bond-containing elastomeric substrates and wrinkled graphene [62]. There was strong interfacial adhesion between the two layers. The composite films exhibited good mechanical and electrical healing, and they could be used in strain sensors to monitor human motion.

Various self-healing polymers and nanocomposite have been applied in TENGs, in order to maintain the stable output performance and provide the devices with multiple functions. Xu et al. firstly combined healable polymer and magnetic-assisted

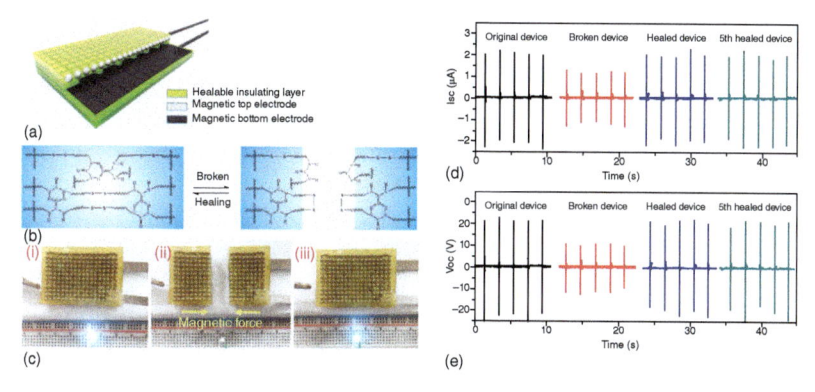

Figure 7.6 (a) Schematic of the self-healing TENG. (b) Self-healing mechanism of as-prepared PDMS–PU film. (c) Optical images of the electric healing process of the top electrode. (d) Short-circuit current and (e) open-circuit voltage of the as-prepared TENG at original, broken, and healed state. Source: Reproduced with permission Xu et al. [65]. Copyright 2017, Elsevier.

electrode to fabricate fully self-healing TENGs (Figure 7.6) [65]. Disulfide bonds were introduced into PDMS–PU polymer to obtain the healable insulating layer. The top and bottom electrodes were made of nickel-plated magnetic balls and magnetic cubes arrays, respectively. The top electrode was embedded in healable PDMS–PU, on the undersurface of which there was pattern to form gaps, and the bottom electrode was attached on another PDMS–PU as substrate. When the broken TENG was brought to contact, on the one hand, the mechanical performance of insulating layer could be restored 97% after healing for 2 hours at 65 °C, owing to the synergistical effect of disulfide and hydrogen bonds; on the other hand, the electric conductivity could be immediately recovered attributed to the magnetic attraction of electrodes, achieving fully self-healing. Therefore, the output performances of the healed TENG were close to the initial state.

Noticeably, the electrode layer with self-healing characteristic can guarantee the recovery of the final electric output performances. The cracked triboelectric layer hardly affects the electric performances since the output of an intact friction layer is equal to that of several cut sections in series with the same total area. Guan et al. devoted to designing a NIR irradiation triggered TENG based on self-healable nanocomposite as electrode layer [66]. CNTs were incorporated in epoxy resin-based polysulfide elastomer to obtain conductive and self-healable nanocomposite. PDMS and dynamic epoxy network served as triboelectric layer and healing-assisting layer, respectively. Driven by the disulfide exchange, the damaged TENG could be efficiently healed under NIR radiation for one minute with the temperature rising to 160 °C. The open-circuit voltage, short-circuit current, and transferred charge almost kept constant after repeatedly healing for 5 times.

Simultaneously realizing transparency, stretchability, and self-healing capability is highly desirable for soft power sources. Sun et al. presented a TENG that satisfied all the aforementioned requirements (Figure 7.7) [67]. A buckled composite electrode composed of Ag nanowires and poly(3,4-ethylenedioxythiophene) (Ag–PEDOT) was sandwiched in the self-healing elastomers based on reversible

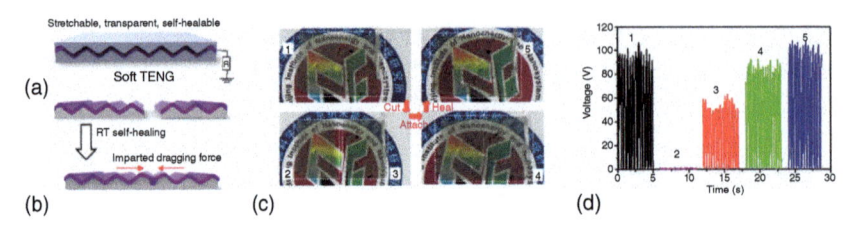

Figure 7.7 (a) Schematic illustration of the soft TENG. (b) Self-healing mechanism of the H-PDMS/Ag-PEDOT film with prestrain. (c) Photos and (d) voltage of a soft TENG at pristine, cut, attached, and healed states. Source: Reproduced with permission Sun et al. [67]. Copyright 2018, American Chemical Society.

imine bonds (H-PDMS). In the obtained H-PDMS/Ag–PEDOT film, the electrode was not intrinsically healable, whereas during the self-healing process of H-PDMS, the broken electrode could be brought into electrical contact under the imparted dragging force. The as-fabricated TENG could almost completely restore its energy-generation function at room temperature.

To further enhance the performance of TENGs, Chen et al. indicated an instantaneous self-healable and highly shape-adaptive triboelectric nanogenerator (SS-TENG) with improved outputs [68]. The healable viscoelastic polymer (called Silly Putty) as triboelectrification layer was synthesized through the condensation reaction between hydroxyl terminated PDMS and boric acid. The self-healing conductive composite as electrode was prepared by filling multiwalled carbon nanotubes (MWCNTs) into the polymer matrix. The sandwich-like TENG could fully recovered in three minutes without extra stimuli after mechanical damage, owing to the dynamic hydrogen bonds and dative bonds between boron atoms and oxygen. The viscoelastic polymer enabled the TENG adapt to arbitrary irregular surfaces. Moreover, due to the self-healing polymer with a certain degree of liquid-like viscous mechanical properties, the ideal soft contact was achieved at the solid–solid interfaces, which enhanced output performances.

Most of the healable TENGs are restricted by external triggers such as magnetism, heat, and light, while the automatically self-healing systems at room temperature may be unstable or slow to repair. Taking advantage of infrared (IR) radiation from the human body, Dai et al. designed an entirely self-healing and flexible TENG as a self-powered sensor to monitor human motion (Figure 7.8) [69]. Human skin serving as natural IR emitter could provide a favorable condition for the device to realize timely self-healing, based on thermal effect of IR radiation. The imine bonds and quadruple hydrogen bonding (2-ureido-4[1H]-pyrimidinone [UPy]) units were introduced into the cross-linking networks to form self-healing polymers as electrification layer. The UPy-functionalized MWCNTs were well dispersed in the polymer matrix to obtain self-healing nanocomposite as electrode layer. Driven by the reversibility of imine and UPy, there was robust interface binding between the overlapped layers and the device exhibited outstanding self-healing properties. The device could undergo repeated healing cycles and maintain the original output performances.

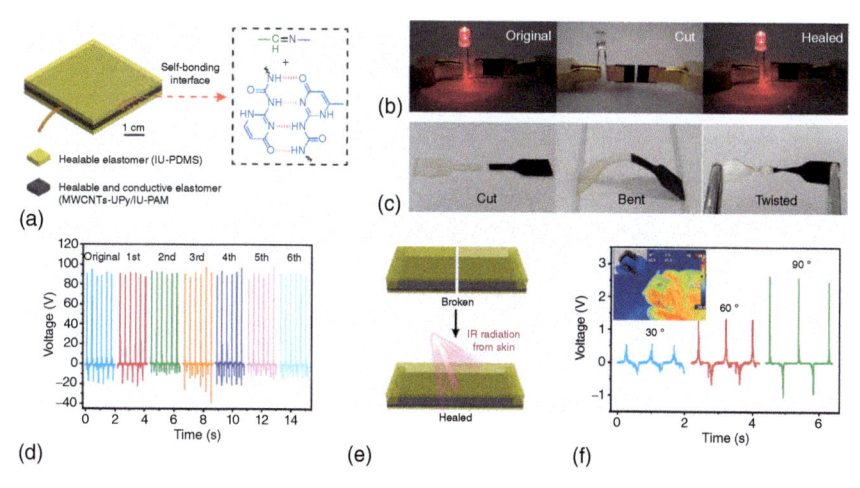

Figure 7.8 (a) Schematic illustration of self-healing TENG. (b) Optical images of self-healing nanocomposite in a circuit. (c) Self-healing enabled welding of cut dumbbell between the polymer matrix and nanocomposite. (d) Open-circuit voltage of the original and self-healed TENG after one to six cycles. (e) Schematic diagrams of the self-healing TENG via IR radiation from skin. (f) Open-circuit voltage responses when bending the finger joint at different angles with inset thermal and optical images of bent TENG on the finger. Source: Reproduced with permission Dai et al. [69]. Copyright 2020, Wiley-VCH.

7.5 Shape Memory Triboelectric Nanogenerators

Shape memory polymer (SMP) is a stimulus-responsive material, which can maintain a temporary shape and return to its original shape under external stimulus such as temperature, light, electricity, and magnetic field [70]. SMP is composed of switch units and net-points, which have great influence on the performance of SMP. The net-points refer to the chemical or physical cross-linking structures, determining the permanent shape. It's the entropic elasticity of the cross-linked polymer networks that drives the recovery of the deformed SMP. The switch units are capable of fixing the temporary shape and inducing the recovery, with the reversible change of the structures when exposure to external stimulus. Importantly, the key point of shape memory effect lies in switch units. The switching mechanism of SMP can be investigated on both phase level and molecular level.

At a phase level, there are three switches that can be introduced into polymer networks to construct SMP, including amorphous phase with glass transition temperature (T_g), semi-crystalline phase with melting temperature (T_m), and liquid crystal phase with isotropic temperature (T_i) [71]. As for the T_g-type SMP, when the temperature rises to above T_g, polymer transforms into highly elastic state in which the movement of polymer chains are improved, and the shape is easy to be deformed under external force. Upon cooling down without releasing the force, the polymer reverts to glass state and the deformed network is fixed. After releasing the force, the temporary shape can be kept. Once heating again, the frozen segments are activated to move. Meanwhile, driven by the force originating from the entropic elasticity of polymer, the network was rearranged to the initial state, with the recovery of original shape. Similarly, for T_m-type SMP, through controlling the temperature above or below the T_m of crystal, the semi-crystalline phase can be disappeared or regenerated, thus accomplishing the shape memory function. For the T_i-type SMP, there is a transition from an anisotropic to an isotropic phase in liquid crystalline elastomers when heating up to T_i. At a molecular level, the reversible units can be utilized to construct SMP. Supramolecular interactions such as hydrogen bonding [72] and reversible covalent bonds such as disulfide bond [73] have been applied in SMP to lock the temporary shape. The reversible system can accommodate its structure in response to external environments.

On the basis of SMP, functional nanomaterial can be incorporated into SMP matrix to improve the shape memory properties such as the fixity ratio and the strain recovery ratio. Noteworthily, SMP nanocomposites can be imparted with different responsiveness to broaden the actuation methods. Heating is a most direct way to actuate SMP. However, some indirect actuation approaches may be more attractive for their controllability and efficiency. The indirectly induced SMP nanocomposite can be categorized into electro-sensitive, light-sensitive, magnetic-sensitive, and microwave-sensitive types. As for electrically sensitive SMP nanocomposite, electrical conductivity is very important, which can be provided by conductive fillers such graphene, CNTs, and metal nanoparticles. Under the electrical field, electrical current passes through the conductive network of SMP nanocomposite. Due to "Joule heating," the generated heat can trigger the shape memory process. The electrical actuation method is more precise, convenient, and efficient than heat-induced method. As for light-sensitive SMP nanocomposite, functional fillers such as graphene, CNTs, and Au nanoparticles (Au NPs) are capable of absorbing light energy and transforming it into heat energy (photothermal effect). Under light radiation, a remotely, spatially, and locally controllable shape memory function can be realized. As for magnetic-sensitive SMP nanocomposites, under alternating magnetic field, there will generate heat in magnetic fillers such as Fe_3O_4 nanoparticles, attributing to electromagnetic thermal effect. As for microwave-sensitive SMP nanocomposite, carbon, silicon carbide, and ferroelectric nanoparticles can be used as microwave absorbers. Under electromagnetic field, the microwave energy is attenuated, with the generation of heat. The microwave-triggered actuation has the advantages of uniform and rapid heating.

Comparing with the TENG with flat surface, the devices with surface morphologies such as micro/nano pattern arrays have significantly enhanced output performances. However, under the long-term or strong force, the created surface microstructures could be gradually degraded or even completely destroyed, leading to the degradation of performances and the failure of device. SMP could be a promising solution to overcome above obstacles. The deformed surface morphologies of SMP-based TENG can recover to their initial state after being triggered by appropriate conditions, thus achieving prolonged lifetime and improved reliability. In addition, since SMP is transformable and programmable, the SMP-based TENG is capable to be adaptively attached on various objects with different shapes, which

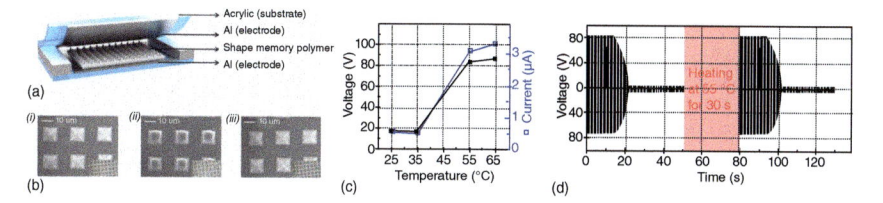

Figure 7.9 (a) Schematic of the TENG structure. (b) SEM images of the SMP micro-pyramid pattern (i) before degradation, (ii) after degradation, and (iii) after recovery. (c) Open-circuit voltage and short-circuit current of a healed SMP–TENG depending on the healing temperature. (d) Open-circuit voltage outputs from the SMP–TENG compressed by a strong force of 12 kg and healed. Source: Reproduced with permission Lee et al. [74]. Copyright 2015, Royal Society of Chemistry.

can widely broaden the application occasions of TENG and render the device effectively function.

Several efforts have been devoted to develop SMP-based TENG. Currently, the research mainly focuses on the TENG based on thermal-induced SMP. Lee et al. firstly used shape memory polyurethane (SMPU) with micropatterns to fabricate the TENG (SMP–TENG) with extending lifetime and enhanced performances (Figure 7.9) [74]. A thin PU film with micro pyramid patterns and Al were utilized as triboelectric layers, respectively. During the operation, the micropatterns degraded over time. And when applying too strong force, the micropatterns would be destroyed, which gave rise to the dramatical reduction of the electrical output performances. The output open-circuit voltage and short-circuit current decreased from 100 to 10 V and from 15 to 1 µA, respectively. However, after raising the temperature above the glass transition temperature of SMPU (T_g = 55 °C) for 30 seconds, the micropatterns recovered to their original shapes, leading to the recovery of the output performances of the TENG.

Apart from the recovery of microscopic morphologies, the macroscopic shape transformation of SMP is valuable for TENGs. Liu et al. reported a transformable TENG using SMP for body motion energy harvesting and self-powered mechanosensing [75]. The semicrystalline polycaprolactone (PCL) was introduced into the soft elastomer to form the SMP. Then, the SMP-based TENG was fabricated by encapsulating a conductive liquid electrode in the synthesized SMP. The TENG is capable of transforming its shape according to different requirements through tuning the temperature above the T_m of PCL. When cooling down to room temperature, the new shape could be fixed. Therefore, the SMP-based TENG was provided with transformable, programmable, and skin-compatible features to adapt to multiple surface configurations through a thermal process, thus effectively harvesting energy and sensing motion.

Combined with the self-recovery advantages of SMP in both micro morphology and macro shape, the performances of TENGs device can be improved and their application can be further developed. Xiong et al. utilized SMP to fabricate a self-restoring TENG with tunable microstructure for self-powered water temperature sensor [76]. An ether-based thermoplastic thermally triggered SMPU was served as triboelectric layer with various microstructures by electrospinning.

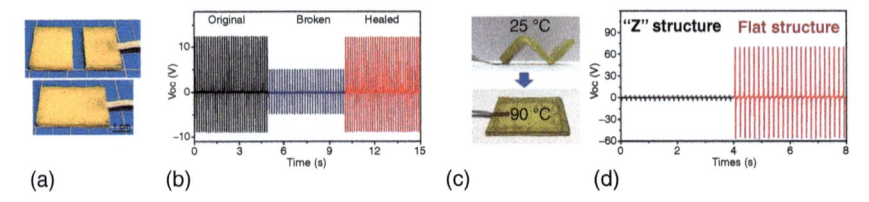

Figure 7.10 (a) Photograph of the healing process of the TENG. (b) Healing process in open-circuit voltage of the TENG. Response of (c) the shape and (d) the electric output of the TENG with the PDMS layer to the ambient temperature. Source: Reproduced with permission Xu et al. [77]. Copyright 2019, Royal Society of Chemistry.

The TENGs based on tunable microarchitecture were provided with enhanced output performances ($\sim$150–320 V, $\sim$2.5–4 μA cm^{-2}). Under repetitive mechanical compression at room temperature, the microstructures on the surface of SMPU were degraded to a flat film due to plastic deformation, leading to the decreased outputs (100 V, 1 μA cm^{-2}). After heating at 75 °C for 40 seconds, the molecular chains rearranged under elastic state, so that the surface morphology recovered to the initial state and the output performances were restored. Higher temperature could accelerate the recovery process. Therefore, the TENG with deformed surface structure could be used as a hot water temperature sensor. In addition, due to the shape memory function, the device could be processed into various shapes by heating–cooling for adaptive attachment on human body.

A healable and shape memory dual functional polymer was designed by Xu et al. to fabricate a reliable and multipurpose TENG (Figure 7.10) [77]. The semicrystalline poly(1,4-butylene adipate) (PBA) segments and disulfide bonds were simultaneously introduced into the polymeric networks, so that the polymer was provided with shape memory and self-healing properties. The TENG assembled by the polymer and conductive layer was endowed with corresponding features, showing remarkable reliability. Both the open-circuit voltage and short-circuit current of the healed device could recover to the initial value without obvious change. When the shape of TENG was changed into "Z" structure, the output signal was weak, due to the low contact area. After the TENG achieved a flat structure under heating owing to the shape memory effect, the electric signal became strong. Therefore, the SMP-based TENG could be used as a self-powered fire alarm for its thermal responsive ability.

7.6 Biodegradable Triboelectric Nanogenerators

Biodegradation can be realized by being physically dissolved in solvent, chemically decomposed into small molecules triggered by heat, light, pH, enzyme, or moisture, and biologically decomposed by bacteria, without leaving hazardous substances to human body and environment. Biodegradable polymers can be divided into two categories: natural and synthetic polymers according to the source. Natural biodegradable polymers are abundant on earth, including chitosan, gelatin, cellulose, silk protein, and starch. Synthetic biodegradable

7.6 Biodegradable Triboelectric Nanogenerators

polymers, by contrast, are more versatile, controllable, and easily modified, containing PCL, polylactic acid (PLA), poly(lactic-co-glycolic acid) (PLGA), PVA, poly(3-hydroxybutyrate-co-3-hydroxyvalerate) (PHBV), and alginate. By incorporating nanofillers into biodegradable polymer matrix, biodegradable polymer nanocomposites can be imparted with enhanced mechanical performance and various functional properties such as thermal, electrical conductivity, and responsiveness, which is beneficial to tuning the devices performances. The nanofillers should be biocompatible and non-toxic, such as cellulose nanowhiskers, nanoclays, graphene, CNTs, polyhedral oligomeric silsesquioxanes (POSS), and Au NPs [78, 79]. Aside from metals that can react with water, such as magnesium (Mg), aluminum (Al), lithium (Li), zinc (Zn), iron (Fe), and molybdenum (Mo) used as conductors in biodegradable devices [80], the biodegradable polymer nanocomposite with excellent conductivity is an alternative approach for the electrode.

With the rapid development of electronic devices and the increasing demands for electric devices in modern society, the disused devices have caused a large amount of electronic waste and increased the environmental burden. Although the commonly used materials in TENGs, such as PDMS, PET, PTFE, and Kapton are biocompatible, they are not biodegradable, since they cannot be completely degraded and are prone to generating harmful chemicals. The concern about waste disposal has become more serious all over the world. To address the challenge, fabricating devices by using biodegradable materials is a sound strategy. Remarkably, biodegradable materials are ecofriendly. Consequently, from an environmental perspective, developing the environmentally friendly TENGs with biodegradability is of great urgency.

Additionally, the biodegradable materials play crucial roles in biomedical field. Currently, the power sources for a majority of implantable medical electronics such as cardiac pacemakers, insulin pump, deep brain stimulators, and cochlear implants are batteries, but the traditional batteries have the demerits of bulky volume and limited lifetime. For the development of biomedical and implanted devices, the biocompatible and biodegradable material-based TENGs are of greatly significance, which is closely related to human healthcare and practical societal needs. The implantable TENGs can effectively convert different kinds of biomechanical energy such as heartbeat, breathing, vasodilation, and vasoconstriction, into electrical energy [81]. Moreover, biodegradability has been proposed as a crucial qualification to avoid secondary surgery for the removal of implanted device. Therefore, biodegradable TENGs have promising prospects in biomechanical energy harvesting, biomedical sensors, and environmental sustainability.

A great deal of research has suggested that biodegradable TENGs can be applied in vitro and vivo. The basic requirement for biodegradable TENGs is the ability to be degraded in the natural environment without pollution after completing the working period. Yang et al. deposited aluminum-doped zinc oxide (AZO) on cellulose nanofibril (CNF) to assemble the TENG with flexibility, high transparence, and good degradability [82]. The AZO–CNF paper could be completely dissolved in warm

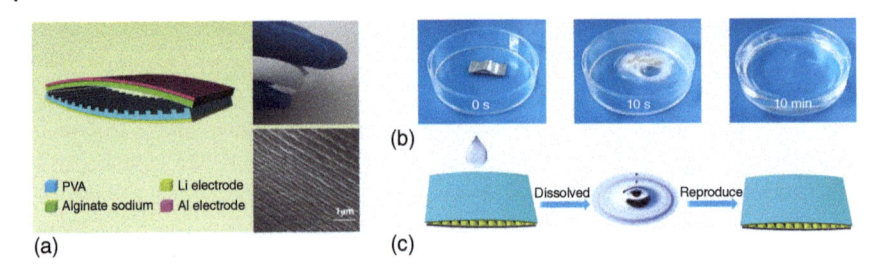

Figure 7.11 (a) Schematic diagram of the TENG, snapshot and SEM image of sodium alginate film. (b) Device dissolved in water completely within 10 minutes. (c) Schematic illustration of reproductivity. Source: Reproduced with permission Liang et al. [83]. Copyright 2017, Wiley-VCH.

water within one hour and the entire device was water degradable. Furthermore, the CNF film and TENG could be reproduced by concentrating the generated solution. The output signals of the TENG based on AZO–CNF paper reached about 7 V and 0.7 μA. Liang et al. demonstrated a fast soluble and recyclable TENG made of PVA and sodium alginate (SA) (Figure 7.11) [83]. Li and Al served as conductors. The whole device could completely degrade into environmentally benign products in water within 10 minutes. Moreover, the TENG could be refabricated through the generated liquid, leaving no waste in environment. The recycled TENG maintained the good performance and the power density of TENG was 3.8 mW m^{-2}.

Most of the biodegradable TENGs presented relatively low output performance and low energy conversion of device. To enhance the output performances of biodegradable TENGs based on cellulose, Zhang et al. chemically modified the CNFs, developing a high-performance flexible positive friction material for TENG [84]. CNFs were functionalized with amino groups to enhance the positive triboelectricity of cellulose, and Ag NPs was coated onto the surface to form a nanostructure, which could further boost the output performance of CNF-based TENG from 68 to 100 V. Besides, with the design of a three-dimensional gear-like structure, the maximum open-circuit voltage could be increased to 286 V. In addition, Pan et al. reported a fully biodegradable TENG with a high power density, which was composed of electrospun PLA, nanostructured gelatin, and Mg electrode [85]. The device could be entirely dissolved in water in about 40 days, and generated a voltage over 900 V and a power density more than 5 W m^{-2}.

3D-printing technology was adopted by Chen et al. to fabricate a single integrated elastic and biodegradable TENG for wearable electronics [86]. The mixture of poly(glycerol sebacate) (PGS) and CNTs was used to print the TENG in one-step direct ink writing (DIW) process without further assembling procedure. Meanwhile, hierarchical porous structure was created in the TENG to improve the output performances. Noteworthily, PGS synthesized from bio-based sebacic acid and glycerol is an excellent sustainable and biodegradable material, thereby allowing for the recycle of expensive CNTs and the refabrication of TENG based on the biodegradability of PGS matrix.

The biodegradable TENGs implanted in vivo should be completely vanished upon exceeding working cycles and resorbed by living systems without adverse effects. The encapsulation is necessary, to avoid liquids permeate into the device and guarantee an ideal operating environment. Zheng et al. built the first fully biodegradable TENG as an implantable power source in vivo [81]. The TENG was fabricated by assembling two of the selected biodegradable polymers (PLGA, PCL, PVA, and PHBV) as triboelectric layers, on one side of which was deposited with thin Mg films as electrode. The device was then encapsulated in the biodegradable polymer. The electrical outputs and degradation features can be changed, depending on the characteristic of the used biodegradable polymers. The whole device could be hydrolyzed and absorbed in animal body without residue. The TENG didn't induce inflammation, and could be applied for electric field-assisted neuron cell orientation.

Natural polymers are more attractive for their advantages of low-cost and easy availability, compared with synthetic polymers. Jiang et al. developed a fully bioabsorbable natural-materials-based triboelectric nanogenerator (BN-TENG) in vivo [87]. Cellulose, chitin, silk fibroin (SF), rice paper (RP), and egg white (EW) as friction materials and ultrathin Mg films as electrodes were used to prepare the BN-TENGs. The devices were encapsulated by SF films. The operation time of the BN-TENG could last from days to weeks through modifying the SF encapsulation films. The BN-TENG could be completely degraded and resorbed in the rats. The BN-TENG as a power source was used to accelerate the beating rates of dysfunctional cardiomyocyte clusters and improve the consistency of cell contraction.

The controllable biodegradability of the device is an ideal capability. Li et al. developed a biodegradable and implantable triboelectric nanogenerator (BD–iTENG) with the photothermally tunable biodegradability for tissue repairing (Figure 7.12) [79]. Due to the NIR photothermal effect based on Au nanorods (Au NRs), the BD–iTENG containing Au-doped PLGA nanocomposite film can respond to NIR light, thus accomplishing the rationally controllable biodegradation process. Under the NIR irradiation, the output of the BD–iTENG containing Au NRs was rapidly decreased to 0 within 24 hours, and the device was largely degraded in 14 days in vivo. By contrast, the BD–iTENG without Au NRs or NIR treatment could operate properly for more than 28 days in vivo.

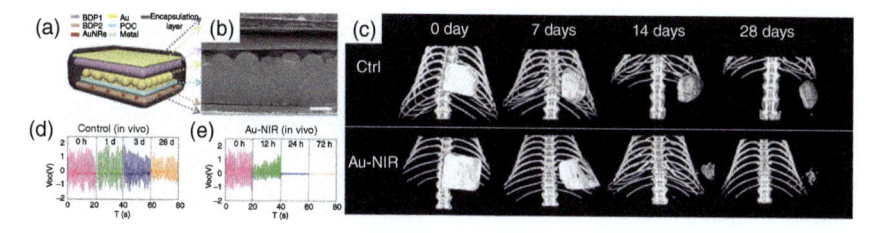

Figure 7.12 (a) Structure design of the photothermal-controlled BD-iTENG. (b) Section view of BD-iTENG (scale bar: 500 μm). (c) Micro-CT image of the implanted PLGA-iTENGs at various time points. (d) Electrical output of the PLGA-iTENG without AuNRs or NIR. (e) Electrical output of the PLGA-iTENG with AuNRs and NIR. Source: Reproduced with permission Li et al. [79]. Copyright 2018, Elsevier.

7.7 Conclusion

As the intensive investigation of the functional materials and the electronic devices, various TENGs with self-healing, shape memory, or biodegradable properties have been explored to satisfy the requirements of reliability, durability, and environmental friendliness for further sustainable development. On the basis of the principle, the appropriate TENGs can be constructed. Functional polymer nanocomposite has great influence on the performances and applications of TENGs. Self-healing TENGs are capable to recover the mechanical and electrical performances after damage, thus extending lifetime and enabling reliability. Shape memory TENGs have the characteristic of restoring the deformation under external stimulus from both micro and macroscope, resulting in the recovery of outputs and shape adaptability. Biodegradable TENGs can be degraded in the natural environment without pollution after completing the work period, contributing to human healthcare and environmental friendliness. This chapter presents the current research process of functional polymer nanocomposite-based TENGs, which is of comprehensive guiding significance for the development of electronic devices.

References

1. Traverse, C.J., Pandey, R., Barr, M.C., and Lunt, R.R. (2017). *Nat. Energy* 2: 849.
2. Fana, F.-R., Tian, Z.-Q., and Wang, Z.L. (2012). *Nano Energy* 1: 328.
3. Wang, Z.L., Chen, J., and Lin, L. (2015). *Energy Environ. Sci.* 8: 2250.
4. Fan, F.R., Tang, W., and Wang, Z.L. (2016). *Adv. Mater.* 28: 4283.
5. Shi, B., Li, Z., and Fan, Y. (2018). *Adv. Mater.* 30: e1801511.
6. Blaiszik, B.J., Kramer, S.L.B., Olugebefola, S.C. et al. (2010). *Annu. Rev. Mater. Res.* 40: 179.
7. Xie, T. (2011). *Polymer* 52: 4985.
8. Ojijo, V. and Sinha Ray, S. (2013). *Prog. Polym. Sci.* 38: 1543.
9. Lin, Z.H., Cheng, G., Lee, S. et al. (2014). *Adv. Mater.* 26: 4690.
10. Huang, L.-B., Xu, W., Bai, G. et al. (2016). *Nano Energy* 30: 36.
11. Zhang, L., Xue, F., Du, W. et al. (2014). *Nano Res.* 7: 1215.
12. Wang, X., Liu, Z., and Zhang, T. (2017). *Small* 13: 1602790.
13. Zhu, G., Pan, C., Guo, W. et al. (2012). *Nano Lett.* 12: 4960.
14. Wang, S., Lin, L., Xie, Y. et al. (2013). *Nano Lett.* 13: 2226.
15. Yang, Y., Zhang, H., Lin, Z.H. et al. (2013). *ACS Nano* 7: 9213.
16. Wang, S., Xie, Y., Niu, S. et al. (2014). *Adv. Mater.* 26: 2818.
17. Zhu, G., Chen, J., Liu, Y. et al. (2013). *Nano Lett.* 13: 2282.
18. Lin, L., Wang, S., Xie, Y. et al. (2013). *Nano Lett.* 13: 2916.
19. Yang, Y., Zhang, H., Chen, J. et al. (2013). *ACS Nano* 7: 7342.
20. Xie, Y., Wang, S., Niu, S. et al. (2014). *Adv. Mater.* 26: 6599.
21. Jie, Y., Ma, J., Chen, Y. et al. (2018). *Adv. Energy Mater.* 8: 1802084.
22. Zhang, Z., Bai, Y., Xu, L. et al. (2019). *Nano Energy* 66: 104169.
23. Wang, Z.L. (2013). *ACS Nano* 7: 9533.

24 Li, H.Y., Su, L., Kuang, S.Y. et al. (2015). *Adv. Funct. Mater.* 25: 5691.
25 Feng, Y., Zheng, Y., Ma, S. et al. (2016). *Nano Energy* 19: 48.
26 Kwon, Y.H., Shin, S.-H., Kim, Y.-H. et al. (2016). *Nano Energy* 25: 225.
27 Chen, J., Guo, H., He, X. et al. (2016). *ACS Appl. Mater. Interfaces* 8: 736.
28 Wu, C., Kim, T.W., Park, J.H. et al. (2017). *ACS Nano* 11: 8356.
29 Cui, N., Gu, L., Lei, Y. et al. (2016). *ACS Nano* 10: 6131.
30 Wu, C., Kim, T.W., and Choi, H.Y. (2017). *Nano Energy* 32: 542.
31 Xie, Y., Wang, S., Lin, L. et al. (2013). *ACS Nano* 7: 7119.
32 Jeong, C.K., Baek, K.M., Niu, S. et al. (2014). *Nano Lett.* 14: 7031.
33 Cheng, G., Lin, Z.H., Du, Z.L., and Wang, Z.L. (2014). *ACS Nano* 8: 1932.
34 Jie, Y., Wang, N., Cao, X. et al. (2015). *ACS Nano* 9: 8376.
35 Fan, F.R., Lin, L., Zhu, G. et al. (2012). *Nano Lett.* 12: 3109.
36 Lee, K.Y., Yoon, H.-J., Jiang, T. et al. (2016). *Adv. Energy Mater.* 6: 1502566.
37 Tang, W., Jiang, T., Fan, F.R. et al. (2015). *Adv. Funct. Mater.* 25: 3718.
38 Pu, X., Guo, H., Tang, Q. et al. (2018). *Nano Energy* 54: 453.
39 Chen, S., Gao, C., Tang, W. et al. (2015). *Nano Energy* 14: 217.
40 Wen, X., Yang, W., Jing, Q., and Wang, Z.L. (2014). *ACS Nano* 8: 7405.
41 Xiao, X., Zhang, X., Wang, S. et al. (2019). *Adv. Energy Mater.* 9: 1902460.
42 Yang, W., Chen, J., Zhu, G. et al. (2013). *ACS Nano* 7: 11317.
43 Wang, G., Xi, Y., Xuan, H. et al. (2015). *Nano Energy* 18: 28.
44 Huang, L.-B., Bai, G., Wong, M.-C. et al. (2016). *Adv. Mater.* 28: 2744.
45 Fan, Y.J., Meng, X.S., Li, H.Y. et al. (2017). *Adv. Mater.* 29: 1603115.
46 Cao, V.A., Lee, S., Kim, M. et al. (2020). *Nano Energy* 67: 104300.
47 Jiang, C. Wu, C., Li, X. et al. (2019). *Nano Energy* 59: 268.
48 Zheng, J., Wang, Y., Yu, Z. et al. (2019). *Nano Energy* 64: 103957.
49 Yan, S., Song, W., Lu, J. et al. (2019). *Nano Energy* 59: 697.
50 Wang, S., Tai, H., Liu, B. et al. (2019). *Nano Energy* 58: 312.
51 Bian, J., Wang, N., Ma, J. et al. (2018). *Nano Energy* 47: 442.
52 Chen, J., Chen, B., Han, K. et al. (2019). *Adv. Mater. Technol.* 4: 1900337.
53 Song, P., Qin, H., Gao, H.L. et al. (2018). *Nat. Commun.* 9: 2786.
54 Song, Y., Chen, H., Su, Z. et al. (2017). *Small* 13: 1702091.
55 Kim, W.-G., Kim, D., Jeon, S.-B. et al. (2018). *Adv. Energy Mater.* 8: 1800654.
56 Huynh, T.P., Sonar, P., and Haick, H. (2017). *Adv. Mater.* 29: 1604973.
57 Yang, Y. and Urban, M.W. (2013). *Chem. Soc. Rev.* 42: 7446.
58 Dai, X., Du, Y., Yang, J. et al. (2019). *Compos. Sci. Technol.* 174: 27.
59 Deng, G., Tang, C., Li, F. et al. (2010). *Macromolecules* 43: 1191.
60 Rekondo, A., Martin, R., Ruiz de Luzuriaga, A. et al. (2014). *Mater. Horiz.* 1: 237.
61 Wu, S., Li, J., Zhang, G. et al. (2017). *ACS Appl. Mater. Interfaces* 9: 3040.
62 Yan, S., Zhang, G., Jiang, H. et al. (2019). *ACS Appl. Mater. Interfaces* 11: 10736.
63 Weng, G., Thanneeru, S., and He, J. (2018). *Adv. Mater.* 30: 1706526.
64 Miyamae, K., Nakahata, M., Takashima, Y., and Harada, A. (2015). *Angew. Chem. Int. Ed.* 54: 8984.
65 Xu, W., Huang, L.-B., and Hao, J.H. (2017). *Nano Energy* 40: 399.
66 Guan, Q.B., Dai, Y.H., Yang, Y.Q. et al. (2018). *Nano Energy* 51: 333.
67 Sun, J.M., Pu, X., Liu, M.M. et al. (2018). *ACS Nano* 12: 6147.

68 Chen, Y., Pu, X., Liu, M. et al. (2019). *ACS Nano* 13: 8936.
69 Dai, X., Huang, L.-B., Du, Y. et al. (2020). *Adv. Funct. Mater.* 30: 1910723.
70 Huang, J., Cao, L., Yuan, D., and Chen, Y. (2018). *ACS Appl. Mater. Interfaces* 10: 40996.
71 Hu, J., Zhu, Y., Huang, H., and Lu, J. (2012). *Prog. Polym. Sci.* 37: 1720.
72 Zhang, G., Zhao, Q., Zou, W. et al. (2016). *Adv. Funct. Mater.* 26: 931.
73 Aoki, D., Teramoto, Y., and Nishio, Y. (2007). *Biomacromolecules* 8: 3749.
74 Lee, J.H., Hinchet, R., Kim, S.K. et al. (2015). *Energy Environ. Sci.* 8: 3605.
75 Liu, R., Kuang, X., Deng, J. et al. (2018). *Adv. Mater.* 30: 1705195.
76 Xiong, J., Luo, H., Gao, D. et al. (2019). *Nano Energy* 61: 584.
77 Xu, W., Wong, M.-C., Guo, Q. et al. (2019). *J. Mater. Chem. A* 7: 16267.
78 Reddy, M.M., Vivekanandhan, S., Misra, M. et al. (2013). *Prog. Polym. Sci.* 38: 1653.
79 Li, Z., Feng, H., Zheng, Q. et al. (2018). *Nano Energy* 54: 390.
80 Huang, X., Wang, L., Wang, H. et al. (2019). *Small* 16: 1902827.
81 Zheng, Q., Zou, Y., Zhang, Y.L. et al. (2016). *Sci. Adv.* 2: e1501478.
82 Yang, B., Yao, C.H., Yu, Y.H. et al. (2017). *Sci. Rep.* 7: 8.
83 Liang, Q., Zhang, Q., Yan, X. et al. (2017). *Adv. Mater.* 29: 1604961.
84 Zhang, C., Lin, X., Zhang, N. et al. (2019). *Nano Energy* 66: 104126.
85 Pan, R., Xuan, W., Chen, J. et al. (2018). *Nano Energy* 45: 193.
86 Chen, S., Huang, T., Zuo, H. et al. (2018). *Adv. Funct. Mater.* 28: 1805108.
87 Jiang, W., Li, H., Liu, Z. et al. (2018). *Adv. Mater.* 30: 1801895.

8

Polymer Nanocomposites for Resistive Switching Memory

Qazi Muhammad Saqib, Muhammad Umair Khan, and Jinho Bae

Jeju National University, Department of Ocean System Engineering, 102 Jejudaehakro, Jeju 63243, Republic of Korea

8.1 Introduction

Over the past decade striking explosion of big data witnessed a striking explosion in information and technology with exponentially growing speed in different fields of real life, which includes consumer electronic gadgets, aerospace operations, and civil defense [1]. It is expected that, in year 2020, trillions of gigabyte data will be produced and 5200 GB will be shared by each person on this planet [2]. In order to meet future challenges for data storage, a new universal memory device is required, which is capable to operate at higher speed with nonvolatile memory. Especially, it can act as a single unit to meet the respective merits of the hard-disk drive (HDD) [3, 4]. In order to meet the recent evolution in the field internet of things with novel storage units, the memory device should be highly flexible and bendable, so that it can be easily deformed on human skin surface to monitor personal healthcare using smart medical equipment [5]. However, the conventionally used silicon based devices have physical constraints of heat problems, quantum uncertainties, and economic issues for new fabrication costs of the CMOS technology [6, 7]. M. M. Waldrop predicted that the CMOS technology will not be able to follow Moore's law anymore [8]. Hence, many researchers are investigating scalable nonvolatile memory techniques that can replace the flash memories to achieve small size, high density, and less power consumption [9]. For this purpose, new storage device is required to be introduced with a novel design that are built from new emerging materials and operate at different mechanisms. Therefore, it is critical task for industrial and academic communities to meet the future requirements.

In 1960s, Moore's law made a prediction of transistor size shrinkage. After that, the researchers focused on new type of memories, namely, as magnetic random access memory (RAM), phase-change RAM, ferroelectric RAM, and resistive RAM [10–13]. In all the aforementioned memory devices particularly, resistive random access memory (RRAM) consists of simple device structure in which a resistance switching layer is sandwiched between two metallic electrodes [14]. In 2008, Hewlett

Polymer Nanocomposite Materials: Applications in Integrated Electronic Devices, First Edition.
Edited by Ye Zhou and Guanglong Ding.
© 2021 WILEY-VCH GmbH. Published 2021 by WILEY-VCH GmbH.

Packard laboratories reported resistive switching memory (RSM) called memristor for the fourth fundamental passive circuit element, and it was postulated by Leon O. Chua and coworkers in the year 1971 [15–18]. The RSM devices have been attracting global attention as an emerging next-generation nonvolatile memory (NVM) and it is scalable down to $4F^2$ in single layer, and this device dimension could be further reduced to $4F^2/n$, where n is number of stacked layers of memory elements [19, 20]. Restive memory behavior is observed with versatile ranges of materials for flexible implementation. The working mechanism of resistive memory device totally depends on the type of material sandwiched between two electrodes [20–24]. Utilizing unique resistance switching behaviors, many single bit memristor-based NVMs have been proposed, which can store two distinctive resistance states (high resistance state (HRS) or low resistance state (LRS)) [9]. Here, resistive switching function can be classified by current–voltage (I–V) curve as symmetric type and asymmetric type [25–28]. The RRAM based resistive switching has been extensively studied, because of its advantages over conventional memories, thus considered as the NVM technology [25, 28–30]. Currently researchers are trying to improve the performance of resistive switching devices to operate at low voltage with fast switching speed in a range of few nanoseconds with fast writing, reading, and erasing speed with the low power consumption. The presence of these RSM devices is considered as an appealing technique for universal memory.

The RSMs are utilizing for physiological activities based on neural synapses using the consecutive modulation of device resistance to execute logic-in-memory function. This property of resistive memory device is considered as emerging application for the designing of a neuromorphic computing to resolve von Neumann bottleneck complications. On the other hand, a crossbar array has been presented for highly dense nonvolatile RSM device applications. The crossbar memory architecture is very simple, as the insulating layer is sandwiched between two crossbar metallic electrodes. However, in crossbar arrays, there are several immanent drawbacks such as cell to cell interference know by sneak current [31, 32] along with power utilization. Especially, a peripheral circuit is required along a reading cycle that can read a single bit at one time. Thus, an extra power is dissipated due to leakage current of nearby cells. The situation further worsens if resistance in crossbar structure becomes low [20]. The power usage of the RSM device increases with the device size, as a result more sneak current is generated. The crosstalk problem usually occurs during the memory data access, when RSM carries lower threshold voltage, hence the RSM device can switch from one state to another state. In the conventionally used crossbar arrays, the desired cell can be reached through rows and columns by sensing the current of particular element. The similar path is acquired to write the bit. Hence, very complex peripheral circuits are needed for data accessing. To overcome these drawbacks, RSM device should have sneak current protection, low operating current, high on/off resistance ratio, sufficient detection margin for writing and reading of logic data, and simple parallel-voltage readable architecture.

To realize these functions, many researchers are studying an RRAM based on polymer nanocomposite (PNC) in recent. PNC consists of nanoparticles/nanofillers blended in the polymers [33]. The hybrid PNC based RSM devices offer excellent performance, high flexibility, low fabrication cost, and simple fabrication [34]. The hybrid device combines the properties of both organic polymers and inorganic nanocomposites, thus it leads to ensure tremendous electrical performance, such as excellent switching speed, high on/off ratio, and non-volatility [35]. Furthermore, the hybrid PNC based RSM devices have several advantages over the inorganic and organic memories. First, these RSM devices have an immense edge of easy processing and simple structure in terms of large-scale manufacturing. Secondly, the remarkable flexibility of hybrid PNC materials opens the doors for wearable and foldable applications [36]. On large-scale device, researchers are designing the application of PNC to target a wide range of the future technology along with the synthesis process.

8.2 Resistive Switching Memory for Polymer Nanocomposite

8.2.1 Resistive Switching

Among the printed electronics, flexible memory devices are necessary element for realization of all flexible systems. In order to meet future data storage demand, the nonvolatile resistive memory is presented with confidence of latest data storage device. RRAM is known as symmetric type switching, when $R_{FL} = R_{RL}$ in LRS and $R_{FH} = R_{RH}$ in HRS (Figure 8.1a). On the other hand, RRAM is called as asymmetric type resistive switching, when $R_{FL} \neq R_{RL}$ and $R_{FH} \neq R_{RH}$. Several RSM devices operate as write once and read many (WORM) [37]. The crossbar array having single bit RSM cells is presented for high density integration of NVMs. However, the crossbar arrays faced a severe problem of sneak current (I_{Sneak}), as a result read/write errors occurred frequently. Also, extra power dissipation occurs due to sneak current problem [20]. A particular row-column paired passive crossbar array connected with RSM cell will be in HRS and LRS by depending upon the stored logic value. In a passive crossbar array, a specific row–column pair is connected by a memory cell that is either in the HRS or the LRS depending on the logic value stored in it. The sneak path problem will occur when memory cell in HRS will be surrounded by other cells in LRS is read, causing a reading (Figure 8.1b).

The sneak current problems are more noticeable for devices for large size devices. Various approaches have been suggested to overcome this problem. A rectifying element was connected in series with a memristor at each cross-point and anti-serial memristors complementary resistive switches. Although these approaches minimized the problem to some extent; however further advances are needed to present a practical solution. The stability of resistive memory device is strongly dependent upon the active layer material like organic materials [34], soft materials [38], metal oxide [39], composite of two different materials [35], 2D materials [40], phase change material [41], and PNC [42]. Due to remarkable properties of PNC based hybrid materials, the printed electronics are extensively

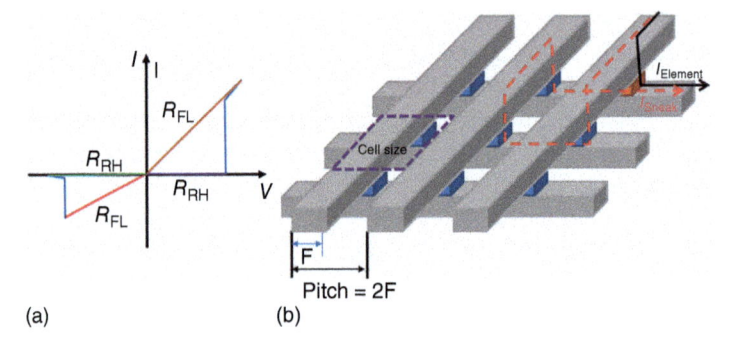

Figure 8.1 (a) Current–voltage (*I*–*V*) characteristics curve of RSM. (b) Schematics of the device structure, presenting the sneak current problem in the RSM devices.

researched because of low cost and easy fabrication. Most importantly, the inherent softness achieved through carbon to carbon polymer long chain increased its importance for flexible and printed electronics. The resistive switching memories fabricated using PNC materials are used to produce e-textiles, which open a gate way to bring future electronic system to our everyday outfits to achieve higher comfort level in human life.

8.2.2 Resistive Switching Memory Operating Mechanism

8.2.2.1 Formation and Rupture of Conductive Filaments

One of the possible working mechanisms in PNC based RSM is the formation and rupture of the conductive filaments between the electrodes. The conduction mechanism is originated due to the oxygen ions migration. Under the influence of external positive voltage, the oxygen ions are pulled out from the PNC layer. As a result, the oxygen vacancies are generated in the lattice structure. The ions acted as a p-type dopant will be attracted toward the opposite polarity electrode. The tiny conducting filaments are formed due to oxygen vacancies in the PNC lattice structure. When the set voltage reaches, the conductive filaments come nearer to each other, thus making a strong filamentary pathway between the top/bottom electrodes (Figure 8.2a). When the opposite bias voltage is applied across the electrodes, the oxygen ions are pushed toward the oxygen vacancies, which will be neutralized. The process will support to rupture the pathway of conductive filaments between the electrodes. The device switches to the HRS, also known as erase process [43–48]. For instance, zinc oxide (ZnO) used as nanoparticles blended in poly(vinyl alcohol) (PVOH) was studied on the flexible polyethylene terephthalate (PET) substrate. A polymer poly(3,4-ethylenedioxythiophene)polystyrene sulfonate (PEDOT:PSS) layer was spin-coated on flexible electrode. Under the application of positive voltage, the oxygen ions were pulled toward the PEDOT:PSS, creating the oxygen vacancies in the lattice structure. As a result, the tiny conductive filaments were created in a PNC layer. When the voltage reached the set level, the conductive filaments

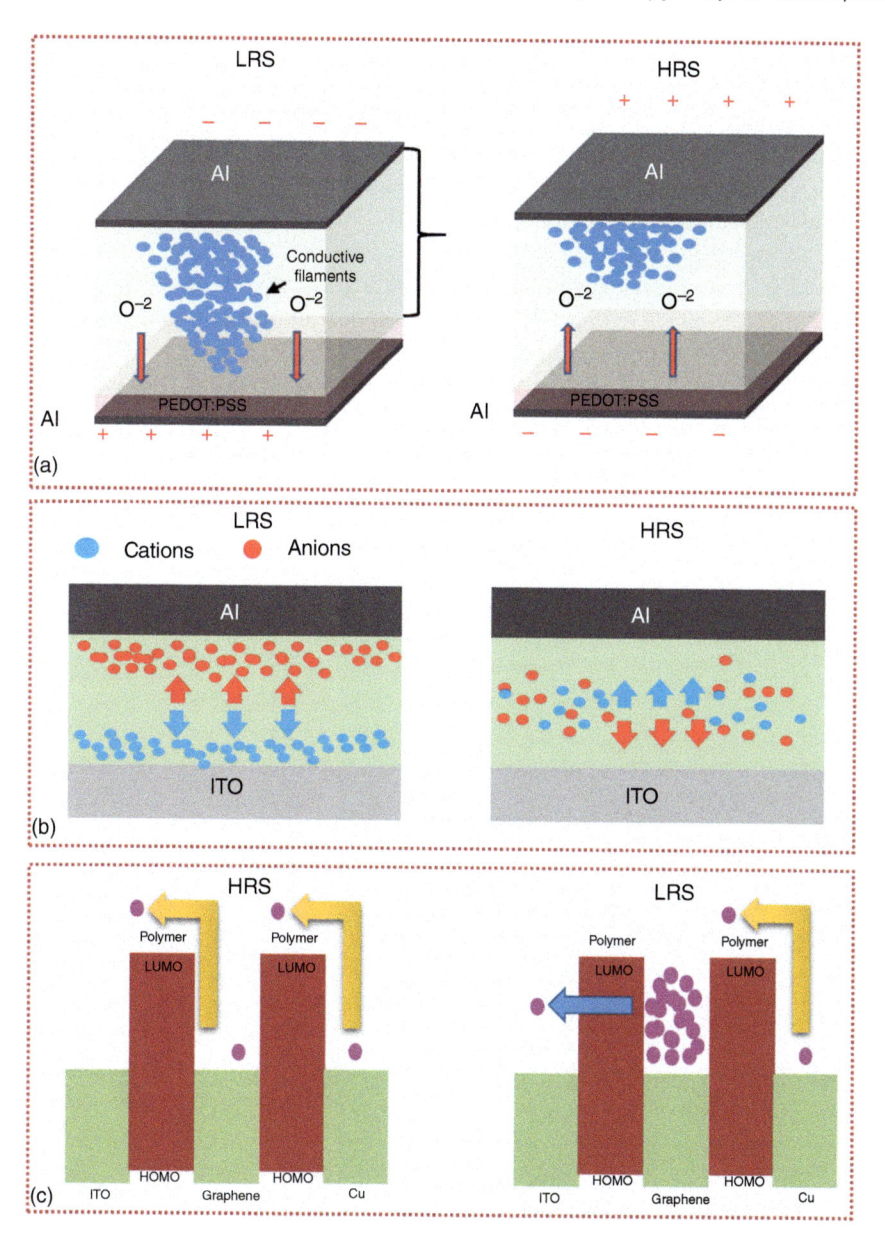

Figure 8.2 Schematics of different switching mechanisms. (a) Formation and rapture of conductive filaments. (b) Cations and anions migration. (c) Trapping and de-trapping of electrons.

gathered to form a pathway between top/bottom electrodes. The device then shifted to the LRS. On the other side, when negative voltage was applied across the RSM device, the oxygen ions were pulled back toward the oxygen vacancies. The conductive filamentary path was diminished. As a result, the device moved back to the HRS [47].

8.2.2.2 Cations and Anions Migration

When the positive sweep is applied across the memory device, more cations and anions are generated in the PNC layer. These cations and anions are migrated toward the opposite polarities due to the high ionic conductivity of polymer material. The PNC layer near the electrodes heavily doped with the cations and anions. As a result, the barrier between electrode and PNC layer are shrunk. These effects enhance the tunneling of ions to cross the barrier. The device switches to the LRS (Figure 8.2b). When the negative sweep is applied across the RSM device, both the cations and anions move in opposite direction. The barrier between the electrodes and PNC recovered again. As a result, the device switches to HRS [49–54]. For instance, in TiS_2-PVP PNC, sulfur vacancies (V_s) are generated during positive sweep. These vacancies may be cationic and anionic. The cations move toward the indium tin oxide (ITO) electrode while the anions move toward the aluminum electrode. When the voltage becomes significantly high, the barrier diminished and the device switched to LRS. During the negative voltage application, the cations and anions move in the opposite direction. The barrier between the PNC layer and electrodes recovered again. As a result, the device switched back to HRS [53].

8.2.2.3 Electrons Trapping and De-tapping

One of the possible mechanisms in PNC based RSM devices is the trapping and de-trapping of the electrons. The higher work function of nanocomposites in PNC RSM makes the possibility of electron trapping and transportation (Figure 8.2c). When the potential bias is increased, a strong SCLC represents the presence of trapped charges in the forbidden energy gap. When low potential bias, the trapped charges do not have enough energy to escape from the deep forbidden energy gap. When the applied potential is increased, trapping density increases. As a result, the whole traps are occupied, and the conductivity also increases. The device switches to LRS. These trapped charges will remain in the deep energy gap. When a sufficiently high reverse bias is applied across the RSM, the trapped charges will be pulled out and device switches to HRS [48, 55–57]. For example, the nonvolatile behavior of poly(methyl methacrylate) (PMMA) embedded graphene and MoS_2 composite was investigated. The graphene and molybdenum disulfide nanocomposites acted as charge trapping sites. Higher work function (4.88 eV) of graphene makes it excellent candidate for charge trapping and as an electron transporter. When higher bias was applied to the Cu electrode, the traps were filled and electron injection started. Consequently, the device switched to LRS. When the power was turned off, the nanocomposites still captured the trapped charges and device remained in the LRS until the whole trapped charges were fully extracted from the deep energy gap [57].

8.2.2.4 Other Conduction Mechanisms

Ohmic conduction is another possible mechanism in PNC based resistive switching devices. In ohmic conduction, a linear curve is obtained between current density and electric field in HRS, having the slope nearly equal to one [58, 59]. The carrier transport follows the ohmic model mechanism given by [60]

$$I \propto V \exp\left(\frac{-\Delta E_{ae}}{kT}\right) \tag{8.1}$$

where $V, I, \Delta E_{ae}, T$, and k represented the applied voltage, current, activation energy, temperature, and Boltzmann constant, respectively [61]. Space charge limited current (SCLC) mechanism contains two regions of trap controlled SCLC region and trap filled SCLC region. The slope of I–V curve defines the trap controlled and trap filled SCLC regions in the RSM devices. When the slope of the curve between I and V is nearly equal to two, it shows the trap controlled SCLC [62]. When the slope of the curve is increased exponentially (I–V^m), it presents the trap filled SCLC region in the PNC RSM devices [63, 64]. The linear curve between ln (I) and $V^{\frac{1}{2}}$ expresses the thermal emission mechanism in the low bias. The linear behavior shows that thermal emission process dominated the charge transportation in HRS. The transport equation for the thermionic emission at low applied potential [65] is given by

$$\ln(J) \propto \ln(AT^2) - \frac{q\varphi}{kT} + q\left(\frac{q^3 V}{4\pi\varepsilon}\right)^{1/2} \tag{8.2}$$

Here, A, U, q, φ, and ε represent the Richardson constant, the Schottky barrier height, electronic charge, and dielectric constant, respectively [66].

8.2.3 Fabrication Techniques

The solution processing has drawn a considerable attention as an alternative of vacuum deposition and other thin film fabrication techniques because of low fabrication cost, large area manufacturing, and simplicity. Both single layer and multilayer PNC based RSM devices have been prepared using simple spin-coating methods. The PNCs are usually formed by blending the inorganic nanocomposite materials with a polymer matrix. Two types of nanoparticles are generally used as nanocomposite materials: semiconductor and metal based. The thickness and uniformity of the active layer can be controlled by varying the spinning conditions. Various low-price solution processing techniques, viz., spin-coating, spray-coating, dip-coating, blade-coating, and roller-coating, are generally used to fabricate the PNC memory device [36]. Many researchers employed the electro hydrodynamic (EHD) technique for deposition of active layer in PNC based RSM devices. The EHD thin film fabrication can be carried out in ambient pressure and temperature conditions, which make it a suitable candidate for active layer fabrication in memory device applications. Furthermore, EHD technique is cost effective, rabid, and convenient for mass production. It consumes less power as compared with other conventional methods and very simple to use. The PNC material is moved by applying electric field. When a high voltage is applied, the ions started moving from nozzle to the substrate. The EHD setup consists of X–Y stage, CCD camera and a light source, ink and supply part, high voltage source, a nozzle, and a display PC [39]. The thermal evaporation is commonly used technique for electrode deposition in PNC based RSM devices. The material is heated in a vacuum chamber that produces vapor cloud in the chamber. This evaporated cloud continues to deposit on the substrate and make a thin film. The process has a capability to produce highly pure films along with high deposition rate and minimum penetration [66–68].

8.2.4 Polymer Nanocomposite Materials

Various polymers are employed as an active layer in PNC based RSM by making blends with nanocomposites used as electroactive elements. The resultant polymer composites give the advantages of improving the electronic behavior and mechanical flexibility the device [66, 68]. PEDOT:PSS is the famous polymer used in RSM, while PSS is used to enhance the conductivity of PEDOT [69]. Some of commonly used polymers in PNC based RSM devices are polyvinylpyrrolidone (PVP), PVA, polyvinylidene difluoride (PVDF), PMMA, and polyimide [42, 70–72]. The nanoparticles are randomly dispersed in the insulating polymer matrix. These polymers were introduced to interact with the inorganic and semiconductor nanoparticles. The nanoparticles were employed as trapping sites by blending in polymer active layer. The metal based nanoparticles used in PNC based RSM devices are FeNi, Au, and Ag [73–76]. Researchers have reported Si, ZnO, CuO, and CdSe as semiconductor nanoparticles in PNC RSM devices. By using a polymer with inorganic nanoparticles, a reliable bipolar behavior was observed with better on/off current ratio [65, 67, 72–74]. Various quantum dots based materials were also developed as a charge trapping sites in blend with the different materials. $MoSe_2$, WS_2 with a polymer, showed excellent performance in PNC based RSM [77–79]. Similarly, two-dimensional (2D) material is another inorganic nanocomposite used as a trapping agent in PNCs based RSM. The most famous 2D materials introduced in PNC based RSM are graphene and MoS_2 [56, 80].

8.3 Polymer Nanocomposite Based RSM Devices

8.3.1 Oxide Based Polymer Nanocomposite RSM

The nanocomposite of metal oxides like ZnO and TiO_2 is extensively studied because of its high holes/electrons diffusion. Furthermore, these RSM devices carried excellent optical absorption functionality. The oxide based RSM devices contain a wide application in electronic devices due to its hybrid, optical, magnetic, and electronic properties [9, 81–83]. Bhattacharjee et al. prepared PNC based resistive RSM device using ionized oxygen rich ZnO nanorods with PMMA (Figure 8.3a) [48]. Only 0.5 wt% nanorods were used as nanocomposite material. Increasing the dopant concentration caused the cluster formation, which connects the neighbors with each other. The memory effect was controlled by the oxygen vacancies present in the ZnO nanorods. The transition from HRS to LRS took place at +1 V (Figure 8.3b), while the device showed the high on/off ratio of 5×10^5. The retention time of the device measured as 10^3 seconds. In low applied bias, a linear relationship was obtained between ln (I) and $V^{\frac{1}{2}}$, which shows that the current

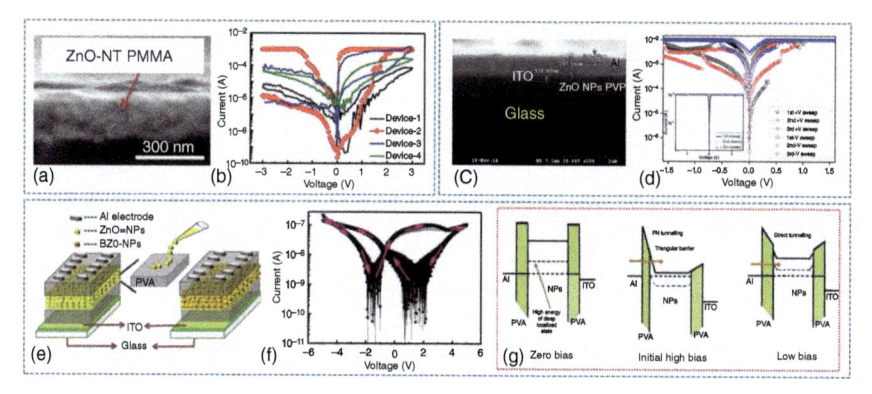

Figure 8.3 (a) Cross sectional SEM image of Al/(0.5%) ZnO-NR: PMMA/ITO/Glass. (b) *I–V* curves of the Al/ZnO NR: PMMA/ITO/Glass structured device at various nanoparticle concentrations. Source: Bhattacharjee et al. [48]. Reproduced with permission. Copyright © 2016 Elsevier Ltd. (c) Cross-sectional SEM of PVP active layer blended with 0.125 wt% of 14 nm ZnO nanoparticles. (d) *I–V* curves of PVP active layer (1.0 wt%) blended with 0.125 wt% of 14 nm ZnO nanoparticles. Source: Jung et al. [84]. Reproduced with permission. Copyright © 2018 Elsevier Ltd. (e) The schematic illustration of tri-layered organic bi-stable device fabrication having the device structure of Al/PVA/BZO/PVA/ITO. (f) *I–V* characteristics of Al/PVA/(0%) BZO nanoparticles PVA/ITO, presenting scan course. (g) Schematic illustration of Al/PVA/BZO/PVA/ITO device presenting current conduction mechanism having different energy bands. Source: Bhattacharjee et al. [85].

in the HRS is conducted by thermionic emission. Jung et al. studied the zinc oxide nanoparticles based RSM embedded in PVP (Figure 8.3c) [84]. A multistate adaptive memory behavior was examined depending upon the device thickness, nanoparticles (NPs) concentration, and format of the device (Figure 8.3d). Similarly, Bhattacharjee et al. designed Boron doped ZnO nanoparticles (BZO) sandwiched between two PVA polymer layers [85]. The bandgap of BZO showed an increment in the bandgap till 3% concentration. The electrical behavior of ZnO improved by doping with group III elements like Al, B, etc. The fabricated device structure was Al/PVA/BZO/PVA/ITO (Figure 8.3e). The transition from HRS to LRS took place at 1 V (Figure 8.3f). The device on/off ratio was >10^2, while the retention time calculated was 10^4 seconds. The electrons were injected through FN tunneling at the initial high negative voltage. When the voltage deceased, the electrons started to move by direct electron tunneling process (Figure 8.3g).

Multiwall carbon nanotubes (MWCNT) blended in PMMA were studied by Li and Wen [86]. The device structure was Ni/MWCNTs (2 wt%): PMMA/ITO. The set voltage of the fabricated device was −2.25 V, while the on/off ratio measured was >10^7. The device depicted excellent retention time of more than 10^6 seconds. The memory device followed the ohmic, trap controlled, and trap filled conduction mechanisms in HRS. In LRS, ohmic conduction mechanism was followed. A novel RSM device based on the Schottky diode was designed by Khan et al. to reduce the sneak current problem using ZnO/PEDOT:PSS heterojunction [87]. The device structure was kept Ag/PEDOT:PSS/ZnO/ITO. The cross-sectional SEM image of the fabricated device is shown in Figure 8.4a. The ZnO (n-type) with PEDOT:PSS (p-type)

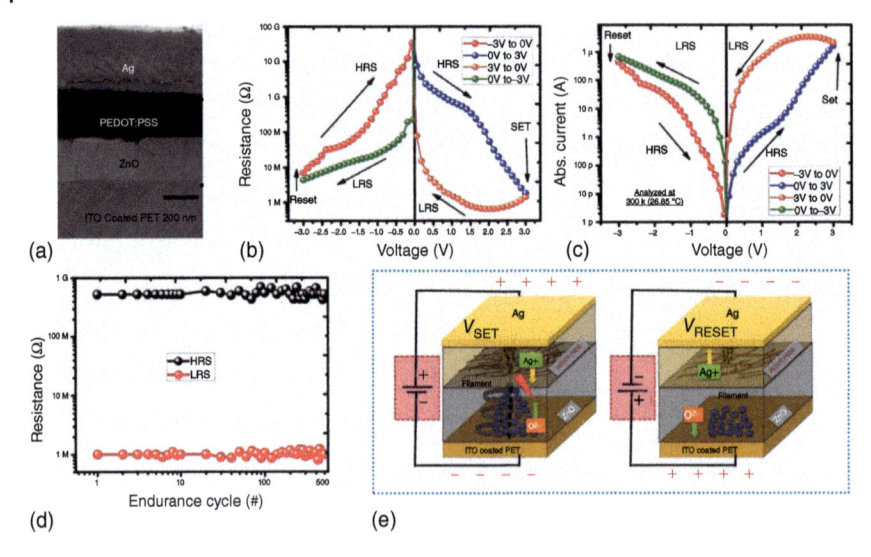

Figure 8.4 (a) Cross-sectional SEM image of the fabricated device. (b) The resistance-voltage characteristics curve of ITO/ZnO/PEDOT: PSS/Ag PNC device. (c) *I–V* characteristics curve of ITO/ZnO/PEDOT: PSS/Ag PNC device. (d) The LRS and HRS resistive switching endurance stability for above 500 cycles. (e) The proposed ITO/ZnO/PEDOT: PSS/Ag proposed TCSCLC mechanism. It shows that Ag charge ions are migrated for the SET and RSET process. Source: (a) Khan et al. [87]. Reproduced with permission. Copyright © 2019 Springer Nature Limited, (b–e) Khan et al. [87].

heterojunction was used to study the memory behavior by utilizing the Schottky diode effect. The ZnO and PEDOT:PSS showed the high forward as well as reverse current ratio. The high electrons concentration in zinc oxide make it feasible candidate for the Schottky behavior in heterojunction based RSM device. Also, the rectification ratio was high due to the barrier high between ZnO and ITO. Similarly, the Schottky function was enhanced significantly due to high band-gap difference between ZnO and PEDOT:PSS. The R_{off}/R_{on} ratio was measured to be 530, while the device presented the retention time of >30 days. The double logarithmic *I–V* analysis curve of PEDOT:PSS and ZnO heterojunction is shown in Figure 8.4b. The *I–V* characteristics curve of ITO/ZnO/PEDOT: PSS/Ag PNC device is shown in Figure 8.4c. Figure 8.4d shows that the device has excellent stability for >500 endurance cycles. The device followed the ohmic and SCLC conduction mechanism in HRS and LRS. Figure 8.4e shows the silver ion migration during LRS and HRS set process.

Kim et al. studied ferroelectric mediated P(VDF-TrFE)/ZnO nanocomposite film for RSM device [88]. The device structure was Au/P(VDF-TrFE)+ZnO/Si. The blending ratio of ZnO was kept 20%. The device achieved the high on/off ratio of 2×10^7, while the retention time was measured to be $\sim 10^4$. In the ferroelectric polymer semiconductor based RSM device, the semiconductor regime used to read while ferroelectric regime was used to write the information. However, a high blending ratio of semiconductor suppresses the ferroelectric properties. ZnO nanoparticles were used to control the resistive switching behavior due to its spontaneous polarization. The device followed the ohmic, trap controlled, and trap filled conduction mechanism. Hmar studied RSM device based on ZnO blended and PVA & PEDOT: PSS on the PET substrate [47]. The device structure was selected to be Al/ZnO-PVA/PEDOT: PSS/Al/flexible PET. The PEDOT: PSS was selected, as it enhances the migration of oxygen ions. Moreover, PEDOT: PSS provides the protection to oxygen vacancies. As a result, the leakage current will be reduced and the on/off ratio increases significantly. The switching voltage, on/off ratio of the fabricated device, was measured to be 3.6 V and 3×10^5, respectively. The PNC based resistive switching device worked on the cation and anions migration conduction mechanism. The oxygen vacancies acted as n-type, while the oxygen ions worked as p-type dopant. When the positive voltage was applied, oxygen ions were pulled out of ZnO and migrated toward the PEDOT: PSS and the depletion region will be reduced. Kaur and Tripathi studied the effect of silver doping in CdSe-PVP nanocomposite [89]. Three different concentrations of Ag (0.1%, 0.2%, and 0.3%) were used. The RSM device showed the on/off ratio of 4×10^2 with 0.1% of Ag doping. The space charges near the PNC layer provided the memory effect in the RSM device. Khan et al. studied PNC RSM device based on zirconium dioxide (ZrO_2) with PVP [90]. The device structure was Ag/ZrO_2: PVP/ITO on the PET substrate. Figure 8.5a shows the detailed fabrication process, 2D and 3D nano profile of top electrode, and cross-sectional image of the RSM device. In Figure 8.5b, the fabricated RSM device showed the maximum R_{off}/R_{on} at the blending ratio of 1 : 0.5 (ZrO_2:PVP). The device showed the transition form off state to on stat at ±1.5 V, while the R_{off}/R_{on} ratio was measured to be 44. Figure 8.5c presents the I–V curve analysis of PNC, showing the symmetric behavior of the fabricated RSM device. The ZrO_2 was used owing to its bio-polar RSM characteristics. PVP was used due to insulating nature, film uniformity, and flexibility. The temperature dependency of fabricated memory device can be seen in Figure 8.5d. The device followed the ohmic and trap controlled and trap-filled conduction mechanism in HRS and LRS. Figure 8.5e shows the energy band diagram of fabricated RSM device.

Jyoti et al. introduced titanium oxide based PNC for RSM device application [91]. Fluorine doped tin oxide (FTO) was used as a substrate. The device structure was FTO/TiO_2-PVA/Ag. The PVA used as a polymer is soluble in water and contained as large number of hydroxyl groups. When the concentration of nanoparticles was increased, the charge storage capability of RSM device also increased significantly. The device worked on filamentary conduction mechanism. When voltage sweep is applied, the oxygen vacancies were repelled toward the top electrode. As the TiO_2 concentration was increased, more oxygen vacancies were formed. When negative voltage was applied, the bottom electrode attracted the oxygen vacancies. As a result, the device switched to HRS. Ukakimaparn et al. suggested P-25 TiO_2 nanoparticles with PVP [92]. The device structure was ITO/PVP: TiO_2 NPs/Al. The device showed the on/off ratio of 10^5 at 1 V. The voltage range of RSM device was 7 to −7 V. The P-25 TiO_2 nanoparticles possessed rutile and anatase phases at the ratio of 3 : 1 having the 3.2 and 3.0 eV. As low bias voltage thermionic emission was observed. When the trap sites were filled, the device switched to LRS. Hence the device worked on trapping and de-trapping conduction mechanism. Similarly, Siddiqui et al. explored

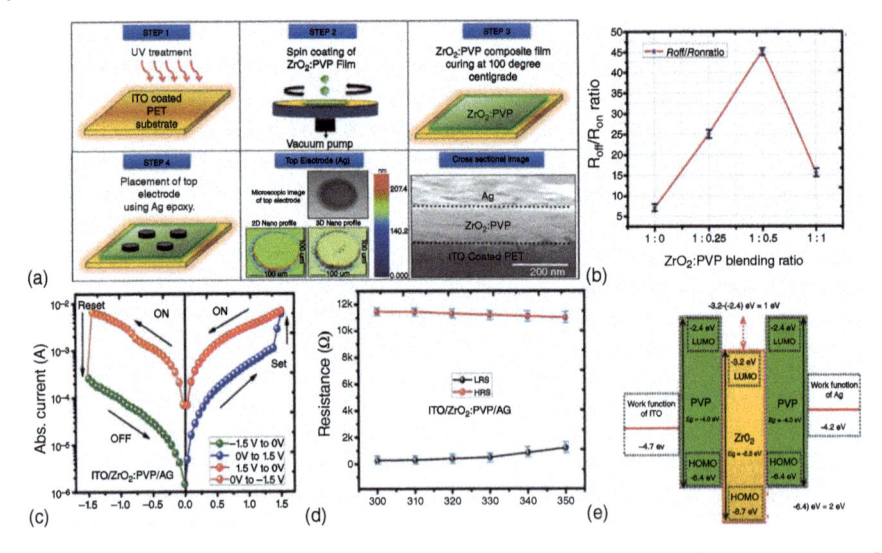

Figure 8.5 (a) The complete fabrication process of the RSM device having the structure of ITO/ZrO$_2$: PVP/Ag. (b) R_{off}/R_{on} ratio for various blending ratios of PVP and ZrO$_2$. (c) I–V characteristics curve of ITO/ZrO$_2$: PVP/Ag PNC device, presenting bi-stable switching behavior. (d) Showing the effect of temperature on the fabricated ITO/ZrO$_2$: PVP/Ag PNC device. (e) The PNC RSM ITO/ZrO$_2$: PVP/Ag energy band diagram. Source: Khan et al. [90]. Reproduced with permission. Copyright © 2019 Springer Nature Limited.

PVOH-ZnSnO$_3$ nanocomposite based RSM on a PET substrate [93]. The active layer of device was deposited by using EHD technique. The fabricated device structure was Ag/PVOH-ZnSnO$_3$/Ag. The transition of fabricated device from off state to ON state took place at 1.5 V, while the off/on ratio was 1.65×10^2. The device showed the high retention time of 10^5 seconds. The memory device followed the same conduction mechanism of TCSCLC in HRS while, ohmic conduction in LRS.

8.3.2 Metal Based Nanoparticles for Polymer Nanocomposite RSM

The hybrid nanocomposites based on inorganic NPs blended in the organic polymers have been largely studied in several domains such as solar cells, optoelectronic devices, and RSM applications [94–97]. The properties of inorganic NPs are combined with the polymers in the hybrid nanocomposites. These hybrid nanocomposite based devices offered advantages of light weight, simple device structure, large area production, flexibility, and less power consumption [98–100]. Wu et al. reported biodegradable RSM device based on Au nanoparticles embedded in alkali lignin [101]. Lignin is a biopolymer found in plants having insulating nature. It can be easily dissolved in water, hence can be easily processable. The fabricated device structure was Al/Au NPs: lignin/Al. The complete device fabrication is illustrated in Figure 8.6a. The AFM of the nanoparticles embedded lignin layer that presented the smooth roughness with the 0.6 nm RMS (Figure 8.6b). The process of peeling the device from SiO$_2$ substrate is shown in Figure 8.6c. Figure 8.6d presents the flexible nature of the device, as it is pasted on the arm. The device switched from HRS

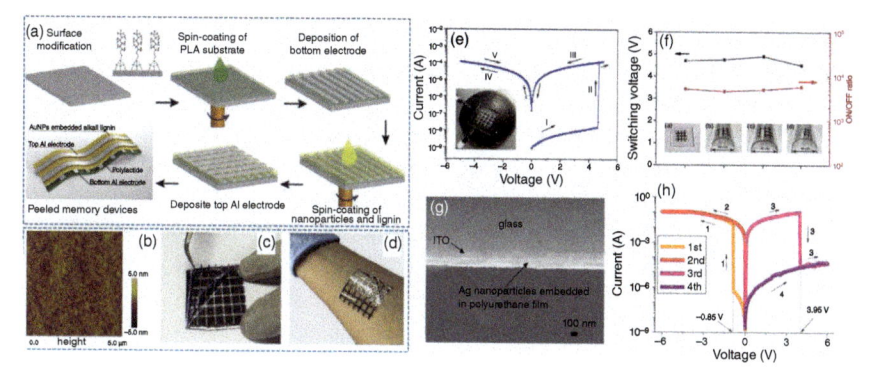

Figure 8.6 (a) The complete schematic illustration of the alkali-lignin based RSM depicting the step by step device fabrication. (b) The digital image of fabricated memory device ensures the flexible nature by pasting on the human arm. (c) The digital image shows the peeling-off the device from substrate. (d) The AFM of the active layer based on Au blended alkali-lignin having the RSM value of 0.6 nm. (e) I–V characteristics showing the resistive switching behavior having the device structure of Al/Au: lignin/Al. (f) the switching voltage and the on/off ratio of Al/Au: lignin/Al device at various bending diameters of 0, 30, 40, and 50 nm, respectively. Source: Wu et al. [101]. Reproduced with permission. Copyright © 2018, Elsevier Ltd. (g) Presented the Cross-section image of indium-tin-oxide/Ag nanoparticles embedded in polyurethane film/Al based RSM device. (h) I–V characteristics of the RSM device having the structure of ITO/Ag nanoparticles embedded in polyurethane film/Al. Source: Liu et al. [102]. Reproduced with permission. Copyright © 2018, Elsevier Ltd.

to LRS at 4.7 V (Figure 8.6e). The on/off ratio of fabricated RSM device was $>10^4$, while the retention time was measured to be 10^3. The switching voltage and on/off ratio showed that the device exhibits excellent stability at various bending diameters (Figure 8.6f). In HRS, the device followed the thermionic emission conduction mechanism at low bias voltage, while at higher bias voltage, trapped controlled, and trap filled SCLC model was followed. In LRS, the memory device followed the ohmic conduction model.

Liu et al. demonstrated the Ag nanoparticles embedded in polyurethane film [102]. The PEDOT: PSS modified layer was used on the ITO. The device structure was Al/polyurethane + Ag/ITO/glass (Figure 8.6g). The device switched to set state at −0.85 V, while the on/off current ratio was measured to be 10^3–10^5 (Figure 8.6h). The retention time of the fabricated device was $>1.8 \times 10^4$. The polyurethane was selected due to high strength, lower cost, and high resistance to moisture. The ohmic, trap controlled, and trap filled models were used to elaborate the conduction mechanism of fabricated device. Kaur et al. demonstrated the effect of Ag doping and PVA as additional dielectric buffer layer in the CdSe nanocomposite based RSM [103]. The device structure was kept Ag/Ag: CdSe nanocrystal (NC)/PVA buffer layer/Al on the aluminum (Al) substrate. The memory device was prepared by chemical synthesis, while the characterization of prepared device was done by photoluminescence and UV–Vis techniques. The fabrication of CdSe NC involved the mixing of solutions of cadmium acetate and sodium selenosulfate at the temperature of 60 °C while the solution of $AgNO_3$ was prepared in the deionized water. The solution of CdSe NC was mixed with the solution of silver nitrate for Ag doping.

The transition from off state to on state was occurred at +4 V. The device showed the on/off ratio of 6.6×10^3. It is shown that the Ag doping enhances the memory effect by increasing on/off ratio. The dopant atoms work as traps for charge carriers and make the trap concentration more uniform resulting in enhancing the charge trapping efficiency. The PVA buffer layer restricts the recombination of electrons and holes. As a result, space charge region is formed due to the accumulation of electrons and holes in nanocomposites. The device presented good reliability and stability characteristics. In HRS, the device followed the ohmic conduction mechanism, while in LRS, SCLC conduction mechanism was followed.

Tripathi et al. studied cadmium sulfide (CdS) nanocomposite with polystyrene (PS) as a polymer in RSM device application [104]. PS was used as an insulating material to prevent the leakage of charges stored in the memory device. An extra copper phthalocyanine (CuPc) layer was inserted between bottom electrode and the nanocomposite layer. The device structure was Al/CuPc/CdS NC/Ag. The transition from off state to on state occurred at 1 V. The device showed the high on/off ratio of 1.4×10^4. P-type organic semiconductor was utilized as hole trapping layer while CdS was used as electron trapping site. Once the electron is trapped, the higher energy level of CuPc prevents the tunneling of trapped electrons back to aluminum electrodes. To understand the conduction mechanism, I–V characteristics have been fitted with different current conduction models. At low applied bias, the memory device followed the Schottky emission model, whereas above 1 V, it follows SCLC. The PNC containing CdSe nanoparticles in poly(N-vinyl carbazole) (PVK) was studied by Kaur et al. [105]. The PNC was synthesized by using ex-situ chemical process. The author selected PVK due to its high resistance from moisture. The device structure was Al/(CdSe/PVK)/Ag. The set voltage of RSM device was 1 V while the on/off ratio measured was $\sim 10^4$. The I–V curve showed symmetrical behavior along different polarities. The operating mechanism of the RSM device is based on charge confinement in PNCs.

8.3.3 Graphene Based Polymer Nanocomposite RSM

Graphene with its derivatives plays a key role in memory devices because of simple fabrication process, and excellent electronic properties. Graphene oxide (GO) is a famous insulating material used in the RSM, which is obtained by modifying the graphene [106]. The graphene derivatives are functionalized by aromatic group present in the polymers. The modified graphene derivatives with the polymers could be promising materials in PNC based RSM devices [107]. Kim et al. studied the GO-PVP nanocomposite for RSM device. PEDOT: PSS was used as a modified layer on ITO [108]. The device structure was Al/GO: PVP/PEDOT: PSS/ITO/PEN (Figure 8.7a). The transition from off to on state took place at −0.2 to −0.7 V, while on/off ratio was found to be 1×10^2 (Figure 8.7b). The device showed the retention time of $>1 \times 10^4$. Since GO is soluble in water so high density of fabrication was achieved. PVP was used because of excellent hydrophilic nature for enhanced dispersion. Also, the performance enhanced significantly because of existence of hydrogen bonding between PVP and GO. As the polymer contained poor wettability

8.3 Polymer Nanocomposite Based RSM Devices | 225

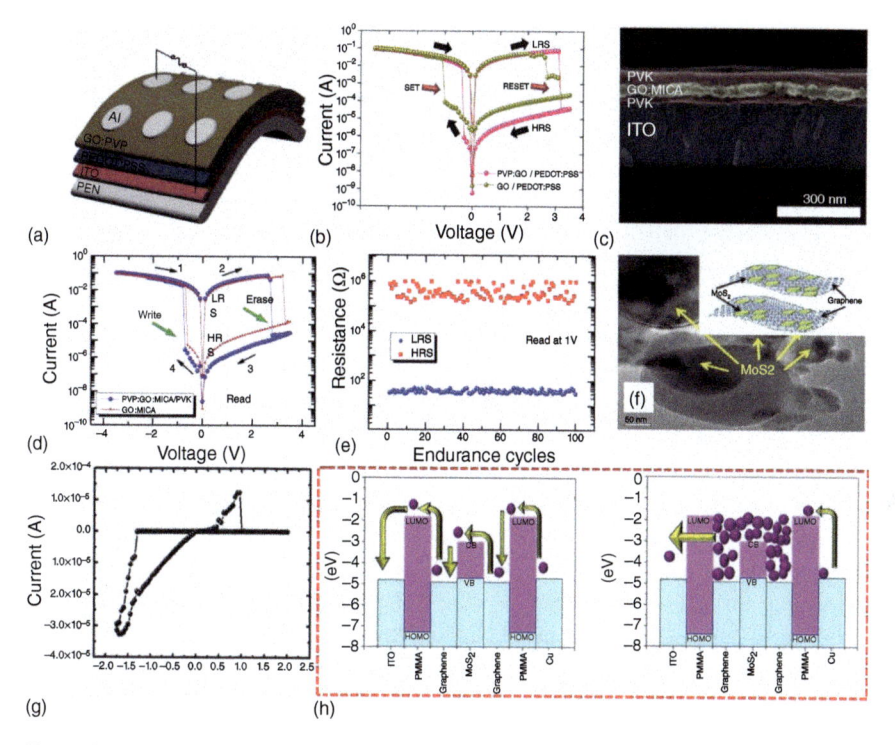

Figure 8.7 (a) The schematic illustration of RSM device fabrication having the device structure of Al/graphene oxide (GO): PVP/PEDOT: PSS/ITO. (b) *I–V* curves of the Al/GO: PVP/PEDOT: PSS/ITO and Al/GO/PEDOT: PSS/ITO on the PEN substrate. Source: Kim et al. [108]. (c) Cross-sectional SEM of PVK/graphene oxide: mica/PVK/ITO structure. (d) *I–V* curves of Al/PVK/graphene oxide (GO): mica/PVK/ITO in blue spherical shapes depicting high on/off ratio as compared with the device having structure of Al/GO: mica/ITO in red triangular shapes. (e) The HRS and LRS of the device with the structure of PVK/graphene oxide: mica/PVK/ITO 100 cycles at 1 V showing excellent stability of the device. Source: Kim et al. [109]. Reproduced with permission. Copyright © 2018, Elsevier Ltd. (f) The TEM image of hybrid structure based composite of graphene and MoS$_2$ presenting the MoS$_2$ active modes. The inset shows the schematic illustration of graphene and MoS$_2$ based nano-assembly. (g) The linear scanned illustration at the first voltage scan of the device having the structure Cu/g-MoS$_2$/PMMA/ITO. (h) The transport mechanism at HRS and LRS of Cu/g-MoS$_2$/PMMA/ITO device at various energy bands. Source: Bhattacharjee et al. [57]. Reproduced with permission. Copyright © 2018, Elsevier Ltd.

property, the modification layer was used to increase the mechanical quality of RSM device. The conductive filament model based on ohmic and SCLC was used for the conduction in fabricated RSM device.

A macromolecular bi-stable memory device was investigated by Huang and coworkers, which includes the composites of insulating PVA and graphene nano flakes [110]. The composites of GNF/PVA is sandwiched between ITO coated PET substrate and Ag to study the bi-stable switching behavior and rewritable memory effects of the fabricated device. The thin film solution processing technique was used for the fabrication of RSM device. The device structure was Ag/GNF: PVA/ITO. All the characterization of this device was carried out at room temperature. The fabrication process of this device involved the spin-coating of GNF/PVA composites on ITO substrate for 60 seconds at 4000 rpm followed by the temperature of 180 °C. The residual solvent was removed by placing the composites in air for 10 minutes. The area of typical memory device was observed as 3 mm^2. The device exhibits the retention time of about 10^4 seconds without significant degradation and exhibits the on/off ratio of $>10^2$ at 0.2 V, which is distinct in memory applications. It is also observed that fabricated device showed bi-stable switching behavior and nonvolatile rewritable memory effects. The conduction mechanism dominating in this device is ohmic, TLSCLC, as well as thermionic emission. The macromolecular devices showed intense attraction toward the scientists due to simple configuration and low temperature simple processing.

Choi et al. fabricated a flexible memory switching device, which included graphene and poly(4-venylphenol) nanocomposites [111]. Electro-hydrodynamic atomization (EDHA) technique was used for the fabrication of device because of non-contact and efficient material printing. The nanocomposites of graphene/PVP were sandwiched between Ag and ITO substrate as an active layer for studying the resistive switching behavior of the device. The fabrication process of this device involved the breakdown of graphene nano flakes into ultra-small nano flakes of 20–200 nm thickness in the presence of N-methyl-pyrrolidone by ultrasonication. These nano flakes were successfully deposited on the ITO coated PET substrate as a thin film of 140 ± 7 nm thickness. The fabricated device showed transition of high resistance state to low resistance state (off to on) characteristic at low voltages especially when operated at ±3 V. The fabricated device indicated a stable memory device due to Ag doping. The device showed the retention time of about one hour. The application of the memory devices is efficient and high density memory devices. Similarly, Kim et al used mica blended with GO and PVK [109]. The device structure was Al/PVK/GO: mica/PVK/ITO (Figure 8.7c). The device switches from HRS to LRS at −0.5 V (Figure 8.7d). The on/off ratio was measured to be 2×10^4, while retention time was 1×10^4. The LRS and HRS of the fabricated device at read voltage of 1 V for 100 cycles showed a stable behavior (Figure 8.7e). Sun et al. studied the composite of PS and GO as a PNC resistive switching device [112]. The device structure was ITO/PS+GO/Al, while the devices showed two set voltages of −0.85 and −1.55 V, respectively. The device presented high retention time of 2×10^5 seconds. The device worked on the three different conduction mechanisms. In HRS, the device worked on ohmic conduction followed by trap controlled and trap filled SCLC. In LRS, ohmic model was followed.

The PNC based on graphene oxide-polyvinyl imidazole-polypyrrole (GO-Pvim-Ppy) was observed by Pang and Ni [113]. The device structure was Al/GO-Pvim-Ppy/ITO. The transition from HRS to HRS takes place at 4.44 V while the on/off ratio measured was 10^6. The GO was used as photoinitiator to polymerize the monomer solution. The hydrophilic nature of imidazole allows the composite to be dispersed in the water. The device followed ohmic and SCLC model for conduction. Bhattacharjee et al. suggested a composite of graphene and MoS$_2$ blended in the PMMA [57]. The device structure was Cu/gMoS$_2$-PMMA/ITO (Figure 8.7f). The device switched from HRS to LRS at 2 V (Figure 8.7g), and the on/off ratio was measured to be $>10^4$. The device showed the excellent retention time of 10 days. The graphene was added to increase the conductivity of MoS_2, as the graphene possess high conductivity. MoS_2 was used as charge trapping points in RSM device. The resistive switching device worked on the trapping and de-trapping conduction mechanism (Figure 8.7h). A stretchable reduced GO with conjugated polymer based RSM device was demonstrated by Ban et al. [114]. The switching voltage from reset to set was 1.1 V, while the on/off ratio was measured to be 10^4. The device showed the retention time >12 000 seconds, and the stretchability was up to 30%. The device structure was r-rGO/PVK/Al. The device was fabricated by direct deposition on PDMS used as substrate. The conjugated polymer was sandwiched between aluminum and reduced graphene oxide used as bottom electrode. The reduced graphene was then transferred to PDMS. The aluminum was thermally deposited as top electrode. At low applied voltage, the RSM device followed ohmic model due to thermally generated charge carriers. When the voltage was increased, the injected charges increased, and the device followed the charge limited model. At higher bias voltage, the conductive channels were developed, known by carbon rich filaments. As a result, the device switched to LRS. At opposite applied voltage, the rupturing of the conductive path takes place.

GO blended in PMMA for PNC based RSM device was explored by Gogoi and Chowdhury [115]. The defect states in GO acted as an energy barrier. The device structure was ITO/PMMA-GOs/Al. The device offered the on/off ratio of 10^4 at 1.87 V. The worked successfully achieved the liquid exfoliation of graphene oxide flakes in few nanometers layer. The GO acted as electron trapping sites because of rich defect centers. At low bias voltage, the injected electrons were blocked at the interface. As threshold bias voltage, the electrons cross the barrier and got trapped by graphene oxide. Upon filling the traps, the conductive path was formed, and the device switched to LRS. When opposite voltage was applied, the trapped electrons was not released due to deep energy sites. Thus, the device remained at LRS, at 0 bias voltage, and at low opposite applied voltage.

8.3.4 Quantum Dot Based Polymer Nanocomposite RSM

Currently quantum dot based RSM devices are largely studied for various memory device applications. These RSM devices provide the advantages of fast switching along with nanoscale device area. The memory function ability of QD based RSM is achieved through the charge storage. The intrinsic energy levels in quantum dots are responsible to store the electrons in quantum dot based PNC RSM device [116–119]. Ali et al. studied the graphene quantum dots (G-QDs) blended with PVP for PNC based RSM device [77]. The RSM devices were papered in 3 × 3 crossbar array, while each device was accessed randomly without any cross talk. The fabricated device structure was Ag/G-QDs: PVP/Ag. EHD technique was used for deposition of electrodes and active layer. EHD is a useful deposition method because it is a cost effective, rabid, convenient for mass production, and deposited at ambient temperature conditions. The schematic diagram of EHD technique with multiple jetting modes is shown Figure 8.8a. The electric field is applied on the nozzle. Figure 8.8b shows

228 | *8 Polymer Nanocomposites for Resistive Switching Memory*

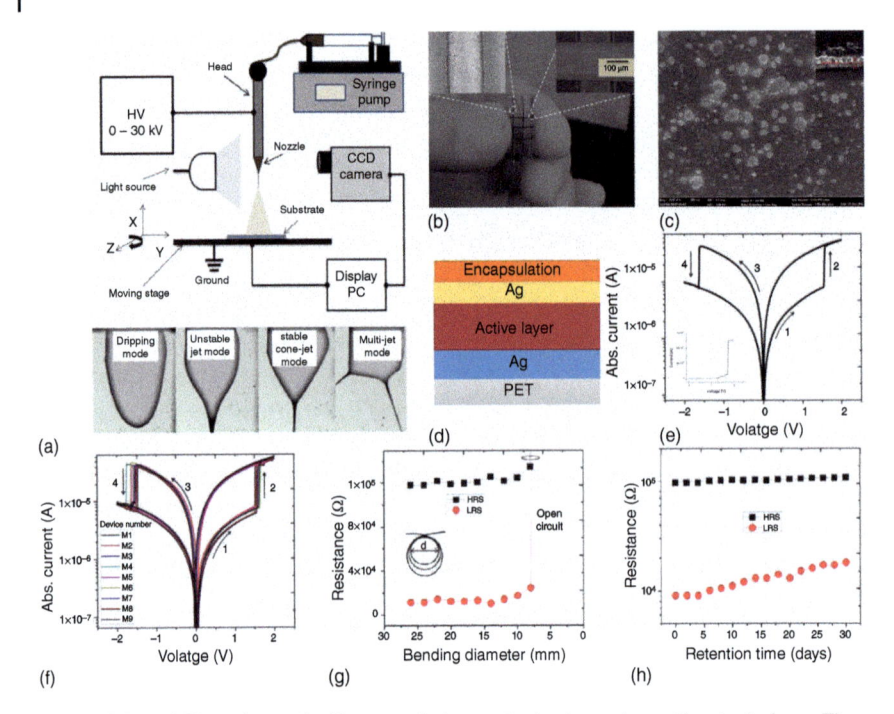

Figure 8.8 (a) The schematic diagram of electro-hydrodynamic coating technique. The different modes, i.e. dripping mode, unstable jet mode, stable cone-jet mode, and multi-jet mode are shown in the figure. (b) The photograph of 3 × 3 RSM based on graphene quantum dots blended with PVP. The left top image shows the zoomed image of silver electrode, while the right top image presents the cross-bar points. (c) SEM of G-QDs and PVP blended layer, while the cross-sectional image is shown in the inset. (d) Cross-sectional view of GQDs/PVP layers based RSM. (e) The absolute *I–V* curve for dual sweeping of voltage at various transition states. (f) Shows the absolute *I–V* analysis for all the memristor devices. (g) The LRS and HRS of fabricated device at various diameters using rods of different diameters ranging from 26 to 5 mm. (h) The retention time of fifth device for about 30 days. The fabricated device presented near stable behavior. The fabricated device presented near stable behavior. Source: Ali et al. [77]. Reproduced with permission. Copyright © 2015, Elsevier Ltd.

the 3 × 3 RSM device, having transparent active layer and visible silver electrodes. Here, inset Figure shows the zoomed image of the silver electrode and crossbar point, respectively. Figure 8.8c shows the SEM image of the active layer. Here, inset Figure shows the cross-sectional view of the RSM device. Figure 8.8d depicted the cross-sectional view of the fabricated device. The device switched from HRS to LRS at 1.8 V, while the on/off ratio as measured to be 14. The RSM device presented the retention time of >30 days. The device showed the symmetric behavior, when the blending ratio of G-QDs and PVP was kept 1 : ½. All other ratios gave the asymmetric behavior. Figure 8.8e shows the *I–V* curve between −2 to +2 voltage sweep showing the HRS (1), LRS (3), transition from HRS to LRS (2), and transition from LRS to HRS (4). Figure 8.8f shows the *I–V* analysis curves for all nine RSM devices in 3 × 3 crossbar array. Figure 8.8g shows the bending diameter of the RSM device. It can be seen that the device was open circuited at 7.5 mm bending diameter. The retention time of RSM device is Figure 8.8h. The device followed the ohmic conduction mechanism in LRS and HRS.

Zhou et al. presented a flexible and transient RSM device based on citric acid quantum dot (CA-QD)-polyvinyl pyrrolidone (PVP) PNC, fabricated by full solution process method [120]. The fabrication methods involve the mixing of 5 mg (CA QDs) and 1 ml PVP solution. The mixture was placed into AgNW-PVP electrode, followed by the spin coating process for 60 seconds and 3000 rpm. For the removal of deionized water, the obtained film was placed in oven for 30 minutes at the temperature of 60 °C. The fabricated device showed DRAM storage memory, reversible switching, and symmetrical behavior of I–V curve. The RSM device structure was AgNWs (top electrode)/CA QDs-PVP/AgNWs/PVP (bottom electrode). The device showed the on/off current ratio of $>2 \times 10^4$. The first transition was occurred at +1 V from off state to on state. The retention time of the device was $>10^4$. The proposed device also showed the high reproducibility and flexibility, which indicates excellent memory characteristic. The working mechanism presented was charge trapping and de-trapping model. Upon the application of positive bias, at low voltage, the thermally generated charge carriers were more dominant at low voltage. When the injected charge carriers dominated the thermally generated charge carriers, trap-controlled SCLC was achieved. When all the traps were filled, current sharply increased, and device switched to LRS. These types of devices find its intensive applications in transient electronics and implanted electronics.

Ooi et al. suggested a mixture of graphene QDs and MoS_2 blended with PVDF for PNC based RSM device [121]. The PVDF was selected due to its high flexibility and light weight. The device structure was kept AgNWs/PVDF/GMP-NC/PVDF/ITO/GLASS. The three stacking layers were sandwiched between metal electrodes. The solution processing technique was used for coating the layers on the electrode. Initially the device was in low conduction state. The device switched to intermediate high conductivity at −0.6 V. Finally, the device switched to high conductivity state at −0.8 V, and the two high conduction states depicted the formation of heterostructures in nanoparticles. On the other hand, the changes occurred at two different voltages, 0.4 and 0.68 V. Thus, the RSM device showed the tristable resistive switching behavior. The device followed the charge trapping and de-trapping conduction mechanism.

8.3.5 Polymer Based Nanocomposites for RSM

The hybrid composite of the various polymers got immense attention due to their hybrid properties in the RSM devices [122, 123]. The researchers explored various composites by using different polymers for hybrid composite based RSM devices to enhance the on/off ratio, flexibility, and stability [35]. Resistive switching properties of RSM using PEDOT: PSS doped PVA were suggested by Huang and Ma [124]. It was demonstrated that the resistive switching characteristics were strongly dependent upon UV treatment of polymer blend. The fabrication of the device involved the UV-zone treatment for 15 minutes followed by cleaning and drying. After that, the aqueous solution of PEDOT: PSS and PVA dispersion was

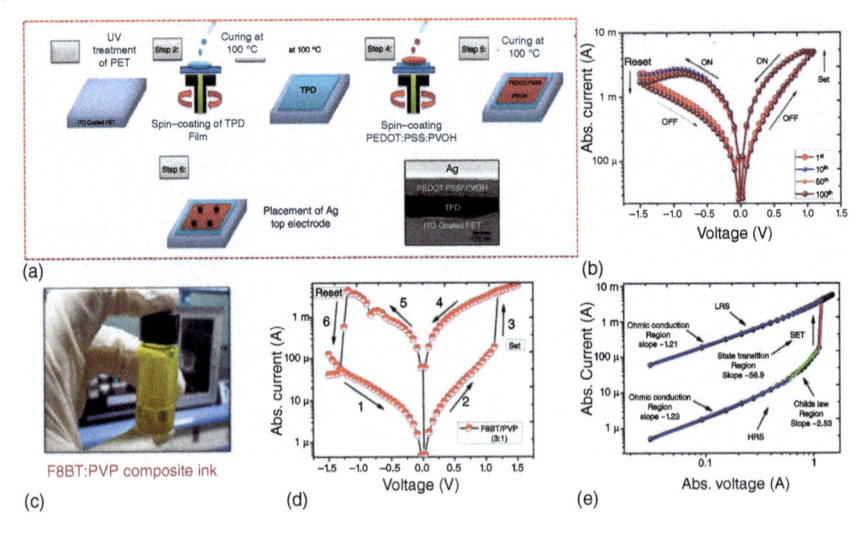

Figure 8.9 (a) The fabrication of the ITO/TPD/(PEDOT:PSS/PVOH)/Ag device using spin-coating technique, and the cross-sectional SEM image (inset figure). (b) I-V curve of ITO/TPD/(PEDOT:PSS/PVOH)/Ag device, showing that bio-polar behavior is observed for 100 different cycles. Source: Khan et al. [125]. Reproduced with permission. Copyright © 2018, Springer Nature Limited. (c) The ink based on PVP-F8BT polymers composite in the THF used as solvent. (d) The electrical curve for I-V measurement with the blending ratio of 3 : 1 (F8BT:PVP). (e) The double logarithmic curve for I-V showing the HRS and LRS following the SCLC conduction mechanism. Source: Hassan et al. [35]. Reproduced with permission. Copyright © 2018, Springer Nature Limited.

mixed at the room temperature and stirred the solution for six hours. The obtained solution was coated upon ITO substrate to form a thin film. It was observed that the treatment of UV-ozone is necessary for better switching performance. The device structure was ITO/PEDOT: PSS: PVA/Al. The device showed the on/off current ratio of about 10^2 and long life time of about 96 hours without any deterioration. The transition from off to on state was occurred from 1.1 to 1.8 V, respectively. The conduction mechanism dominated in this device in on state was ohmic conduction while the Frenkel–Poole emission model described the conduction mechanism in off state. The memory devices based upon polymer blends films have a variety of applications in storage devices due to excellent storage performance and simple structures. N,N′-bis(3-methylphenyl)-N,N′-diphenylbenzidine (TPD) and PEDOT: PSS/PVOH polymer composite was studied by Khan et al. [125]. Spin-coating technique was used to fabricate the RSM device. The fabrication process is illustrated in Figure 8.9a, while the cross-sectional SEM image is shown in the inset figure. The device structure was Ag/(PEDOT: PSS/PVOH)/TPD/ITO. The device switched from HRS to LRS at low voltage of ±1.5 V. The I–V curve showed the bi-polar resistive switching behavior of the fabricated device (Figure 8.9b). TPD carried organic photo conducting property with semiconducting capability. The high bandgap of PVOH provided high insulating and dielectric properties in the active layer of fabricated RSM device. The device presented the R_{off}/R_{on} ratio = 28.7, while the retention time was measured to be 10^4 seconds.

Choi et al. introduced the RSM device having the structure of Al/PI-mica/PEDOT: PSS/ITO [126]. The device showed the transition from off to on state at −0.3 V, while the on/off ratio was 4.28×10^3. The device presented the retention time of 1×10^4 seconds. The PEDOT: PSS was used as a buffer layer. Mica was used owing to its larger bandgap and insulating nature. The bandgap can be decreased by decreasing mica layers. The mica layers acted as a charge trapping sites. Above 1%, the on/off ratio of the device was reduced due to aggregation. The device followed the charge trapping and de-trapping conduction mechanism for switching the RSM device between HRS and LRS. Hassan et al. reported poly[(9,9-di-*n*-octylfluorenyl-2,7-diyl)-*alt*-(benzo[2,1,3]thiadiazol-4,8-diyl)] (F8BT) and (PVP) based PNC for RSM (Figure 8.9c) [35]. The *I–V* curve in Figure 8.9d shows that the device presented excellent on/off ratio with the blending ratio of 3 : 1 (F8BT and PVP). The device followed the SCLC conduction mechanism as explained in Figure 8.9e.

8.3.6 2D Material Based Polymer Nanocomposites RSM

The two-dimensional (2D) materials are largely studied because of unique functionalities of ultrathin thickness, large surface area, excellent transparency, and 2D morphology [127]. These materials are comprised of two-dimensional sheets. There is a strong covalent bond between the interlayer atoms present in the stacked layers. The van der Waals forces are responsible to hold the stacked layers with each other. The conduction in 2D material based RSM devices is usually carried by metallic ion penetration and intrinsic species migration [127–129]. The RSM device characteristics of a hybrid PNC synthesized by coating semi-conductive molybdenum disulfide (MoS_2) flakes embedded with dielectric PVA polymer was introduced by Rehman et al. [80]. The device structure was PET/Ag/MoS_2-PVA/Ag. Hybrid MoS_2-PVA nanocomposite, bottom electrode (Ag), and top electrode (Ag) are deposited using electrohydrodynamic (EHD) atomization, reverse offset, and EHD patterning, respectively. The set voltage for the device was 3 V, while the on/off ratio was 1.28×10^2. The retention time of the fabricated device was measured to be 10^5. The device presents the excellent properties with considerable deviation. Initially, the conduction band of MoS_2 was partially filled with electrons. As the applied field was increased, electrons jump to conduction band, resulting in the higher conductivity of device.

The resistive switching characteristics of boron nitride (BN) and PVOH were suggested by Siddiqui et al. [130]. Liquid based method was used to exfoliate the *h*BN in IPA. The device structure was Ag/*h*BN-PVOH/ITO/PET. The device switched at low bias voltage of 0.78 V from the off state to the on state, and the on/off ratio measured was 4.8×10^2. The BN was selected due to its unique thermal, mechanical, chemical, and electrical characteristics. There is strong covalent bonding between boron and nitrogen, which enhances the flexibility of memory device. Furthermore, large band gap in BN restricts the charge carriers to return in initial energy states when the biasing voltage is removed. The device works on conduction mechanism of metallic filaments formation. The LRS did not show any change upon changing the device size, which ensures the metallic filament formation in RSM device. A 2D material, TiS$_2$ nano-flakes was first time reported by Lyu et al. with PVP for nonvolatile bipolar RSM device [53]. The structure of fabricated device was Al/TiS$_2$-PVP/ITO/PET. The processing was carried in low boiling state of solvent. The HRS and LRS were carried out at <2 V, while the retention time was measured to be 10^4. The TiS$_2$ was adopted due to the following reasons. Due to quantum confinement effect, the TiS$_2$ acted as trap rich site for charges. The TiS$_2$ possess sulfur vacancies, which are important for RSM behavior. During the contact with metal, TiS$_2$ makes the Schottky barrier. This RSM device having PVP and TiS$_2$ presented good bending endurance and retention time. The RSM device worked on the Schottky emission in HRS. In LRS, Poole–Frenkel effect was followed. During the positive sweep, the cations are anions were generated. The anion moved toward the aluminum, while the cations moved toward the ITO. Since TiS$_2$ is trap rich site, the localized traps dominated the carrier transport. The existed Schottky barrier becomes thinned at high voltage, and device shifted to LRS. When negative sweep was applied, the cations and anions moved in opposite direction. The Schottky barrier recovered, and the device switched to HRS.

8.3.7 Other Materials Used for Polymer Nanocomposite Based RSM

Similarly, various other materials are also studied for PNC based RSM devices. Haoqun An et al. reported Cesium-lead-chlorine (CsPbCl$_3$) based perovskite quantum dots blended with PMMA [131]. The device structure was Al/PQDs-PMMA/ITO. The transition takes place at −0.3 V from HRS to LRS. The device showed the high on/off ratio of 2×10^4 and the retention time measured was 1×10^4. The device was operated on charge trapping/de-trapping mechanism. The perovskite quantum dots acted as a charge trapping sites, which form conductive filaments in the polymer layer. When the opposite bias voltage was applied, the captured electrons were released. As a result, the device switched to HRS. Rajan et al. observed the RS properties of polyvinylidene fluoride-hexafluoropropylene (PVDF-HFP), PMMA, PEO, and mixture of PEO and PVDF-HFP [132]. AgNO$_3$ was used as switching matrix along with ionic liquid. The ionic liquid was introduced due to its capability of enhancing charge transport in the polymer matrix. The various fabricated device structures were Ag/Ti (adhesion layer)/AgNO$_3$_IL_PVDF-HFP/Ta (adhesion layer)/Pt/SiO$_2$, Ag/Ti (adhesion layer)/AgNO$_3$-IL_PEO/Ta (adhesion layer)/Pt/SiO$_2$, Ag/Ti (adhesion layer)/AgNO$_3$_IL_PMMA/Ta (adhesion layer)/Pt/SiO$_2$, and Ag/Ti (adhesion layer)/AgNO$_3$_IL_PVDF-HFP+PEO/Ta (adhesion layer)/Pt/SiO$_2$. Planar asymmetric switching was observed in all these PNC resistive switching devices. The device having PVDF-HFP used as switching matrix presented 1200 switching cycles, while the PEO based switching matrix presented large memory properties.

Chaudhary et al. explored P3HT and carbon nanotubes composite thin film for PNC based resistive switching device [133]. The fabricated device structure was FTO/P3HT-CNT/Al. When 4% of CNT was embedded with the polymer, the devices switched at the set voltage of 1.8 V from HRS to LRS, while the on/off ratio war measured to be >10^2. Carbon rich filaments (CRF) are responsible for resistive switching. When enough bias voltage is applied across the device, CRF are formed due to pyrolyzation, which switches the device to the set state. When 8% of CNT was blended with polymer, the carbon rich filaments may come in contact with each other. As a result there will be no resistive behavior in the device. When negative bias was applied across the device, the CRF dissolved. Consequently, the device switched to reset state. The natural polymers are easily processable, non-toxic, and biodegradable. Liu et al. introduced a biodegradable cellulose based graphene oxide memory device [134]. The device structure was Al/CMC-GO/Al/SiO$_2$. The aluminum was thermally evaporated on SiO$_2$ as a bottom electrode. The PNC material was spin-coated on the bottom electrode. The device presented the high on/off ratio of 6×10^5 at the low switching voltage of 2.22 V because of hybridization of GO and carboxymethyl cellulose. The device followed ohmic, trap controlled, and trap filled conduction mechanism in HRS. At LRS, ohmic model was followed. When the voltage was removed, the trapped charges were not discharged due to insulting behavior in the surroundings.

8.4 Concluding Remarks

In 1960's, Moore's law has precisely predicted the performance guideline for the development of the microelectronic technology. The downscaling of CMOS technology beyond the few nanometers (2–3 nm) will not be possible, due to dominant quantum effects, which leads to leakage current through thinner gate insulators. In a single piece of integrated circuit, when more and more electronic components and circuits get jammed and electrons move faster between the processor and memory, the large amount of heat is generated in electronic devices. Hence, it makes too hot for consumers to sustain and it affects the device performance on large-scale. On the other hand, many researchers have been developing new materials for RSM devices to solve this law's limitation with new solutions, which can bring low-power, flexible, and multifunctional devices to meet challenges of future electronics. The conduction mechanism of the RSM devices usually depends upon the movements of ions and electrons between two conductive electrodes to store the data. This data storage is a novel method as compared with previously reported technologies. RRAM brings hope to overcome the quantum effect caused by the size shrinking to few nanometers and helps to reduce leakage current and replace entire memory heritage and achieve nonvolatile storage with ultrahigh density. RSM also enable the gate way to improve power consumption and computation efficiency. To realize these functions, PNC based RSMs were developed. The hybrid memory devices usually operate on charge transfer mechanism with the separation of holes and electrons.

In this chapter, previously reported PNCs based hybrid RSM devices are discussed. The oxide based ferroelectric mediated P(VDF-TrFE)/ZnO nanocomposite presented the highest on/off ratio of 2×10^7. Similarly, the blended mixture of graphene QDs and MoS$_2$ with PVDF showed the high on/off ratio of 10^7. The nanocomposite of ZnSnO$_3$ blended with the PVOH polymer offered the excellent retention time of 10^5 seconds; however the on/off ratio of the fabricated device is low as compared with other reported memory devices. The GO embedded PVP based PNC memory device switched from HRS to LRS at the low transition voltage of −0.2 to −0.7 V. The PEDOT: PSS was used a modified layer in the fabricated device. However, the device presented the low on/off ratio and retention time of 1×10^2 and 1×10^4, respectively. Similarly, $CsPbCl_3$–PMMA embedded RSM device switched from HRS to LRS at the transition voltage of −0.3 V. Hybridizing the organic polymers with other materials have solved stability problem, wherein organic polymers with metal composites were also confirmed to be highly stable and flexible RSMs. With the invention of 2D materials, it was opening a new hope to fabricate thinnest resistive memory device, which could show a stable performance under high mechanical deformation. 2D materials based on few layer or mono layer structure improve mechanical flexibility and deformability of RSM devices. In these expectations, many researchers were mainly forcing on different exfoliations techniques to fabricate thinnest monolayer of 2D materials.

To achieve flexibility for soft memory devices, the hybrid polymer composites based on organic polymers and inorganic materials could be tuned through coordination bond angle by engineering in molecular design. The polymers nanocomposite RSM device presented simple fabrication process along with high flexibility and less fabrication cost. As the hybrid PNC memory device merges the properties of both organic and inorganic materials, it presents the high on/off current ratio, excellent retention time, and non-volatile nature. The flexible nature of hybrid RSM device makes it a remarkable option for wearable and printable electronic device. Vast versatility in the properties of organic polymers and hybrid nanocomposites rapidly becomes part of emergent electronic and optoelectronic devices to improve the daily life of human beings. The various polymers nanocomposite for RSM were still required lot more attention to formed into a complete set of system until. Current situation opens an infinite possibilities and opportunities for all researchers to develop a novel electronic system based on RSM polymers nanocomposites. For summering the RSMs based on PNCs, all performance comparisons according to device material and structure are described in Appendix 8.A.

Acknowledgments

The National Research Foundation of Korea (NRF) grant funded by the Korean government (MSIP) (2020R1A2C1011433).

References

1 Chen, C.L.P. and Zhang, C.Y. (2014). Data-intensive applications, challenges, techniques and technologies: a survey on big data. *Inf. Sci. (NY)* 275: 314–347.
2 Vergouw, B., Nagel, H., Bondt, G., and Custers, B. (2016). *The Future of Drone Use*, vol. 27, 21–46. Asser Press.
3 Forrest, S.R. (2004). Electronic appliances on plastic. *Nature* 428 (6986): 911–918.

4 Sekitani, T., Zaitsu, K., Noguchi, Y. et al. (2009). Printed nonvolatile memory for a sheet-type communication system. *IEEE Trans. Electron Devices* 56 (5): 1027–1035.

5 Cheng, C.H., Yeh, F.S., and Chin, A. (2011). Low-power high-performance non-volatile memory on a flexible substrate with excellent endurance. *Adv. Mater.* 23 (7): 902–905.

6 Haron, N.Z. and Hamdioui, S. (2008). Why is CMOS scaling coming to an END? Proceedings of the 2008 3rd International Design and Test Workshop, IDT 2008, pp. 98–103.

7 Isaac, R.D. (1998). Reaching the limits of CMOS technology. IEEE 7th Topical Meeting on Electrical Performance of Electronic Packaging (Cat. No. 98TH8370), 3.

8 Heath, J.R., Kuekes, P.J., Snider, G.S., and Williams, R.S. (1998). A defect-tolerant computer architecture: opportunities for nanotechnology. *Science* 280 (5370): 1716–1721.

9 Strukov, D.B., Snider, G.S., Stewart, D.R., and Williams, R.S. (2008). The missing memristor found. *Nature* 453 (7191): 80–83.

10 Kwon, D.H., Kim, K.M., Jang, J.H. et al. (2010). Atomic structure of conducting nanofilaments in TiO_2 resistive switching memory. *Nat. Nanotechnol.* 5 (2): 148–153.

11 Strukov, D.B., Stewart, D.R., Borghetti, J. et al. (2010). Hybrid CMOS/memristor circuits. 2010 IEEE International Symposium on Circuits and Systems: Nano-Bio Circuit Fabrics and Systems, ISCAS 2010, pp. 1967–1970.

12 Kumar, A., Rawal, Y., and Baghini, M.S. (2012). Fabrication and characterization of the ZnO-based memristor. 2012 International Conference on Emerging Electronics, ICEE 2012, pp. 1–3.

13 Li, B., Shan, Y., Hu, M. et al. (2013). Memristor-based approximated computation. Proceedings International Symposium on Low Power Electronics and Design, pp. 242–247.

14 Eshraghian, K., Cho, K.R., Kavehei, O. et al. (2011). Memristor MOS content addressable memory (MCAM): hybrid architecture for future high performance search engines. *IEEE Trans. Very Large Scale Integr. Syst.* 19 (8): 1407–1417.

15 Di Ventra, M., Pershin, Y.V., and Chua, L.O. (2009). Circuit elements with memory: and meminductors. *Proc. IEEE* 97 (10): 1717–1724.

16 Adhikari, S.P., Sah, M.P., Kim, H., and Chua, L.O. (2013). Three fingerprints of memristor. *IEEE Trans. Circuits Syst. Regul. Pap.* 60 (11): 3008–3021.

17 Corinto, F., Member, S., Civalleri, P.P. et al. (2015). A theoretical approach to memristor devices. *IEEE J. Emerg. Sel. Top. Circuits Syst.* 5 (2): 123–132.

18 Ascoli, A., Tetzlaff, R., Chua, L.O. et al. (2016). History erase effect in a non-volatile memristor. *IEEE Trans. Circuits Syst. Regul. Pap.* 63 (3): 389–400.

19 Cavin, R.K., Zhirnov, V.V., Herr, D.J.C. et al. (2006). Research directions and challenges in nanoelectronics. *J. Nanopart. Res.* 8 (6): 841–858.

20 Ali, S., Bae, J., Lee, C.H. et al. (2018). Resistive switching device with highly asymmetric current-voltage characteristics: a solution to backward sneak current in passive crossbar arrays. *Nanotechnology* 29 (45): 455201.

21 Sawa, A. (2008). Resistive switching in transition metal oxides. *Mater. Today* 11 (6): 28–36.

22 Terabe, K., Hasegawa, T., Nakayama, T., and Aono, M. (2005). Quantized conductance atomic switch. *Nature* 433 (7021): 47–50.

23 Waser, R., Dittmann, R., Staikov, C., and Szot, K. (2009). Redox-based resistive switching memories nanoionic mechanisms, prospects, and challenges. *Adv. Mater.* 21 (25–26): 2632–2663.

24 Yang, J.J., Pickett, M.D., Li, X. et al. (2008). Memristive switching mechanism for metal/oxide/metal nanodevices. *Nat. Nanotechnol.* 3 (7): 429–433.

25 Jeong, H.Y., Kim, Y.I., Lee, J.Y., and Choi, S.Y. (2010). A low-temperature-grown TiO_2-based device for the flexible stacked RRAM application. *Nanotechnology* 21 (11): 115203.

26 Jo, S.H., Kim, K.-H., and Lu, W. (2009). High-density crossbar arrays based on a Si memristive system. *Nano Lett.* 9 (2): 870–874.

27 Lee, M.J., Kim, S.I., Lee, C.B. et al. (2009). Low-temperature-grown transition metal oxide based storage materials and oxide transistors for high- density non-volatile memory. *Adv. Funct. Mater.* 19 (10): 1587–1593.

28 Torrezan, A.C., Strachan, J.P., Medeiros-Ribeiro, G., and Williams, R.S. (2011). Sub-nanosecond switching of a tantalum oxide memristor. *Nanotechnology* 22 (48): 485203.

29 Strachan, J.P., Torrezan, A.C., Medeiros-Ribeiro, G., and Williams, R.S. (2011). Measuring the switching dynamics and energy efficiency of tantalum oxide memristors. *Nanotechnology* 22 (50): 505402.

30 Miao, F., Strachan, J.P., Yang, J.J. et al. (2011). Anatomy of a nanoscale conduction channel reveals the mechanism of a high-performance memristor. *Adv. Mater.* 23 (47): 5633–5640.

31 Cho, B., Kim, T.W., Song, S. et al. (2010). Rewritable switching of one diode-one resistor nonvolatile organic memory devices. *Adv. Mater.* 22 (11): 1228–1232.

32 Kim, T.W., Choi, H., Oh, S.H. et al. (2009). One transistor-one resistor devices for polymer non-volatile memory applications. *Adv. Mater.* 21 (24): 2497–2500.

33 Harito, C., Bavykin, D.V., Yuliarto, B. et al. (2019). Polymer nanocomposites having a high filler content: synthesis, structures, properties, and applications. *Nanoscale* 11 (11): 4653–4682.

34 Cho, B., Song, S., Ji, Y. et al. (2011). Organic resistive memory devices: performance enhancement, integration, and advanced architectures. *Adv. Funct. Mater.* 21 (15): 2806–2829.

35 Hassan, G., Khan, M.U., and Bae, J. (2019). Solution-processed flexible non-volatile resistive switching device based on poly[(9,9-di-n-octylfluorenyl-2,7-diyl)-alt-(benzo[2,1,3]thiadiazol-4, 8-diyl)]: polyvinylpyrrolidone composite and its conduction mechanism. *Appl. Phys. A Mater. Sci. Process.* 125 (1): **18**.

36 Shan, Y., Lyu, Z., Guan, X. et al. (2018). Solution-processed resistive switching memory devices based on hybrid organic-inorganic materials and composites. *Phys. Chem. Chem. Phys.* 20 (37): 23837–23846.

References | 237

37 Rajan, K., Bocchini, S., Chiappone, A. et al. (2017). WORM and bipolar inkjet printed resistive switching devices based on silver nanocomposites. *Flex. Print. Electron.* 2 (2): 024002.
38 Koo, H.J., So, J.H., Dickey, M.D., and Velev, O.D. (2011). Towards all-soft matter circuits: prototypes of quasi-liquid devices with memristor characteristics. *Adv. Mater.* 23 (31): 3559–3564.
39 Ali, S., Bae, J., Choi, K.H. et al. (2015). Organic non-volatile memory cell based on resistive elements through electro-hydrodynamic technique. *Org. Electron.* 17: 121–128.
40 Cheng, P., Sun, K., and Hu, Y.H. (2016). Memristive behavior and ideal memristor of 1T phase MoS_2 nanosheets. *Nano Lett.* 16 (1): 572–576.
41 Cassinerio, M., Ciocchini, N., and Ielmini, D. (2013). Logic computation in phase change materials by threshold and memory switching. *Adv. Mater.* 25 (41): 5975–5980.
42 Kim, W.T., Jung, J.H., Kim, T.W., and Son, D.I. (2010). Current bistability and carrier transport mechanisms of organic bistable devices based on hybrid Ag nanoparticle-polymethyl methacrylate polymer nanocomposites. *Appl. Phys. Lett.* 96 (25): 94–97.
43 Lee, K.W., Kim, K.M., Lee, J. et al. (2011). A two-dimensional DNA lattice implanted polymer solar cell. *Nanotechnology* 22 (37): 375202.
44 Mondal, S.P., Reddy, V.S., Das, S. et al. (2008). Memory effect in a junction-like CdS nanocomposite/conducting polymer poly[2-methoxy-5-(2-ethylhexyloxy)1,4-phenylene-vinylene] heterostructure. *Nanotechnology* 19 (21): 215306.
45 Zhang, N., Tang, W., Wang, P. et al. (2013). In situ enhancement of NBE emission of Au-ZnO composite nanowires by SPR. *CrystEngComm* 15 (17): 3301–3304.
46 Hmar, J.J.L., Majumder, T., Roy, J.N., and Mondal, S.P. (2015). Electrical and photoelectrochemical characteristics of flexible CdS nanocomposite/conducting polymer heterojunction. *Mater. Sci. Semicond. Process.* 40: 145–151.
47 Hmar, J.J.L. (2018). Flexible resistive switching bistable memory devices using ZnO nanoparticles embedded in polyvinyl alcohol (PVA) matrix and poly(3,4-ethylenedioxythiophene) polystyrene sulfonate (PEDOT:PSS). *RSC Adv.* 8 (36): 20423–20433.
48 Bhattacharjee, S., Sarkar, P.K., Roy, N., and Roy, A. (2016). Improvement of reliability of polymer nanocomposite based transparent memory device by oxygen vacancy rich ZnO nanorods. *Microelectron. Eng.* 164: 53–58.
49 Chen, M., Nam, H., Wi, S. et al. (2014). Multibit data storage states formed in plasma-treated MoS_2 transistors. *ACS Nano* 8 (4): 4023–4032.
50 Liu, J., Zeng, Z., Cao, X. et al. (2012). Preparation of MoS_2-polyvinylpyrrolidone nanocomposites for flexible nonvolatile rewritable memory devices with reduced graphene oxide electrodes. *Small* 8 (22): 3517–3522.
51 Li, A., Pan, J., Yang, Z. et al. (2018). Charge and strain induced magnetism in monolayer MoS_2 with S vacancy. *J. Magn. Magn. Mater.* 451: 520–525.

52 Rajeswari, N., Selvasekarapandian, S., Karthikeyan, S. et al. (2011). Conductivity and dielectric properties of polyvinyl alcohol- polyvinylpyrrolidone poly blend film using non-aqueous medium. *J. Non-Cryst. Solids* 357 (22–23): 3751–3756.

53 Lyu, D., Hu, C., Jiang, Y. et al. (2019). Resistive switching behavior and mechanism of room-temperature-fabricated flexible Al/TiS$_2$-PVP/ITO/PET memory devices. *Curr. Appl. Phys.* 19 (4): 458–463.

54 Kibble, T.W.B. (1967). Symmetry breaking in non-Abelian gauge theories. *Phys. Rev.* 155 (5): 1554.

55 Kao, P.C., Liu, C.C., and Li, T.Y. (2015). Nonvolatile memory and opto-electrical characteristics of organic memory devices with zinc oxide nanoparticles embedded in the tris(8-hydroxyquinolinato)aluminum light-emitting layer. *Org. Electron.* 21: 203–209.

56 Zhang, Q., Pan, J., Yi, X. et al. (2012). Nonvolatile memory devices based on electrical conductance tuning in poly(N-vinylcarbazole)-graphene composites. *Org. Electron.* 13 (8): 1289–1295.

57 Bhattacharjee, S., Das, U., Sarkar, P.K., and Roy, A. (2018). Stable charge retention in graphene-MoS$_2$ assemblies for resistive switching effect in ultra-thin super-flexible organic memory devices. *Org. Electron.* 58: 145–152.

58 Hosseini, N.R. and Lee, J.S. (2015). Biocompatible and flexible chitosan-based resistive switching memory with magnesium electrodes. *Adv. Funct. Mater.* 25 (35): 5586–5592.

59 Son, D.I., Kim, T.W., Shim, J.H. et al. (2010). Flexible organic bistable devices based on graphene embedded in an insulating poly(methyl methacrylate) polymer layer. *Nano Lett.* 10 (7): 2441–2447.

60 Chiu, F.C. (2014). Conduction mechanisms in resistance switching memory devices using transparent boron doped zinc oxide films. *Materials (Basel)* 7 (11): 7339–7348.

61 Wu, C., Li, F., and Guo, T. (2014). Efficient tristable resistive memory based on single layer graphene/insulating polymer multi-stacking layer. *Appl. Phys. Lett.* 104 (18): 1–6.

62 Lampert, M.A., Rose, A., and Smith, R.W. (1959). Space-charge-limited currents as a technique for the study of imperfections in pure crystals. *J. Phys. Chem. Solids* 8 (C): 464–466.

63 Wu, C., Li, F., Guo, T. et al. (2011). Efficient nonvolatile rewritable memories based on three-dimensionally confined Au quantum dots embedded in ultrathin polyimide layers. *Jpn. J. Appl. Phys.* 50 (3): 2–5.

64 Chen, X., Hu, W., Li, Y. et al. (2016). Complementary resistive switching behaviors evolved from bipolar TiN/HfO$_2$/Pt device. *Appl. Phys. Lett.* 108 (5): 053504.

65 Wu, C., Li, F., Zhang, Y., and Guo, T. (2012). Recoverable electrical transition in a single graphene sheet for application in nonvolatile memories. *Appl. Phys. Lett.* 100 (4): 2010–2014.

66 Lin, W.P., Liu, S.J., Gong, T. et al. (2014). Polymer-based resistive memory materials and devices. *Adv. Mater.* 26 (4): 570–606.

67 Ling, Q.D., Liaw, D.J., Zhu, C. et al. (2008). Polymer electronic memories: materials, devices and mechanisms. *Prog. Polym. Sci.* 33 (10): 917–978.

68 Paquin, F., Rivnay, J., Salleo, A. et al. (2015). Multi-phase semicrystalline microstructures drive exciton dissociation in neat plastic semiconductors. *J. Mater. Chem. C* 3: 10715–10722.

69 Bhansali, U.S., Khan, M.A., Cha, D. et al. (2013). Metal-free, single-polymer device exhibits resistive memory effect. *ACS Nano* 7 (12): 10518–10524.

70 Paul, S., Kanwal, A., and Chhowalla, M. (2006). Memory effect in thin films of insulating polymer and C_{60} nanocomposites. *Nanotechnology* 17 (1): 145–151.

71 Ji, Y., Lee, S., Cho, B. et al. (2011). Flexible organic memory devices with multilayer graphene electrodes. *ACS Nano* 5 (7): 5995–6000.

72 Ji, Y., Zeigler, D.F., Lee, D.S. et al. (2013). Flexible and twistable non-volatile memory cell array with all-organic one diode-one resistor architecture. *Nat. Commun.* 4 (1): 1–7.

73 Gao, S., Song, C., Chen, C. et al. (2013). Reply to "comment on 'dynamic processes of resistive switching in metallic filament-based organic memory devices'". *J. Phys. Chem. C* 117 (22): 11881–11882.

74 Cho, B., Yun, J.M., Song, S. et al. (2011). Direct observation of Ag filamentary paths in organic resistive memory devices. *Adv. Funct. Mater.* 21 (20): 3976–3981.

75 Gao, S., Song, C., Chen, C. et al. (2013). Formation process of conducting filament in planar organic resistive memory. *Appl. Phys. Lett.* 102 (14): 141606.

76 Krishnan, K., Tsuruoka, T., Mannequin, C., and Aono, M. (2016). Mechanism for conducting filament growth in self-assembled polymer thin films for redox-based atomic switches. *Adv. Mater.* 28 (4): 640–648.

77 Ali, S., Bae, J., Lee, C.H. et al. (2015). All-printed and highly stable organic resistive switching device based on graphene quantum dots and polyvinylpyrrolidone composite. *Org. Electron.* 25: 225–231.

78 Kim, D.H., Kim, W.K., Woo, S.J. et al. (2017). Highly-reproducible nonvolatile memristive devices based on polyvinylpyrrolidone: graphene quantum-dot nanocomposites. *Org. Electron.* 51: 156–161.

79 Zhang, X., Lai, Z., Liu, Z. et al. (2015). A facile and universal top-down method for preparation of monodisperse transition-metal dichalcogenide nanodots. *Angew. Chem.* 127 (18): 5515–5518.

80 Rehman, M.M., Siddiqui, G.U., Gul, J.Z. et al. (2016). Resistive switching in all-printed, flexible and hybrid MoS_2-PVA nanocomposite based memristive device fabricated by reverse offset. *Sci. Rep.* 6: 1–10.

81 Sanchez, C. and Lebeau, B. (1996). Hybrid organic-inorganic materials with second-order optical nonlinearities synthesized via sol-gel chemistry. *Pure Appl. Optics J. Eur. Opt. Soc. A* 5 (5): 689.

82 Wang, C., Zheng, W., Ji, S. et al. (2018). Identity-based fast authentication scheme for smart mobile devices in body area networks. *Wirel. Commun. Mob. Comput.* 2018: 1–6.

83 Wortham, E., Zorko, A., Arcon, D., and Lappas, A. (2002). Organic-inorganic perovskites for magnetic nanocomposites. *Physica B* 318 (4): 387–391.

84 Jung, J., Kwon, D., Jung, H. et al. (2018). Multistate resistive switching characteristics of ZnO nanoparticles embedded polyvinylphenol device. *J. Ind. Eng. Chem.* 64: 85–89.

85 Bhattacharjee, S., Sarkar, P.K., and Roy, A. (2016). Polyvinyl-alcohol based devices with highly conductive, optically active boron-doped ZnO nanoparticles for efficient resistive-switching at ultralow operating voltage. *Superlattices Microstruct.* 100: 1057–1063.

86 Li, L. and Wen, D. (2016). Memory behavior of multi-bit resistive switching based on multiwalled carbon nanotubes. *Org. Electron.* 34: 12–17.

87 Khan, M.U., Hassan, G., Raza, M.A. et al. (2019). Schottky diode based resistive switching device based on ZnO/PEDOT:PSS heterojunction to reduce sneak current problem. *J. Mater. Sci. - Mater. Electron.* 30 (5): 4607–4617.

88 Kim, T.Y., Anoop, G., Son, Y.J. et al. (2018). Ferroelectric-mediated filamentary resistive switching in P(VDF-TrFE)/ZnO nanocomposite films. *Phys. Chem. Chem. Phys.* 20 (23): 16176–16183.

89 Kaur, R. and Tripathi, S.K. (2018). Dopant dependent electrical switching characteristics of a CdSe- poly(vinyl-pyrrolidone) nanocomposite. *Org. Electron.* 61: 235–241.

90 Khan, M.U., Hassan, G., and Bae, J. (2019). Non-volatile resistive switching based on zirconium dioxide: poly(4-vinylphenol) nano-composite. *Appl. Phys. A* 125 (6): 1–11.

91 Jyoti, Kaur, R., Singh, S. et al. (2019). Effect of TiO_2 concentration on the non-volatile memory behavior of TiO_2-PVA polymer nanocomposites. *J. Electron. Mater.* 48 (9): 5995–6002.

92 Ukakimaparn, P., Chantarawong, D., Songkeaw, P. et al. (2019). Electrical bistable properties of P-25 TiO_2 nanoparticles composited with PVP for memory devices. *J. Electron. Mater.* 48 (10): 6792–6796.

93 Siddiqui, G.U., Rehman, M.M., and Choi, K.H. (2016). Enhanced resistive switching in all-printed, hybrid and flexible memory device based on perovskite $ZnSnO_3$ via PVOH polymer. *Polymer (Guildf)* 100: 102–110.

94 Nguyen, T.P. (2011). Polymer-based nanocomposites for organic optoelectronic devices. A review. *Surf. Coat. Technol.* 206 (4): 742–752.

95 Patil, U.V., Ramgir, N.S., Karmakar, N. et al. (2015). Room temperature ammonia sensor based on copper nanoparticleintercalated polyaniline nanocomposite thin films. *Appl. Surf. Sci.* 339 (1): 69–74.

96 Feng, X., Zhang, G., Zhuo, S. et al. (2016). Dual responsive shape memory polymer/clay nanocomposites. *Compos. Sci. Technol.* 129: 53–60.

97 Woo, S., Lee, S.J., Kim, D.H. et al. (2014). Conducting polymer/in-situ generated platinum nanoparticle nanocomposite electrodes for low-cost dye-sensitized solar cells. *Electrochim. Acta* 116: 518–523.

98 Gambardella, A., Prezioso, M., and Cavallini, M. (2014). Tunnel conductivity switching in a single nanoparticle-based nano floating gate memory. *Sci. Rep.* 4: 1–5.

99 Meena, J.S., Sze, S.M., Chand, U., and Tseng, T.Y. (2014). Overview of emerging nonvolatile memory technologies. *Nanoscale Res. Lett.* 9 (1): 1–33.

100 Kim, Y.N., Yun, D.Y., Arul, N.S., and Kim, T.W. (2015). Carrier transport mechanisms of multilevel nonvolatile memory devices with a floating gate consisting of hybrid organic/inorganic nanocomposites. *Org. Electron.* 17 (1): 270–274.

101 Wu, W., Han, S.T., Venkatesh, S. et al. (2018). Biodegradable skin-inspired nonvolatile resistive switching memory based on gold nanoparticles embedded alkali lignin. *Org. Electron.* 59: 382–388.

102 Liu, L., Lu, K., Yan, D. et al. (2018). Enlarged memory margins for resistive switching devices based on polyurethane film due to embedded Ag nanoparticles. *Solid State Electron.* 147: 6–12.

103 Kaur, R., Kaur, J., and Tripathi, S.K. (2015). Effect of Ag doping and insulator buffer layer on the memory mechanism of polymer nanocomposites. *Solid State Electron.* 109: 82–89.

104 Tripathi, S.K., Kaur, R., and Jyoti (2016). Effect of CuPc layer insertion on the memory performance of CdS nanocomposite diodes. *Mater. Sci. Eng., B* 211: 7–12.

105 Kaur, R., Singh, J., and Tripathi, S.K. (2017). Incorporation of inorganic nanoparticles into an organic polymer matrix for data storage application. *Curr. Appl. Phys.* 17 (5): 756–762.

106 Qi, X., Pu, K.Y., Li, H. et al. (2010). Amphiphilic graphene composites. *Angew. Chem. Int. Ed.* 49 (49): 9426–9429.

107 Qi, X., Tan, C., Wei, J., and Zhang, H. (2013). Synthesis of graphene-conjugated polymer nanocomposites for electronic device applications. *Nanoscale* 5 (4): 1440–1451.

108 Kim, W.K., Wu, C., and Kim, T.W. (2018). Effect of a PEDOT:PSS modified layer on the electrical characteristics of flexible memristive devices based on graphene oxide:polyvinylpyrrolidone nanocomposites. *Appl. Surf. Sci.* 444: 65–70.

109 Kim, W.K., Wu, C., Lee, D.U. et al. (2018). Enhancements of the memory margin and the stability of an organic bistable device due to a graphene oxide:mica nanocomposite sandwiched between two polymer (9-vinylcarbazole) buffer layers. *Appl. Surf. Sci.* 429: 231–236.

110 Lai, Y.C., Wang, D.Y., Huang, I.S. et al. (2013). Low operation voltage macromolecular composite memory assisted by graphene nanoflakes. *J. Mater. Chem. C* 1 (3): 552–559.

111 Choi, K.H., Ali, J., and Na, K.H. (2015). Fabrication of graphene-nanoflake/poly(4-vinylphenol) polymer nanocomposite thin film by electrohydrodynamic atomization and its application as flexible resistive switching device. *Physica B* 475: 148–155.

112 Sun, Y., Wen, D., Bai, X. et al. (2017). Ternary resistance switching memory behavior based on graphene oxide embedded in a polystyrene polymer layer. *Sci. Rep.* 7 (1): 1–11.

113 Pang, L. and Ni, X. (2018). The preparation of water-dispersible graphene oxide/conjugated polymer nanocomposite and the nonvolatile resistive switching memory properties. *J. Mater. Sci. - Mater. Electron.* 29 (3): 2007–2015.

114 Ban, C., Wang, X., Zhou, Z. et al. (2019). A universal strategy for stretchable polymer nonvolatile memory via tailoring nanostructured surfaces. *Sci. Rep.* 9 (1): 1–7.

115 Gogoi, K.K. and Chowdhury, A. (2019). Electric field induced tunable memristive characteristics of exfoliated graphene oxide embedded polymer nanocomposites. *J. Appl. Phys.* 126 (2): 025501.

116 Kannan, V., Kim, H.S., and Park, H.C. (2016). Non-volatile resistive memory device fabricated from CdSe quantum dot embedded in thermally grown In_2O_3 nanostructure by oblique angle deposition. *Phys. Lett. A* 380 (44): 3743–3747.

117 Geller, M., Marent, A., Nowozin, T. et al. (2008). A write time of 6 ns for quantum dot-based memory structures. *Appl. Phys. Lett.* 92 (9): 2006–2009.

118 Kannan, E.S., Kim, G.H., and Ritchie, D.A. (2009). Memory characteristics of InAs quantum dots embedded in GaAs quantum well. *Appl. Phys. Lett.* 95 (14): 93–96.

119 Fischbein, M.D. and Drndic, M. (2005). CdSe nanocrystal quantum-dot memory. *Appl. Phys. Lett.* 86 (19): 1–3.

120 Zhou, Z., Mao, H., Wang, X. et al. (2013). Transient and flexible polymer memristors utilizing full-solution processed polymer nanocomposites. *Nanoscale* 10 (31): 14824–14829.

121 Ooi, P.C., Haniff, M.A.S.M., Wee, M.F.M.R. et al. (2019). Electrical transportation mechanisms of molybdenum disulfide flakes-graphene quantum dots heterostructure embedded in polyvinylidene fluoride polymer. *Sci. Rep.* 9 (1): 1–8.

122 Kim, T.W., Yang, Y., Li, F., and Kwan, W.L. (2012). Electrical memory devices based on inorganic/organic nanocomposites. *NPG Asia Mater.* 4 (6): 1–12.

123 Gao, S., Yi, X., Shang, J. et al. (2019). Organic and hybrid resistive switching materials and devices. *Chem. Soc. Rev.* 48 (6): 1531–1565.

124 Huang, J. and Ma, D. (2014). Electrical switching and memory behaviors in organic diodes based on polymer blend films treated by ultraviolet ozone. *Appl. Phys. Lett.* 105 (9): 093303.

125 Khan, M.U., Hassan, G., Raza, M.A., and Bae, J. (2018). Bipolar resistive switching device based on N,N′-bis(3-methylphenyl)-N,N′-diphenylbenzidine and poly(3,4-ethylenedioxythiophene):poly(styrene sulfonate)/poly(vinyl alcohol) bilayer stacked structure. *Appl. Phys. A* 124 (10): 726.

126 Choi, M.K., Kim, W.K., Sung, S. et al. (2018). Flexible memristive devices based on polyimide:mica nanosheet nanocomposites with an embedded PEDOT:PSS layer. *Sci. Rep.* 8 (1): 1–8.

127 Rehman, M.M., Rehman, H.M.M.U., Gul, J.Z. et al. (2020). Decade of 2D-materials-based RRAM devices: a review. *Sci. Technol. Adv. Mater.* 21 (1): 147–186.

128 Huang, X., Zeng, Z., and Zhang, H. (2013). Metal dichalcogenide nanosheets: Preparation, properties and applications. *Chem. Soc. Rev.* 42 (5): 1934–1946.

129 Hwang, B. and Lee, J.S. (2019). Recent advances in memory devices with hybrid materials. *Adv. Electron. Mater.* 5 (1): 1–22.

130 Siddiqui, G.U., Rehman, M.M., Yang, Y.J., and Choi, K.H. (2017). A two-dimensional hexagonal boron nitride/polymer nanocomposite for flexible resistive switching devices. *J. Mater. Chem. C* 5 (4): 862–871.

131 An, H., Kim, W.K., Wu, C., and Kim, T.W. (2018). Highly-stable memristive devices based on poly(methylmethacrylate): CsPbCl3 perovskite quantum dot hybrid nanocomposites. *Org. Electron.* 56: 41–45.

132 Rajan, K., Roppolo, I., Bejtka, K. et al. (2018). Performance comparison of hybrid resistive switching devices based on solution-processable nanocomposites. *Appl. Surf. Sci.* 443: 475–483.

133 Chaudhary, D., Munjal, S., Khare, N., and Vankar, V.D. (2018). Bipolar resistive switching and nonvolatile memory effect in poly (3-hexylthiophene) –carbon nanotube composite films. *Carbon (NY)* 130: 553–558.

134 Liu, T., Wu, W., Liao, K.N. et al. (2019). Fabrication of carboxymethyl cellulose and graphene oxide bio-nanocomposites for flexible nonvolatile resistive switching memory devices. *Carbohydr. Polym.* 214: 213–220.

8.A Performance Comparison According to Device Material and Structure

Year	Device structure	Materials	V_{on}	On/off ratio	Retention time	Mechanism	References
2013	Ag/GNF:PVA/ITO	Mixture of graphene nanoflakes (GNFs) and insulating PVA	+1.4	~10^2	>10^4	TLSCLC-Ohmic	[110]
2018	Ag NWs (TOP)/CA QDs PVP/Ag NWs/PVP(BE)	Citric acid quantum dot (CA QD)-polyvinyl pyrrolidone (PVP) and Ag NWs	1	2×10^4	10^4	Ohmic – SCLC	[120]
2014	ITO/PEDOT:PSS:PVA/Al	PEDOT:PSS doped with PVA	1.1–1.8	~10^2	36×10^4	Ohmic– Frenkel–Poole emission model	[124]
2013	Ag/G-QDs : PVP/Ag	G-QDs and PVP	1.8	—	30	Ohmic – SCLC	[77]
2015	Ag/Ag:CdSe NC/PVA buffer layer Al	CdSe: Ag NC, PVA as buffer layer	—	6.6×10^3	—	Ohmic- SCLC	[103]
2015	Ag/Graphene: PVP NC/ITO	Graphene: PVP NC	+2	—	>3600 seconds	—	[111]
2016	Al/CuPc/CdS NC/Ag	CdS nanocomposite Polystyrene (PS), CuPc layer	1–2	~1.4×10^4	—	Schottky emission - SCLC	[104]
2011	Al/ZnO nanorods : PMMA/ITO	Ionized oxygen rich ZnO nanorods with PMMA	+1	5×10^3	10^3	Thermionic emission - SCLC	[43]
2016	Ni/MWCNTs(2 wt%): PMMA/ITO	Multiwalled carbon nanotubes (MWCNT) blended in PMMA	-2.25	>10^7	>10^6	Ohmic- SCLC	[86]
2013	PET/Ag/MoS$_2$-PVA/Ag	2D material, MoS$_2$ and PVA	3	1.28×10^2	10^5	Ohmic- TCSCLC	[80]
2016	Ag/PVOH-ZnSnO$_3$/Ag	Perovskite ZnSnO$_3$ with PVOH polymer	1.5	>10^2	10^5	Ohmic- TCSCLC	[93]
2016	Al/PVA/BZO/PVA/ITO	BZO sandwiched between two PVA polymer layers	1	10^2	1×10^4	Ohmic conduction	[85]
2011	Ag/Hbn-PVOH/ITO/PET	hBN and polymer PVOH.	0.78	4.8×10^2	—	Ohmic - TCSCLC	[130]
2017	Al /PS +GO/ITO	Polystyrene (PS) and graphene oxide (GO)	-1.55	—	2×10^5	Ohmic-thermionic emission - TCSCLC	[112]

Year	Device structure	Materials	V_{on}	On/off ratio	Retention time	Mechanism	References
2018	Al/GO-Pvim-Ppy/ITO	GO-Polyvinyl imidazole-polypyrrole (GO-Pvim-Ppy)	4.44	10^6	10^4	Ohmic- SCLC	[113]
2017	Al/(CdSe/PVK)/Ag	CdSe with PVK	1	$\sim 10^4$	—	Charge confinement mech.	[105]
2018	Al/PQDs -PMMA/ITO	$CsPbCl_3$ and PMMA	−0.3	2×10^4	1×10^4	Ohmic - SCLC	[131]
2018	Ag/Ti /$AgNO_3$_IL_PVDF-HFP/Ta Pt/SiO_2	$AgNO_3$_IL with PVDF-HFP	2.3	15	—	Filamentary conduction	[132]
2018	Ag/Ti/$AgNO_3$-IL_PEO/Ta/Pt/SiO_2	$AgNO_3$-IL with PEO	8.4	10^4	—	Filamentary conduction	[132]
2018	Ag/Ti/ $AgNO_3$-IL_PMMA /Ta /Pt/SiO_2	$AgNO_3$_IL with PMMA	6.5	25	—	Filamentary conduction	[132]
2018	Ag/Ti/$AgNO_3$_IL_PVDF-HFP+PEO/Ta/Pt/SiO_2	$AgNO_3$_IL with PVDF-HFP+PEO	4.5	10^3	—	Filamentary conduction	[132]
2018	FTO/P3HT-CNT/Al	CNT and P3HT	1.8	$>10^2$	$>10^3$	Ohmic – SCLC	[133]
2018	Ag/Ag: CdSe-PVP/Al	Ag with CdSe-PVP	1.4	4×10^2	—	SCLC	[89]
2018	Al/GO:PVP/PEDOT: PSS/ITO/PEN	GO with polymer PVP & PEDOT: PSS as modified layer	−0.2 to −0.7	10^2	10^4	Ohmic – SCLC	[108]
2018	Al/polyurethane + Ag/ITO/glass	Ag nanoparticles in polyurethane	−0.85	10^5	1.8×10^4	Drift – SCLC	[102]
2018	Au/P(VDF-TrFE) + ZnO/Si	P(VDF-TrFE) with ZnO	−2.74	2×10^7	10^4	Ohmic – SCLC – TCLC	[88]
2018	Al/ZnO-PVA/PEDOT: PSS/Al/PET	ZnO in PVA & PEDOT:PSS on PET	3.6	3×10^5	>1920	Ohmic – SCLC	[47]
2018	Al/PI-mica/PEDOT: PSS/ITO	Mica in PI	−0.3	4.28×10^3	1×10^4	Ohmic – SCLC	[126]

Year	Device structure	Materials	V_{on}	On/off ratio	Retention time	Mechanism	References
2018	Al/PVK/GO:mica/PVK/ITO	GO:mica with PVK	−0.5	2×10^4	1×10^4	Thermal emission – Ohmic conduction	[109]
2018	Al/Au NPs:lignin/Al	Au NP in lignin	4.7	$>10^4$	$>10^3$	Thermionic emission-SCLC – ohmic	[101]
2018	ITO/ZnO NPs + PVP/Al	ZnO NPs in PVP	1.2	—	—	Ohmic – SCLC	[84]
2011	Cu/gMoS$_2$:PMMA/ITO	G MoS$_2$ in PMMA	2	$>10^4$	10 days	Thermionic emission – SCLC	[57]
2019	Al/TiS$_2$-PVP/ITO/PET	2D TiS$_2$ nanoflakes and PVP	<2 V	10^2	10^4	Schottky emission–Poole–Frenkel emission – SCLC	[53]
2019	r-rGO/PVK/Al	PVK and reduced graphene oxide films are used active layer and cathode	1.1 V	10^4	12 000 seconds	Ohmic – SCLC – conducting filaments	[114]
2019	FTO/TiO$_2$-PVA/Ag	TiO$_2$ PNC	<2 V	10^3	—	Conductive filament – oxygen vacancies	[91]
2019	ITO/PMMA-GOs/Al	GO embedded PMMA	1.87 V	10^4	10^4 seconds	Trap charge limited conduction–Ohmic conduction	[115]
2018	ITO/PVP:TiO$_2$ NPs/Al	P-25 TiO$_2$ NPs in PVP	1 V	10^5	>20 000 seconds	Thermionic emission – Ohmic	[57]
2019	Al/CMC-GO/Al/SiO$_2$	Cellulose and its derivatives	2.22 V	6×10^5	>10 000 seconds	SCLC – Ohmic	[134]
2019	Ag NWs/PVDF/GMP-NC/PVDF/ITO/GLASS	GQD-MoS$_2$-PVDF nanocomposites	−0.65/0.8	10^7	10^4	Schottky emission-PF emission-TCLC – Ohmic	[121]
2016	Ag/ZrO$_2$:PVP/ITO PET	Zirconium dioxide (ZrO$_2$) and poly(4-vinylphenol) (PVP)	±1.5	44	30	Ohmic – SCLC	[90]
2018	Ag/PEDOT:PSS/ZnO/ITO	PEDOT:PSS doped with PVA	3	530	30	Ohmic – SCLC	[87]
2019	Ag/F8BT:PVP/ITO PET	(F8BT) and Polyvinylpyrrolidone (PVP) composite	±1.5 V	274.31	60	Ohmic – SCLC	[35]

9

Polymer Nanocomposites for Temperature Sensing and Self-regulating Heating Devices

Yi Liu[1], Han Zhang[2,3], and Emiliano Bilotti[2,3]

[1] Loughborough University, Department of Materials, Loughborough LE11 3TU, UK
[2] Queen Mary University of London, School of Engineering and Materials Science, Mile End Road, London E1 4NS, UK
[3] Queen Mary University of London, Nanoforce Technology Ltd., Joseph Priestley Building, Mile End Road, London E1 4NS, UK

9.1 Introduction

In the development of integrated electronic devices with various functionalities, there is an increasing demand for technological progress in new and intelligent materials and devices able to respond to external stimuli spontaneously. Polymer nanocomposites are one of the most promising materials owing to their lightweight structure, ease of processing, and multifunctionality [1, 2]. By incorporating a filler into a polymer matrix, the characteristics of the polymer matrix, such as low density and flexibility, can be combined with other physical properties of the filler [3]. Benefiting from numerous choices of polymer matrices and conductive fillers, conductive polymer composites (CPCs) can be precisely tuned and controlled to fulfill the complete spectrum of properties for various applications. Specific CPCs can show a pyroresistive behavior ("pyro" is a Greek word meaning "fire" or "heat"), with the electrical resistivity changing with temperature [4]. Pyroresistivity can manifest itself with a positive temperature coefficient (PTC) effect or negative temperature coefficient (NTC) effect. PTC effect refers to an increase in electrical resistivity with increasing temperature, mostly a sharp jump by several orders of magnitude in resistivity at the critical temperature. NTC stands for a decrease in electrical resistivity with increasing temperature [5, 6]. This PTC phenomenon was first discovered by Frydman in 1948 [7]. However, it had not drawn much attention until 1966, when Kohler demonstrated the significance of obvious PTC effect in carbon black (CB) doped high density polyethylene (HDPE), patented under the name of "resistance element" [8].

The change in resistivity is defined by the PTC intensity, which is quantified as the maximum resistivity (ρ_{max}) over the room temperature resistivity (ρ_{RT}) [4]. The temperature at which there is a sudden change in resistivity is defined as the switch-

Polymer Nanocomposite Materials: Applications in Integrated Electronic Devices, First Edition.
Edited by Ye Zhou and Guanglong Ding.
© 2021 WILEY-VCH GmbH. Published 2021 by WILEY-VCH GmbH.

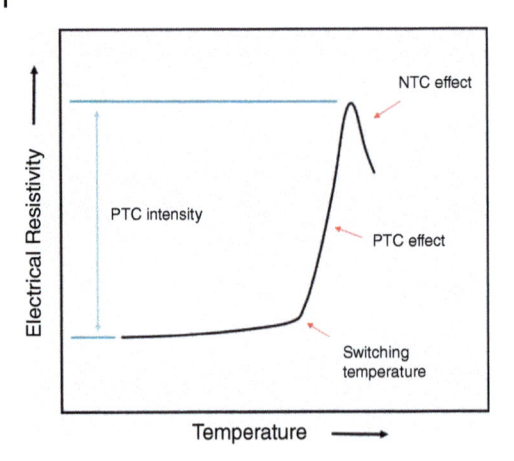

Figure 9.1 Schematic illustration of pyroresistive curve, including PTC effect, NTC effect, PTC intensity, and switching temperature.

ing temperature as illustrated in Figure 9.1. NTC effect in most cases happens after PTC effect.

With the commercial attractive features of PTC effect, pyroresistive CPCs have been utilized in a wide range of industrial application, ranging from temperature sensors, self-regulating heaters, safety fuses/current protection devices, and safety switch for batteries [4, 9]. Recent advances on PTC studies have enabled more flexibility, reproducibility, and sensitivity on polymer nanocomposite materials, especially on integrated electronic devices. This chapter summarizes several theories of PTC effect in CPCs, as well as influencing factors including conductive fillers and polymer matrices. Both traditional use and recent progress of PTC based temperature sensors and self-regulating heaters are discussed, with further exploration of some applications.

9.2 Conducting Mechanism and Percolation Theory

The CPC conductivity dependence on the filler concentration is often described by percolation theory. The electrical conductivity of CPCs experiences a dramatic change when the filler concentration reaches a critical value, from insulating to electrically conductive or semi-conductive when the initial conducting channels are formed [10, 11]. The critical filler concentration needed to form an initial conductive path is called the percolation threshold (φ_c). Two of the early statistical percolation models that have been often referenced were originally proposed by Kirkpatrick (1973) and Zallen (1983) [12, 13]. They used finite arrays of points connected by bonds to predict the percolation threshold. The model followed a power-law equation

$$\sigma = \sigma_0(\varphi - \varphi_c)^t$$

where σ is the conductivity of CPC, σ_0 is a scaling factor, φ is the filler content, φ_c is the percolation threshold, and t is the critical exponent, which is expected to depend on the conductive system dimensionality only.

Figure 9.2 Schematic of percolation behavior of conductive polymer composite. Source: Gulrez et al. [14].

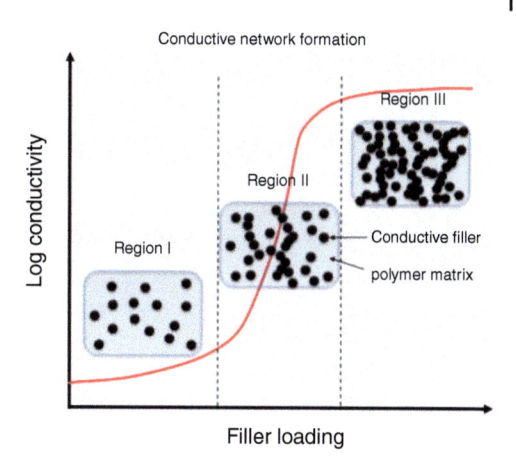

The plot of log of conductivity vs. filler loading presents a typical S-shape as shown in Figure 9.2. There are three typical regions in this S-shape curve, marked as Region I, II, and III. At low filler loadings, before the conductive network forms, the conductivity remains close to that of the pure insulating polymer matrix. The fillers are present only individually or in small clusters throughout the polymer matrix, with no physical contact or electron tunneling between conductive particles to allow charge transport (Region I). With rising in filler concentration, the distance between the fillers reduces. As the critical value of conductive filler is reached, the conductivity of the composite increases dramatically by many orders of magnitudes within a narrow filler loading range (Region II) [15]. It is believed that the fillers form network at φ_c, and electrical field assisted tunneling between adjacent fillers was suggested by Van Beek and Van Pul when the mean particle distance is below 10 nm [16]. Alternatively, Frenkel suggested an electrical field assisted hopping mechanism in his early review [17]. With further increase in the filler loading, the conductivity reaches a plateau and no major change in conductivity can be observed (Region III) [10].

9.3 PTC Theory

It is generally believed that the disruption of the conductive network during heating of the composites causes the PTC effect. Although many models and theories have been proposed in the attempt to explain PTC effect in the CPCs, a model that can exhaustively explain the phenomenon is not yet available. The true mechanism for the PTC effect is not well established since most of these models cannot be verified experimentally. Nevertheless, this section aims to summarize the existing models, to provide a brief overview of the PTC theory for CPCs.

The main models to explain PTC effect include conductive chain and thermal expansion model by Kohler [8], tunneling current model by Ohe and Naito [5], filler congregation and migration model by Klason and Kubát [18], Ohm conductance and phase change model by Allak et al. [19], and internal stress model by Mamunya et al. [20]. The summary of these models is shown in Table 9.1.

Nevertheless, there is no satisfactory theory to explain the PTC and NTC phenomena both theoretically and experimentally. Most of these models suggest that the volume expansion of matrix plays an important role in PTC behavior. The explanation based on tunneling effect is widely accepted that the rapid expansion close to the transition temperature of polymer matrix increases the width of gaps and thus hinders the process of electron tunneling. In other words, the "effective conductive filler loading" is affected by the volume change of the matrix upon heating, resulting in a lower "effective loading" level below the percolation threshold to provide the electron pathways. On the other hand, NTC effect is presumably due to the reaggregation of conductive particles in polymer melting state and therefore reparation of disconnected conductive pathways [9].

9.4 Main Factors Influencing the PTC Effect

The polymer nanocomposites with PTC behavior belong to the family of CPCs, with the polymer matrix as the framework of composites and the filler within the matrix to establish the conductive pathways. Therefore, the electrical conductivity of the nanocomposites is sensitively influenced by the "robustness" of conductive network formed by filler in the matrix. In this section, several influencing factors ranging from fillers to matrices are discussed and analyzed, in order to design and control the PTC behaviors in CPCs.

9.4.1 Effect of Filler Size and Shape

Many researchers have found that the CPCs with larger average filler size exhibit a higher PTC intensity and higher room temperature resistivity than those CPCs with the same filler content but smaller filler size [23]. According to both Ota et al. and Jing et al., the interparticle distance between particles increases with particle size [24, 25]. The number of contact points is lower when larger particles are used as fillers.

It is believed that the PTC behavior of conductive composites is a function of the "robustness" of the conductive networks or pathways formed within the polymer matrix [26, 27]. The reason for this difference in PTC intensity lies in the fact that when the number of conductive pathways is smaller, the resistivity of the composite is more sensitive to small variation of the conductive network. For instance, in CB filled HDPE composites, at a given filler loading, a more robust conductive network is formed with smaller CB fillers due to the increase in the number of CB in contact with each. Therefore, the thermal volume expansion of the polymer matrix has a weaker effect on the conductive network formed by smaller CB filler size and few conductive pathways are broken, leading to a low PTC intensity [28]. Recently, Asare et al. also proved the effect of conductive filler size on PTC behaviors using silver coated glass spheres (AgS) as model conductive fillers with well controlled uniform

Table 9.1 Summary of PTC models, their explanation, and defects.

Model	Explanation	Limitations
Thermal expansion of polymer and conductive chain model by Kohler [8]	At low temperature, the conductive network is presented, and polymer is in semi-crystalline state	The reason that the resistance change is not a function of volume change
	At switching temperature, the polymer phase transits from crystalline to amorphous ⇒ volume is expanded ⇒ conductive chains are broken	Cannot explain the NTC effect after PTC transition
Electron tunneling conductance model by Ohe and Naito [5]	The curve of resistivity and field of PE/CB composites according to the theory equation between tunneling current and field has been calculated and is consistent with experiment	No explanation for the distribution of grain gaps should become more random
	At low temperature, distribution of inter-grain gap is uniform, and gaps are small enough for extensive tunneling to occur	No correlation of resistance drops quite rapidly after crystalline melting
	At switching temperature, distribution becomes random ⇒ gaps between conductive fillers increase to break the electron tunnels	
Electron tunneling conductance advanced model by Meyer [21, 22]	When a crystalline polymer is melted and loaded with finely divided conductive particles, the particles are swept by advancing crystallite fronts into the amorphous regions between crystallites on cooling	It rests on an assumption that thin films of crystalline polymers are highly conductive
	The crystallite change in melting point leading to a sharp reduction of tunneling, which comes from expansion of the previously compressed carbon black masses to form a conductive network stretching through the system	

(continued)

Table 9.1 (Continued.)

Model	Explanation	Limitations
Filler congregation and migration model by Klason and Kubát [18]	At low temperature, CB structure is determined by crystalline phase of polymer, and CB particles distribute in amorphous phase with formed conductive chain	Lack of enough supporting experimental results
	After melting, the structure breaks up, causing a more homogenous particle distribution and a sharp increase in resistivity	
	At even higher temperature, a new CB structure is supposed to form, resulting in a decrease in resistivity	
Ohm conductance and phase change model by Allak et al. [19]	Logarithmic current-voltage plots were found to be linear with unity shapes at temperature above and below melting temperature, indicating good ohmic behavior	Lack of enough supportive experimental results
	It is assumed the conductive chains in the amorphous regions can easily establish throughout the material, and as the temperature close to melting, the crystallites become amorphous with large increasing in their volume. The inter-granular gaps between these newly formed and enlarged amorphous regions are reduced	
Internal stress model by Mamunya et al. [20]	During cooling of thermoplastic matrix, internal stresses appear. These stresses increase the pressure between adjacent particles, give the contact pressure, and decrease the contact resistance	Based on the assumption that particles are in a state of close packing with intimate contact to next neighbors, forming conducting paths throughout the composite
	Instead, the observed strong resistivity change at high temperature is caused by a release of particle contact pressure and a change in gap distance	

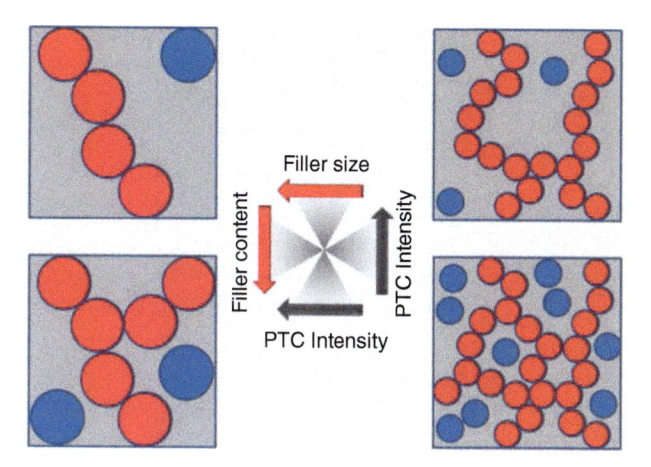

Figure 9.3 Schematic of how a reduction in filler size or an increase in filler content increases the number of conductive pathways in a conductive polymer composite, therefore improving the "robustness" of the network, leading to a smaller PTC intensity. Source: Asare et al. [26].

filler sizes [26]. They demonstrated that the PTC intensity increases with increasing filler size and with decreasing filler content, as illustrated in Figure 9.3.

In terms of filler shape, spherical as well as spiky shaped fillers are expected to be among the most popular candidates for PTC materials as there are less "contacting points" in the conductive network to break. However, the percolation threshold can be drastically reduced for particles with an aspect ratio larger than one, where there can be a huge increase in the connecting paths at relatively low filler contents and less disturbance to the original matrix properties.

9.4.2 Effect of Filler Dispersion and Distribution

The conductivity of the CPCs strongly depends on the dispersion and distribution of filler particles in the matrix, which can greatly influence the PTC behavior. Apart from the intrinsic nature of the materials, the processing method also greatly influences the dispersion state. Special attention should be paid on the filler dispersion and filler-matrix interface when using nanofillers in the composites [29]; the bottleneck is that as-produced fillers tend to be held together in bundles/agglomerates by van der Waals interactions and/or entanglements. Figure 9.4a shows, for instance, both a "bad" distribution and dispersion with no conductive pathway formed in the specific region of the composite investigated. It is of necessity to have a good dispersion of fillers to build a conductive network within polymers at low filler concentration, as a poor dispersion of fillers prohibits networks formation (Figure 9.4b). Figure 9.4c forms a conductive two-dimensional network with the preferential distribution of well dispersed fillers. While Figure 9.4d shows that at extremely low filler concentration, even the perfect distribution of well dispersed fillers cannot form a conductive pathway as the distance between the neighboring fillers inhibiting the conductive network formation [30]. The ideal case to form a good

(a) Bad distribution dispersion (agglomeration problem); no conductive path formation

(b) Good distribution but poor dispersion (agglomerates still there); no conductive path formation

(c) Bad distribution but good dispersion (no agglomerates); continuous conductive path for electron flow

(d) Good distribution and dispersion (no agglomerates); no conductive path formation

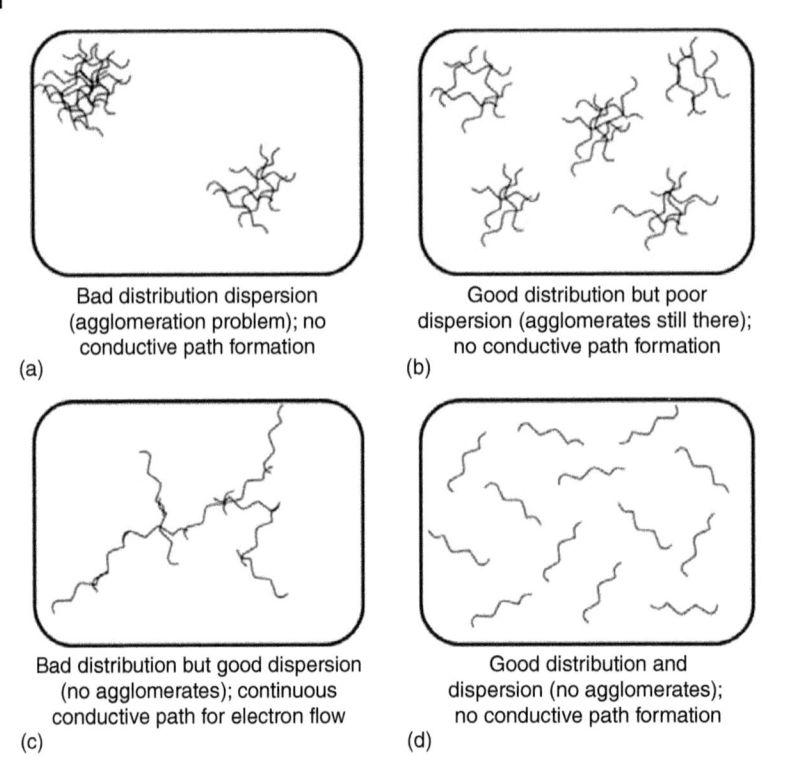

Figure 9.4 Schematic of dispersion and distribution effect of fillers on the conductivity of CPCs. (a) bad distribution and dispersion, (b) good distribution but poor dispersion, (c) bad distribution but good dispersion, (d) good distribution and dispersion. Source: Gulrez et al. [14].

PTC material is to have less contacting points between fillers while preserving high conductivity.

9.4.3 Effect of Mixed Filler

One of the common routes to enhance the room temperature (RT) conductivity and PTC behavior of CPCs is to introduce a second filler into the composites [31, 32]. Secondary fillers with high aspect ratio fillers such as CNT or CF have been explored the most [33]. The purpose of adding these kinds of particles is usually to bridge or connect the gaps between the primary particles, which span across insulating regions and establish conductive pathways more easily [34–36]. For instance, Lee et al. investigated the addition of CNTs on PTC effect of conventional HDPE/CB composites. The obtained results indicated that the PTC intensity and repeatability of the hybrid nanocomposites were dramatically improved by adding a small amount of CNTs. The initial resistivity of the materials decreased with increasing CNT content, while the PTC intensity decreases with further addition of CNT as shown in Figure 9.5a [34]. However, the opposite observation has also been reported. Some researchers claimed that the PTC intensity as well as the repeatability of the composite has been

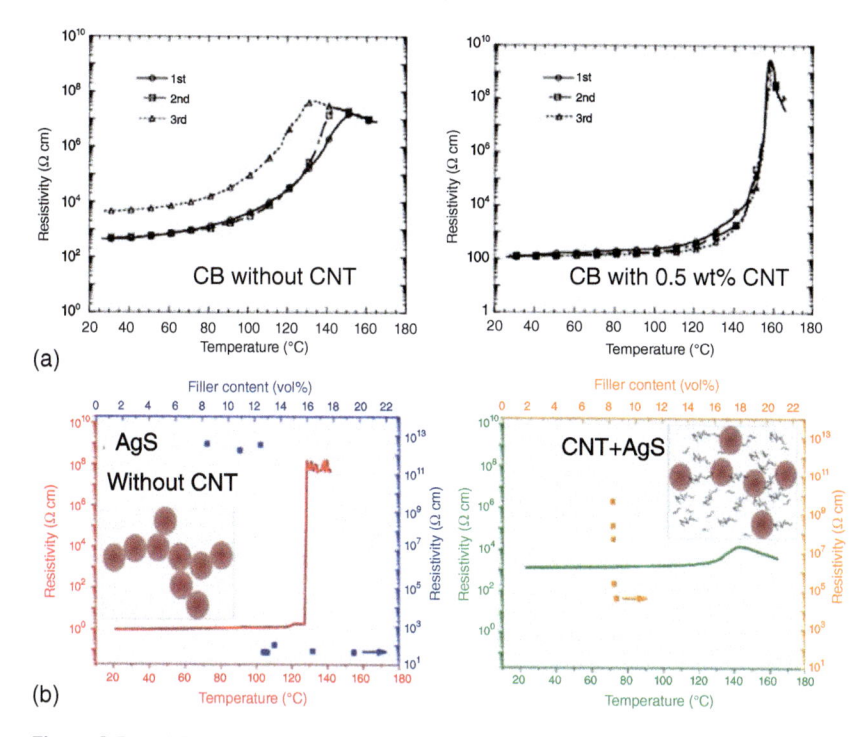

Figure 9.5 (a) Temperature dependence of electrical resistivity of the nanocomposites with CB content of 25 wt% for three heating cycles: without MWCNTs, 0.5 wt% MWNTs and 1.0 wt% MWNTs. (b) Pyroresistive behavior of HDPE/MWCNT composite with low PTC intensity; HDPE/AgS composite with high PTC intensity and mixed filler composites with low PTC intensity similar to HDPE/MWCNT composite. Source: (a) Lee et al. [34], (b) Asare et al. [37]. Licensed under CC-BY-4.0.

enhanced as discussed earlier, while others found out that the second filler had lowered the PTC intensity by several orders of magnitude due to the increased difficulty in breaking the conductive paths [37]. Asare et al. reported a huge decrease in PTC intensity even when a small amount of CNTs was added to HDPE/AgS composites, as shown in Figure 9.5b [37]. They believe that the PTC intensity of mixed filler composites is dominated by the filler with the lowest PTC intensity, even at very low loadings. These different outcomes of mixed fillers CPCs may come from the obvious distinct primary filler size, which might result in CNT having a different degree of influence.

9.4.4 Effect of Polymer Thermal Expansion and Crystallinity

Since it is acknowledged by many researchers that the thermal expansion of the polymer matrix is one of the dominating factors for PTC effect, it is very important to understand how the intrinsic characteristics of various types of polymer matrices are affecting the PTC intensity values. Horibe et al. compared HDPE, PP, and syndiotactic-PS and found that within the three semi-crystalline polymers, HDPE

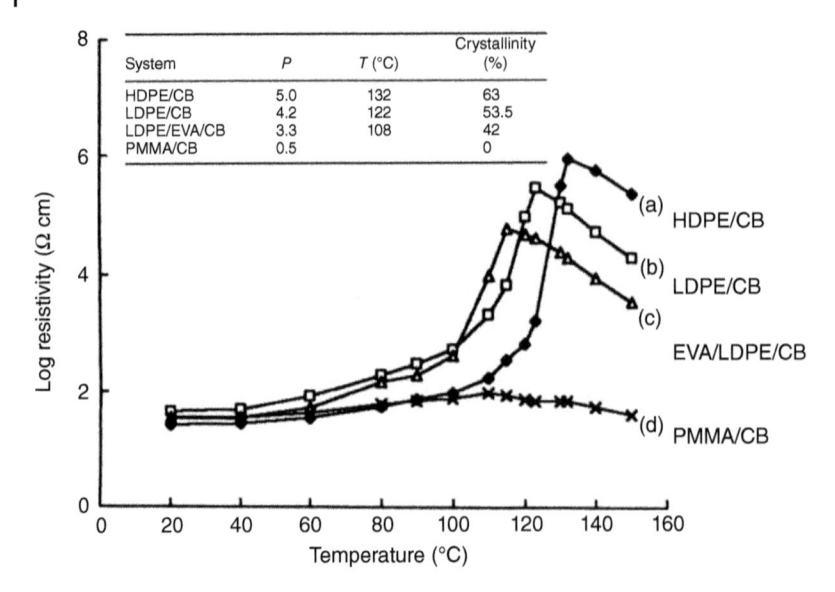

Figure 9.6 The dependence of PTC phenomenon on crystallinity of different systems: (a) HDPE/CB; (b) LDPE/CB; (c) EVA/LDPE/CB; (d) PMMA/CB. Source: Luo et al. [39].

System	P	T (°C)	Crystallinity (%)
HDPE/CB	5.0	132	63
LDPE/CB	4.2	122	53.5
LDPE/EVA/CB	3.3	108	42
PMMA/CB	0.5		0

exhibits the highest PTC intensity (4 order of magnitudes). They believed that this is due to the largest expansion of HDPE in polymer melting, and hence highest increase in distance between the CB filler particles upon polymer melting [38]. It is easy to understand that the larger expansion of the polymer is, the higher the chance to break up the conductive networks during polymer expansion. It is also generally acknowledged that the thermal expansion of semicrystalline polymers is larger than amorphous polymers due to the melt of crystal region. Therefore, it is expected that crystallinity plays an important role in PTC effect; in particular, the PTC intensity of semicrystalline polymer-based composites should be higher when compare with amorphous polymer-based composites.

In early studies, Luo et al. found that the polymer crystalline phase changed with temperature had a great influence on the electrical conductivity of CPCs [39]. They investigated polymer/CB composites with different crystallinity and showed that the higher the crystallinity of the polymer was, the higher the PTC intensity obtained (as shown in Figure 9.6). They associated this behavior on account of the larger expansion accompanying crystalline melting. For the amorphous polymer system (PMMA/CB), the PTC intensity is very weak, usually less than 1 order of magnitude, since it does not have a noticeable expansion phenomenon caused by crystalline melting [39]. This is in agreement with the work on CB/isomeric polymers by Meyer [22].

However, although crystallinity of polymers has been widely believed to be a dominating factor for PTC effect, it is not true for all cases presented. Some experiments show no correlation between PTC intensity and crystallinity [40–42]. For example, large PTC intensities (5 orders of magnitude) have been reported for amorphous CPC by Kar and Khatua (PMMA composite containing 40 wt% Ag coated glass beads) [43]. The same authors also reported that a nickel coated graphite (40 wt%) filled polycarbonate (PC) with a 3 orders of magnitude change in resistance. In simple words, the crystallinity of polymer matrix has a great influence on the PTC effect, but is not the only factor to determine the PTC phenomenon.

9.4.5 Effect of Polymer Transition Temperature

Based on experimental observations, the PTC phenomenon of CPCs occurs around a phase transition temperature (such as T_m, T_g) of polymer matrix, therefore thermal properties of polymers are also important factors, especially the transition temperature of the PTC effect. For semicrystalline polymers, both crystal regions and amorphous regions exist; hence three reversible stages/changes can be found with increased temperature. Those stages include glass transition in the amorphous regions, and crystallization and melting in the crystalline region. Each phase transition is associated with an obvious change in specific volume of polymers, which will affect the PTC phenomenon.

For example, HDPE/CB composites have a transition temperature at around 130 °C while poly(vinylidene fluoride) (PVDF)/CB switches at around 170 °C [39, 44]. For amorphous PC, the switching temperature is close to its glass transition temperature (150 °C) [45]. Based on the models proposed for pyroresistivity, the expansion caused by those thermal transitions will break up the conductive network formed within the composites, leading to a reduction in conductivity of the polymer composites. These results suggested that switch temperature can be predicted by the glass transition point/melting point of the polymer matrices. It is true for most of the cases that PTC effect happens at around melting temperature for semicrystalline polymers, but for amorphous polymers there are some exceptions. For instance, in a composite of PC and nickel coated graphite composite, a PTC effect well below the glass transition temperature was observed when polycaprolactone (PCL) was introduced [46].

9.4.6 Effect of Polymer Blend

Most previous studies on PTC effects were focused on composites containing only one single semi-crystalline polymer filled with CB. The introduction of a secondary polymer phase may influence the percolation threshold as well as the conductivity of the composite, which may subsequently affect the PTC behavior. Binary or ternary polymer blends-based CPCs have shown properties, which cannot be obtained by single polymer CPC. Several advantages have been proved by researchers [47, 48]. For example, localization of conductive particles within one of the phases, or at the interface of an immiscible polymer blend, has demonstrated that the percolation threshold can be substantially reduced [49]. Sufficiently high conductivity can hence be achieved at relatively low filler loading without sacrificing PTC intensity. Double PTC effect has been reported in CPCs with the conductive fillers dispersed at the interface between polymer phases. In the research work of Wei et al., a double PTC

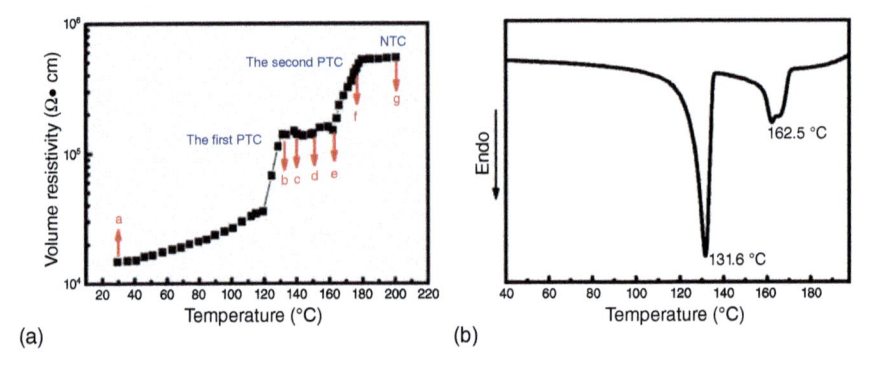

Figure 9.7 (a) Temperature–resistivity relationship of the CB/PP/UHMWPE composite in the heating process. The CB content is 0.94 vol%; (b) DSC curve for CB/PP/UHMWPE composites during the heating process with a CB content of 0.94 vol%. The heating rate was 2 °C min^{-1}. Source: Wei et al. [50].

effect was observed in binary polymer matrices of PP/UHMWPE, which effectively broadened the PTC region, as shown in Figure 9.7 [50]. Generally, in most PTC composites, after the presence of the first PTC peak, the resistivity of the CPC would normally decrease with increasing temperature, exhibiting the NTC effect. In this case, the presence of a second PTC effect at higher temperature can suppress the subsequent NTC effect.

The PTC intensity reported in the literature for polymer blends CPCs are usually higher than the values for a mono matrix CPCs. A reduction in filler content leads to a reduction in conductive pathways, hence resulting in a larger resistivity change when the networks are disrupted. Moreover, polymer blends can also influence the PTC intensity by varying the ratios between polymers in their blends. Some researchers reported that the PTC intensity can increase with an increase in polymer concentration of the matrix carrying the conductive filler. For example, Yu et al. showed that an increase in PTC intensity with increase in LDPE content in a CB/EVA/LDPE composite, where the CB particles preferred to locate in the LDPE phase [47]. They believed that the reduction in CB concentration is the main reason of increasing PTC intensity. However, the opposite has also been reported by other researchers. Xu et al. reported that an increase in the amount of the polymer phase containing the conductive filler reduced the PTC intensity of a mixed polymer composites [51]. Furthermore, for polymer blends, additional benefits are a wider processing window for the production of composite as a higher reproducibility [46, 52]. The composite properties can be adjusted as a function of blending techniques, as expected conventionally. Moreover, careful selection of the second polymer can help to eliminate the NTC effect without other treatment of the polymers [50].

In summary, addition of a second polymer provides new possibilities for tailoring the overall properties of a polymer-based PTC composite. The effect of using polymer blends on the PTC behavior will be discussed in more details in the following section.

9.5 Temperature Sensors

For some specific applications, such as monitoring the human body temperature, a relatively high temperature sensitivity is required. In this case, a switching type PTC thermistor is an ideal candidate. Bao's group at Stanford University has developed a flexible and wireless temperature sensor using nickel micro-particle filled polyethylene oxide (PEO)/PE blends. The PTC switching temperature is adjustable by carefully tuning the molecular weight of PEO (Figure 9.8a) [53]. Ni particles were processed to preferentially locate in the PEO phase of the polymer blend. The sensor showed a strong PTC effect within the temperature range of 35–42 °C according to PEO's transition temperature, which was accurate enough for human body temperature monitoring. A radio frequency identification (RFID) antenna was also integrated in the temperature sensor to enable wireless monitoring [53]. Similarly, Yokota et al. developed an ultra-flexible temperature sensor, in the range of 29.8–37.0 °C. They used semi-crystalline acrylate polymers and graphite, printed through stencil masks (Figure 9.8b). The device exhibited large resistance increase near body temperature with a fast response time and high repeatability (tested over 1800 thermal cycles) [54]. This method can also be used to produce a flexible temperature sensor sheet, allowing for large-area temperature mapping as shown in Figure 9.8c,d [54]. This demonstrates the possibility of very precise and large area temperature monitoring.

Apart from body temperature sensors, other applications require the measurements of relatively high temperatures (exceeding 100 °C) with high sensitivity, especially with the rapid development of electronic devices. For example, Zha et al. recently used modified CB- and MWCNT-filled HDPE to fabricate high performance PTC composites as flexible temperature resistivity sensors with a switching temperature of around 130 °C [55]. As mentioned in Section 9.4.5, the PTC switching temperature can be tuned by choosing appropriate polymer matrices.

9.6 Self-regulating Heating Devices

Another intrinsic and interesting feature of CPCs, which has been widely explored when making heating devices, is resistive heating, also known as the Joule heating effect. When a voltage is applied to a CPC to generate sufficient current flowing through it, the heat generated results in a temperature increase. By combining the PTC effect with the Joule heating effect, a self-regulating heating device can be created, which possesses an integrated safety temperature control system to avoid overheating.

A self-regulating heater is basically an electrical heater, which strives to maintain a constant temperature regardless of how the ambient temperature changes. It was firstly introduced by Chemelex in 1971 and Raychem further discovered it, revolutionizing the trace heating market [56]. This invention counted as revolutionary at that time, due to the ability of a material to control the heating

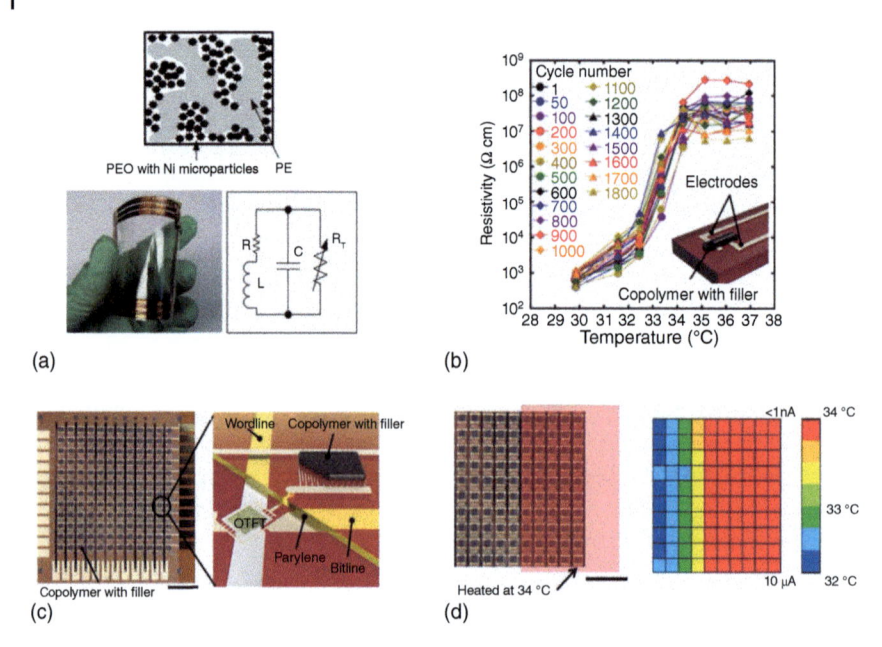

Figure 9.8 (a) A flexible wireless temperature sensor based on Ni microparticle filled PEO/PE blends with a stable and adjustable PTC switching temperature; (b) heat cycling tests for this temperature sensor up to1800 cycles; (c) flexible temperature sensor sheet (scale bar 1 cm) with enlarged cross-sectional illustration of the matrix; (d) area of the temperature sensor sheet heated to 34 C (scale bar 1 cm) and temperature gradient mapping . Source: (a) Jeon et al. [53], (b–d) Yokota et al. [54].

power output autonomously in response to temperature variation, for instance, in pipes. Apart from the self-controlled power output, this type of material has also the advantage of easy design, installation, and maintenance. For example, trace heating cables can provide controllable heat to prevent pipes from freezing or to maintain certain temperature in a container, which are more intelligent than traditional standard heating cables. The cable is constituted by placing a self-regulating heating element between two bus wires (one live and one neutral) and covered with an insulating thermoplastic and an over-jacket [57]. When an alternating current of 220–240 V is applied to the bus wires, the heating element will heat up due to Joule heating effect, but the power output of the heating element falls as the cable gets warmer. Self-regulating heating cables can adjust the output response to temperature all along its length, independently from the length. As the temperature of a given part of the cable increases, the number of conductive pathways decreases, resulting in an increase in resistance, which lowers the output wattage [57]. Conversely, as the temperature decreases, the resistance decreases so more heat can be produced to maintain the temperature.

Although the traditional self-regulating cables have been used in many industrial applications, their performance is still not ideal. For instance, the high filler loading, the relative rigid nature of polymer matrices, or the cross-linking significantly limited mechanical flexibility and prevent recyclability. Hence both industry and academia are still dedicating a great deal of effort to solve these issues and further improve the overall performance of pyroresistive CPC materials. The rise of flexible electronic devices in twentieth century brings new challenges for materials and technologies, as one of the critical requirements for flexible devices is to maintain good performances at relatively large deformations, as well as adaptability in real environmental conditions [58]. It would be very difficult to use traditional self-regulating heating materials as flexible heating devices, due to their limited mechanical flexibility. Material designing and engineering could focus on at least two complementary routes to achieve mechanical flexibility: one is to replace traditional semiconductors with intrinsically flexible materials of good electronic performance, the other one is through engineering the structure.

As mentioned earlier, polymer blend system is an effective way of tuning CPC properties. One strategy to tune the flexibility and PTC effect of a heating device, while preserving the Joule heating properties, is by adding a secondary thermoplastic elastomeric (TPE) phase. A group of researchers from Queen Mary University of London proved that the selection of the TPE phase will greatly affect the morphology of the resulting blends, leading to an immiscible binary blend with either a fine or course droplet morphology or a co-continuous morphology. In the case of the HDPE/TPE filled by graphene nanoplatelet (GNP), the temperature stabilized at the self-regulating temperature of around 120 °C, with the thermal image showing a uniformly heated sample (Figure 9.9a) [48]. The material flexibility was greatly enhanced in these TPE modified composites as demonstrated by the increased bending radii.

The other strategy is to engineer the macro structure of composites. Recently, a versatile sandwich-based design approach has also been presented by connecting different CPCs with distinct pyroresistive behaviors in series, achieving for the first time a high PTC intensity combined with excellent Joule heating as well as mechanical flexibility (Figure 9.9b) [42]. In this study, the sandwich design combines the switching unit [silver coated glass sphere filled thermoplastic polyurethane (TPU) composites] and the heating units (TPU/CNT composite at both ends of the sandwich device) in a desired manner. The use of TPU as the polymer matrix combined with the series configuration and the fillers selected significantly improved the flexibility of such a heating device.

Moreover, advanced use of self-regulating heating materials has also been applied to the manufacturing processes for fiber-reinforced plastics (FRPs). A highly energy efficient and safe out-of-oven curing method is presented by Liu et al. with integrating a pyroresistive surface layer. The novel manufacturing method applied the intrinsic self-regulating heating composite layer onto a composite laminate as shown in Figure 9.9c [59]. This surface layer possesses self-regulating Joule heating capabilities, which can be used to cure epoxy-based composites at a desired temperature without the risk of over-heating. Moreover, the thermoplastic nature of the surface layer enables easy fabrication with good flexibility for complex shapes.

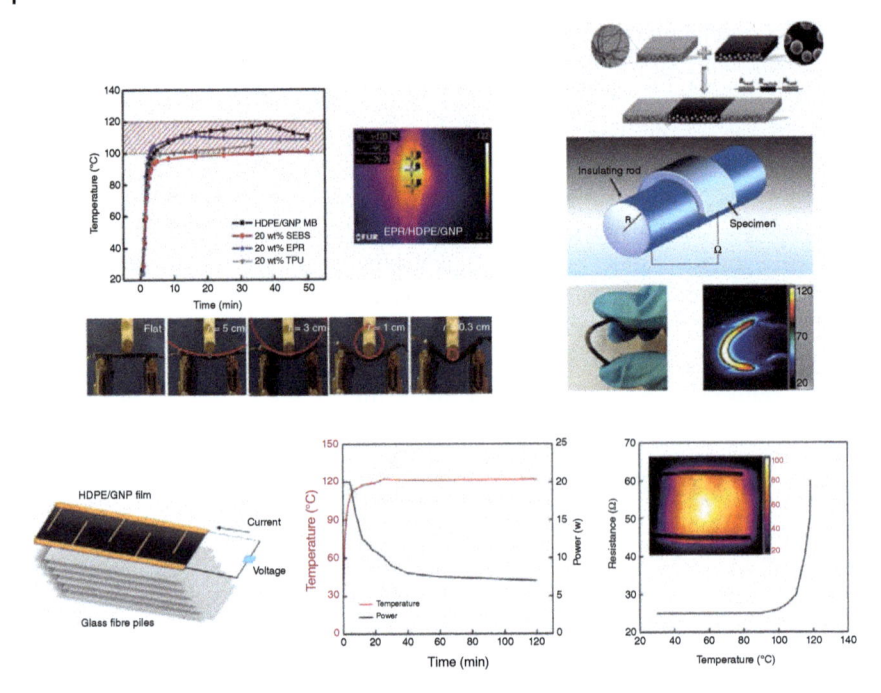

Figure 9.9 (a) Secondary thermoplastic elastomeric (TPE) phase filled HDPE/GNP composites with the temperature stabilized at the self-regulating temperature around 120 °C, and enhanced flexibility; (b) a simple and versatile sandwich-based approach that tackles typical design compromises in smart self-regulating heating devices; (c) a highly energy efficient and safe out-of-oven curing method to cure FRPs with PTC nanocomposites layer on the surface. Source: (a) Liu et al. [4], (b) Liu et al. [42], (c) Liu et al. [59].

9.7 Conclusions

Overall, the insightful understanding of pyroresistive material has brought successful research outcome into industrial applications, such as self-regulating heaters and temperature sensors. The extensive knowledge accumulated in the study of polymer nanocomposites with PTC and NTC effect has enabled the development in the field of integrated electronics. By summarizing the mechanisms/theories of pyroresistive effect and the influencing factors, hypothesis has been made that the PTC effect has been greatly affected by the "robustness" of conductive network. The condition of conductive network is determined by the combination of all the factors, including filler size/shape, filler dispersion/distribution, mixed fillers, polymer matrix type (crystallinity and thermal expansion), and polymer blends. Regarding temperature sensing, the transition temperature of the polymer matrix, such as melting or glass transition temperature, plays an important role in determine the PTC switching temperature.

Though many efforts have been made to achieve the best performance of PTC materials, such as morphological control over conductive networks, many problems are yet to be solved. For instance, the most appropriate way to enhance reproducibilityx and eliminate NTC effect, while reserve high PTC intensity is still worth exploring.

References

1 Deng, H., Lin, L., Ji, M. et al. (2014). Progress on the morphological control of conductive network in conductive polymer composites and the use as electroactive multifunctional materials. *Prog. Polym. Sci.* 39 (4): 627–655.
2 Wan, K., Taroni, P.J., Liu, Z. et al. (2019). Flexible and stretchable self-powered multi-sensors based on the n-type thermoelectric response of polyurethane/nax(Ni-ett)$_n$ composites. *Adv. Electron. Mater.* 5 (12): 1900582.
3 Byrne, M.T. and Gun Ko, Y.K. (2010). Recent advances in research on carbon nanotube-polymer composites. *Adv. Mater.* 22 (15): 1672–1688.
4 Liu, Y., Zhang, H., Porwal, H. et al. (2019). Pyroresistivity in conductive polymer composites: a perspective on recent advances and new applications. *Polym. Int.* 68 (3): 299–305.
5 Ohe, K. and Naito, Y. (1971). A new resistor having an anomalously large positive temperature coefficient. *Japn. J. Appl. Phys.* 10 (1): 99.
6 Xiang, Z.D., Chen, T., Li, Z.M., and Bian, X.C. (2009). Negative temperature coefficient of resistivity in lightweight conductive carbon nanotube/polymer composites. *Macromol. Mater. Eng.* 294 (2): 91–95.
7 E. Frydman (1948). UK patent specification 604695171814s
8 F. Kohler, Resistance element, US Grant US3243753A (1966).
9 Xu, H. (2016). Positive temperature coefficient effect of polymer nanocomposites. In: *Polymer Nanocomposites: Electrical and Thermal Properties* (eds. X. Huang and C. Zhi), 83–110. Cham, Switzerland: Springer International Publishing.
10 Lux, F. (1993). Models proposed to explain the electrical conductivity of mixtures made of conductive and insulating materials. *J. Mater. Sci.* 28 (2): 285–301.
11 Kernin, A., Wan, K., Liu, Y. et al. (2019). The effect of graphene network formation on the electrical, mechanical, and multifunctional properties of graphene/epoxy nanocomposites. *Compos. Sci. Technol.* 169: 224–231.
12 Kirkpatrick, S. (1973). Percolation and conduction. *Rev. Mod. Phys.* 45 (4): 574–588.
13 Zallen, R. (1983). Percolation: a model for all seasons. In: *Percolation structures and processes*, Adam Hilger, Bristol; The Israel Physical Society, Jerusalem; The American Institute of Physics, New York (eds. G. Deutscher, R. Zallen and J. Adler), 3–16.
14 Gulrez, S.K.H., Ali Mohsin, M.E., Shaikh, H. et al. (2014). A review on electrically conductive polypropylene and polyethylene. *Polym. Compos.* 35 (5): 900–914.
15 Stru·mpler, R. and Glatz-Reichenbach, J. (1999). FEATURE ARTICLE conducting polymer composites. *J. Electroceram.* 3 (4): 329–346.
16 Van Beek, L. and Van Pul, B. (1962). Internal field emission in carbon black-loaded natural rubber vulcanizates. *J. Appl. Polym. Sci.* 6 (24): 651–655.

17 Frenkel, J. (1930). On the electrical resistance of contacts between solid conductors. *Phys. Rev.* 36, 1604 (11).

18 Klason, C. and Kubát, J. (1976). Thermal and current noise in carbon black-filled polystyrene and polyethylene in the vicinity of T_g and T_m. *J. Appl. Polym. Sci.* 20 (2): 489–499.

19 Al-Allak, H., Brinkman, A., and Woods, J. (1993). I–V characteristics of carbon black-loaded crystalline polyethylene. *J. Mater. Sci.* 28 (1): 117–120.

20 Mamunya, Y.P., Zois, H., Apekis, L., and Lebedev, E.V. (2004). Influence of pressure on the electrical conductivity of metal powders used as fillers in polymer composites. *Powder Technol.* 140 (1): 49–55.

21 Meyer, J. (1974). Stability of polymer composites as positive-temperature-coefficient resistors. *Polym. Eng. Sci.* 14 (10): 706–716.

22 Meyer, J. (1973). Glass transition temperature as a guide to selection of polymers suitable for PTC materials. *Polym. Eng. Sci.* 13 (6): 462–468.

23 Xu, H., Wu, Y., Yang, D. et al. (2011). Study on theories and influence factors of PTC property in polymer based conductive composites. *Rev. Adv. Mater. Sci* 27: 173-183.

24 Ota, T., Fukushima, M., Ishigure, Y. et al. (1997). Control of percolation curve by filler particle shape in Cu–SBR composites. *J. Mater. Sci. Lett.* 16 (13): 1182–1183.

25 Jing, X., Zhao, W., and Lan, L. (2000). The effect of particle size on electric conducting percolation threshold in polymer/conducting particle composites. *J. Mater. Sci. Lett.* 19 (5): 377–379.

26 Asare, E., Evans, J., Newton, M. et al. (2016). Effect of particle size and shape on positive temperature coefficient (PTC) of conductive polymer composites (CPC) – a model study. *Mater. Des.* 97: 459–463.

27 Liu, Y., Asare, E., Porwal, H. et al. (2020). The effect of conductive network on positive temperature coefficient behaviour in conductive polymer composites. *Composites Part A* 139: 106074.

28 Luo, S. and Wong, C. (2000). Study on effect of carbon black on behavior of conductive polymer composites with positive temperature coefficient. *IEEE Trans. Compon. Packag. Technol.* 23 (1): 151–156.

29 Tkalya, E.E., Ghislandi, M., de With, G., and Koning, C.E. (2012). The use of surfactants for dispersing carbon nanotubes and graphene to make conductive nanocomposites. *Curr. Opin. Colloid Interface Sci.* 17 (4): 225–232.

30 Al-Saleh, M.H. and Sundararaj, U. (2009). A review of vapor grown carbon nanofiber/polymer conductive composites. *Carbon* 47 (1): 2–22.

31 Chen, L., Hou, J., Chen, Y. et al. (2019). Synergistic effect of conductive carbon black and silica particles for improving the pyroresistive properties of high density polyethylene composites. *Composites Part B* 178: 107465.

32 Shi, G., Cai, X., Wang, W., and Wang, G. Improving resistance-temperature characteristic of polyethylene/carbon black composites by poly(3,4-ethylenedioxythiophene)-functionalized multilayer graphene. *Macromol. Chem. Phys.* 221 (14): 2000144.

References

33 Dang, Z.-M., Li, W.-K., and Xu, H.-P. (2009). Origin of remarkable positive temperature coefficient effect in the modified carbon black and carbon fiber cofillled polymer composites. *J. Appl. Phys.* 106 (2): 024913.

34 Lee, J.H., Kim, S.K., and Kim, N.H. (2006). Effects of the addition of multi-walled carbon nanotubes on the positive temperature coefficient characteristics of carbon-black-filled high-density polyethylene nanocomposites. *Scr. Mater.* 55 (12): 1119–1122.

35 Fang, Y., Zhao, J., Zha, J.-W. et al. (2012). Improved stability of volume resistivity in carbon black/ethylene-vinyl acetate copolymer composites by employing multi-walled carbon nanotubes as second filler. *Polymer* 53 (21): 4871–4878.

36 Zha, J.W., Li, W.K., Liao, R.J. et al. (2013). High performance hybrid carbon fillers/binary-polymer nanocomposites with remarkably enhanced positive temperature coefficient effect of resistance. *J. Mater. Chem. A* 1 (3): 843–851.

37 Asare, E., Basir, A., Tu, W. et al. (2016). Effect of mixed fillers on positive temperature coefficient of conductive polymer composites. *Nanocomposites* 2 (2): 58–64.

38 Horibe, H., Kamimura, T., and Yoshida, K. (2005). Electrical conductivity of polymer composites filled with carbon black. *Japn. J. Appl. Phys.* 44, 2025 (4R).

39 Luo, Y., Wang, G., Zhang, B., and Zhang, Z. (1998). The influence of crystalline and aggregate structure on PTC characteristic of conductive polyethylene/carbon black composite. *Eur. Polym. J.* 34 (8): 1221–1227.

40 Xiong, C., Zhou, Z., Xu, W. et al. (2005). Polyurethane/carbon black composites with high positive temperature coefficient and low critical transformation temperature. *Carbon* 43 (8): 1788–1792.

41 Fournier, J., Boiteux, G., Seytre, G., and Marichy, G. (1997). Positive temperature coefficient effect in carbon black/epoxy polymer composites. *J. Mater. Sci. Lett.* 16 (20): 1677–1679.

42 Liu, Y., Zhang, H., Porwal, H. et al. (2017). Universal control on pyroresistive behavior of flexible self-regulating heating devices. *Adv. Funct. Mater.* 27 (39): 1702253.

43 Kar, P. and Khatua, B.B. (2011). Highly reversible and repeatable PTCR characteristics of PMMA/Ag-coated glass bead composites based on CTE mismatch phenomena. *Polym. Eng. Sci.* 51 (9): 1780–1790.

44 Kono, A., Shimizu, K., Nakano, H. et al. (2012). Positive-temperature-coefficient effect of electrical resistivity below melting point of poly(vinylidene fluoride) (PVDF) in Ni particle-dispersed PVDF composites. *Polymer* 53 (8): 1760–1764.

45 Zribi, K., Feller, J.F., Elleuch, K. et al. (2006). Conductive polymer composites obtained from recycled poly(carbonate) and rubber blends for heating and sensing applications. *Polym. Adv. Technol.* 17 (9–10): 727–731.

46 Kar, P. and Khatua, B. (2011). PTCR characteristics of polycarbonate/nickel-coated graphite-based conducting polymeric composites in presence of poly(caprolactone). *Polym. Compos.* 32 (5): 747–755.

47 Yu, G., Zhang, Q., Zeng, H.M. et al. (1999). Conductive polymer blends filled with carbon black: positive temperature coefficient behavior. *Polym. Eng. Sci.* 39 (9): 1678–1688.

48 Liu, Y., Zhang, H., Porwal, H. et al. (2018). Tailored pyroresistive performance and flexibility by introducing a secondary thermoplastic elastomeric phase into graphene nanoplatelet (GNP) filled polymer composites for self-regulating heating devices. *J. Mater. Chem. C* 6 (11): 2760–2768.

49 Feng, J. and Chan, C.-M. (2000). Double positive temperature coefficient effects of carbon black-filled polymer blends containing two semicrystalline polymers. *Polymer* 41 (12): 4559–4565.

50 Wei, Y., Li, Z., Liu, X. et al. (2014). Temperature-resistivity characteristics of a segregated conductive CB/PP/UHMWPE composite. *Colloid. Polym. Sci.* 292 (11): 2891–2898.

51 Xu, H.-P., Dang, Z.-M., Shi, D.-H., and Bai, J.-B. (2008). Remarkable selective localization of modified nanoscaled carbon black and positive temperature coefficient effect in binary-polymer matrix composites. *J. Mater. Chem.* 18 (23): 2685–2690.

52 Xi, Y., Ishikawa, H., Bin, Y., and Matsuo, M. (2004). Positive temperature coefficient effect of LMWPE–UHMWPE blends filled with short carbon fibers. *Carbon* 42 (8–9): 1699–1706.

53 Jeon, J., Lee, H.B.R., and Bao, Z. (2013). Flexible wireless temperature sensors based on Ni microparticle-filled binary polymer composites. *Adv. Mater.* 25 (6): 850–855.

54 Yokota, T., Inoue, Y., Terakawa, Y. et al. (2015). Ultraflexible, large-area, physiological temperature sensors for multipoint measurements. *Proc. Natl. Acad. Sci.* 112 (47): 14533–14538.

55 Zha, J.-W., Wu, D.-H., Yang, Y. et al. (2017). Enhanced positive temperature coefficient behavior of the high-density polyethylene composites with multi-dimensional carbon fillers and their use for temperature-sensing resistors. *RSC Adv.* 7 (19): 11338–11344.

56 Hammack, T.J. and Kucklinca, S.J. (1977). Self-limiting electrical heat tracing: new solution to old problems. *IEEE Trans. Ind. Appl.* 2: 134–138.

57 Pretorius, P., Liang, A., & Mann, P. (2017). U.S. Patent Application No. 15/433,907.

58 Liu, Y., He, K., Chen, G. et al. (2017). Nature-inspired structural materials for flexible electronic devices. *Chem. Rev.* 117 (20): 12893–12941.

59 Liu, Y., van Vliet, T., Tao, Y. et al. (2020). Sustainable and self-regulating out-of-oven manufacturing of FRPs with integrated multifunctional capabilities. *Compos. Sci. Technol.* 190: 108032.

10

Polymer Nanocomposites for EMI Shielding Application

Ajitha A. Ramachandran[1,2] and Sabu Thomas[1,3]

[1]*Mahatma Gandhi University, International and Inter University Centre for Nanoscience and Nanotechnology, Kottayam, Kerala 686 560, India*
[2]*Nirmala College Muvattupuzha, Department of Chemistry, Kerala 686 661, India*
[3]*Mahatma Gandhi University, School of Chemical Sciences, Kottayam, Kerala 686 560, India*

10.1 Introduction

Conducting polymer composites have great importance in our current research since it can find several applications such as sensor, energy storage material, electromagnetic interference (EMI) shielding material, etc. Conducting nanomaterials incorporated polymer nanocomposites have great attention in this field due to their superior electrical and optical properties. The interesting optical and electrical properties of nanomaterials can be explained by their quantum confinement effect. It explains the variation of the electrical, optical, and physical properties of nanomaterials compared with the corresponding bulk materials. The large surface area and high aspect ratio of carbon nanomaterials play a great role in the enhancement of electrical properties of polymers while they are incorporating with carbon nanomaterials. Among the carbon nanomaterials MWCNTs, graphene, graphene oxides, carbon black, etc. have a tremendous interest in the electronic field [1–5].

Conducting polymers or polymers with conductive nanofillers yield high performance conducting polymer nanocomposites and it can be used for the various electronic application, in which EMI shielding application is a current topic and several studies are reported with high shielding effectiveness values for polymer composites. The conducting polymer composites with high dielectric constant, high dielectric loss, and high conductivity have a great interest in the electronic field especially in the field of EMI shielding, sensor, imaging, etc. Polymer composites have acquired great importance as an EMI shielding material since it can attenuate the unwanted electromagnetic radiation coming out from the electronic devices. Due to the increased usage of electronic devices in our daily life, electromagnetic pollution is a major problem facing by our society. Hence, the necessity of fabrication of an effective EMI shielding material is increasing day by

Polymer Nanocomposite Materials: Applications in Integrated Electronic Devices, First Edition.
Edited by Ye Zhou and Guanglong Ding.
© 2021 WILEY-VCH GmbH. Published 2021 by WILEY-VCH GmbH.

day to limit the electromagnetic pollution. By the effective shielding of electronic devices, it can protect not only the surroundings but also it can help parent devices itself for proper functioning from the adverse effect of electromagnetic radiation. Metals are widely used as materials for EMI shielding applications due to its high conductivity and dielectric constants. But, metals have some disadvantages which makes them an undesired choice for an electronic application. According to the recent literature works, it can be said that the current research scenario focused on the development of lightweight and flexible effective EMI shielding materials based on polymers [6–10]. The present chapter mainly discusses the nanofiller incorporated polymer composites and its applicability as EMI shielding material.

10.2 Mechanism of EMI Shielding of Polymer Composites

EMI is a disturbance caused by electro electromagnetic radiation emitted from an external source and that affects surrounding electrical circuits. EMI also called radio frequency interference (RFI). EMI problems increasing day by day due to the increased usage of mobile electronic systems, wireless communication systems, and computer networks [11]. Electromagnetic shielding is the process of minimizing the dispersion of electromagnetic waves into surroundings by shielding the electromagnetic waves with a shield made of a conductive material. Proper functioning of parent electrical instruments or surrounding instruments may be interrupted, or limited due to EMI. The mechanism of EMI shielding of a material mainly takes place by three main mechanisms, which include the reflection of the incident wave, absorption of the incident wave, and multiple reflections caused by internal reflection inside the material. In the case of materials having a low surface area or high absorption dominant shielding effectiveness, multiple reflections can be negligible. In which reflection is the primary mechanism of EMI shielding and for reflection, the material must possess mobile charge carriers such as electrons or holes that interact with electromagnetic radiation. The widely used EMI shielding materials were metals, the available free electrons in metals can interact with the electromagnetic waves. If the material is highly conductive, it can attenuate electromagnetic waves. However, conductivity is not a condition for EMI shielding but it does enhance the reflection mechanism of an EMI shielding material. Absorption is the second mechanism for EMI shielding, which requires the existence of electric or magnetic dipoles to interact with electromagnetic radiation. Materials with high dielectric constant offer electric dipoles and materials with high magnetic permeability provide magnetic dipoles for the EMI shielding by absorption. The third mechanism, multiple reflections, measures the reflections at different surfaces or at the interface of the material. Materials that have large specific internal surfaces or composites with fillers show a multiple reflection mechanism. When electromagnetic waves strike on an object they may

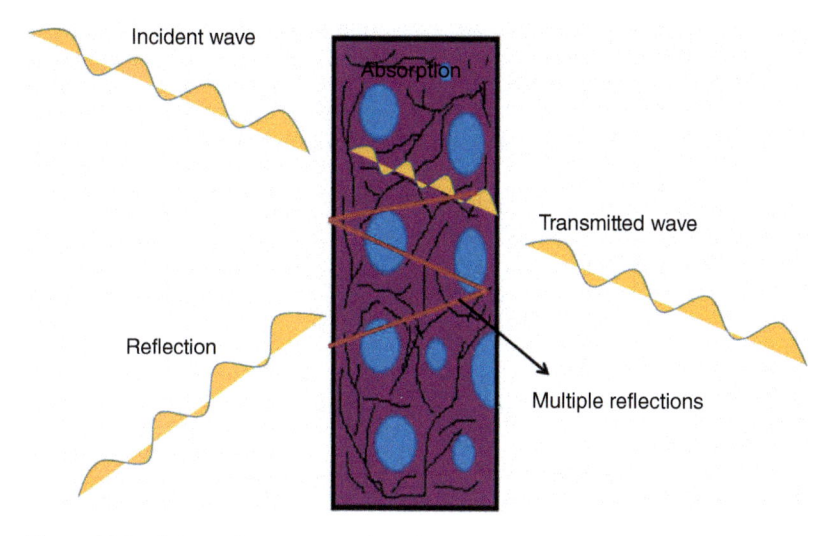

Figure 10.1 Schematic representation of EMI shielding mechanism.

undergo reflection, multiple reflections, absorption, and transmission as shown in Figure 10.1

10.2.1 Materials for EMI Shielding

Due to the increased usage of electronic equipment, shielding such instruments is a major requirement for the parent devices itself, neighboring devices, and human beings from the adverse effect of electromagnetic waves. It is a very serious issue in the current scenario. Conductive polymer nanocomposites have a great interest in the current research field. Today conducting polymer composites have a great deal of interest both academic and industrial field due to their properties such as the cost-effectiveness, easy processability, and their possible applications in many areas including EMI shielding

Metals are commonly used for EMI shielding material in the form of thin sheets or sheathing in automotive applications due to its high conductivity and dielectric constants. But, metal is expensive, prone to corrosion, heavy, and the cost of manufacturing processes is also very high, which makes them an undesired choice for an electronic application. Conductive polymer composites can overcome the problems of metals. It has a great deal of academic and industrial interest by considering the cost-effectiveness, easy processability, and possible applications in many areas including EMI shielding [12]. Polymers incorporated with graphene carbon black (CB), carbon nanotubes (CNTs), metal nanoparticles, nanowires, carbon nanofibers, hybrid fillers, foams, and magnetic nanoparticles show good conductivity and shielding capacity against EM waves. Conductive polymers have also a great interest in this field. Several groups have studied and reported the EMI shielding effectiveness of different materials and mechanisms behind the EMI shielding ability of those materials [12–16].

10.3 Polymer Nanocomposites for EMI Shielding Application

10.3.1 Nanofiller Incorporated Conducting Polymer Composites

The incorporation of conducting fillers into the insulating polymer matrix enhances the conductivity of the polymer matrix. Conducting polymer nanocomposites with high conductivity can be used for EMI shielding applications. The percolation theory can be used to explain the electrically conducting behavior of composites consisting of conducting fillers and insulating polymer matrices [17–19].

When adding conducting nanomaterials (filler) into a polymer matrix, at a critical filler content there will be a chance to increase the electrical conductivity of the insulative polymer matrix sharply by several orders of magnitude. This critical filler content is termed as electrical percolation threshold [20, 21]. At this critical filler content, continuous electron paths or conducting networks will form in the system as a result the system became conductive. So it can be said that the percolation threshold (pc) is the minimum amount of filler to achieve a continuous conducting network in the polymer. Below the percolation threshold, the electrical properties of the composite are mainly due to the hopping mechanism. While above the percolation threshold, continuous electron paths by the direct contact of fillers throughout the system lead to the enhanced electrical conductivity of the composite. At the initial loading of nanofiller, the system shows very low conductivity but after percolation threshold conductivity will rise and often shows a saturation plateau at higher filler loading. Schematic representation of the mechanism is shown in Figure 10.2 [22–25].

The concentration of the conducting filler must be above the percolation threshold to attain efficient conducting networks in the composite [26]. The percolation threshold mainly depends on the characteristics of nanomaterials such as aspect

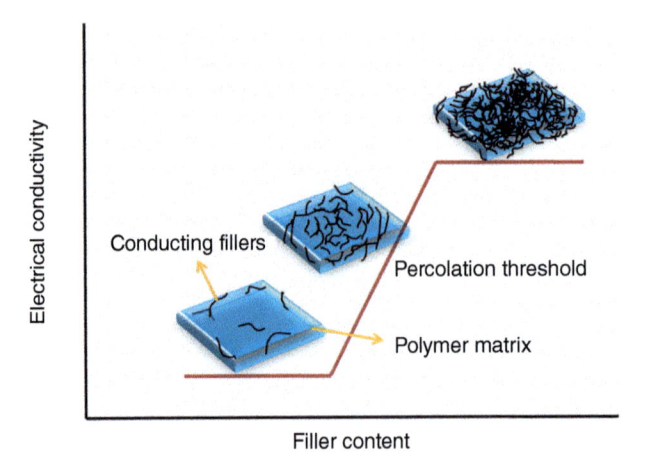

Figure 10.2 Schematic representation of the percolation phenomenon in polymer composites. Source: Ponnamma et al. [22].

ratio, dispersion, and alignments. In addition to that filler, polymer interaction has also a major role in predicting the electrical properties of the fabricated composites. Generally, the percolation threshold decreases with increases rapidly with an increase in aspect ratio. Similarly, well-dispersed nanomaterials can reduce the aggregation and possess a high aspect ratio and low percolation threshold [10, 15, 23, 27–32].

10.3.2 Polymer Blend Nanocomposites for Electromagnetic Interference (EMI) Shielding

Polymer blends are mixtures of two or more polymers with the combined properties of each polymer [33, 34]. Incorporation of conducting fillers into polymer blend yields conducting polymer blend nanocomposites, and it can be used for several electronic applications. Polymer blend nanocomposites with high conductivity have a great attention for EMI shielding applications. The usage of conducting polymers will also be used to fabricate conducting polymer nanocomposites [35, 36].

In addition to single polymer composites, several studies are going on with EMI shielding applicability of polymer blend nanocomposites. Polymer blending is an attractive way to build up new materials for specific applications since it is a very cost-effective and simple method. But most of the polymer blends are immiscible and in some cases, it shows poor properties than the individual polymer components. Even though most of the polymer blends are immiscible, polymer blend nanocomposites with selectively localized conducting nanofillers can achieve an improved conductivity than a single polymer phase composite. This is due to the formation of continuous conducting pathways at a low percolation threshold and this phenomenon termed as double percolation effect [1, 3, 37]. The selective localization of conducting fillers in any one of the phases and the continuation of that phase can be designated as a double percolation effect and is a key factor to achieve good conductivity in a phase-separated system. Immiscible polymer blends have several types of morphology such as co-continuous, domain or droplet morphology, and lamellar morphology. There are many studies related to the double percolation effect observed in the case of droplet morphology and co-continuous morphology [1, 38–40].

According to many reports available in the literature, it can be said that in the case of immiscible polymer blends, phase separation and the selective localization of conducting fillers play a major role in the formation of conducting networks. Selective localization of filler leads to a double percolation effect, which makes them more conductive than a single polymer composite with the same amount of conducting fillers [41–46]. Selective localization of fillers in immiscible polymer blends can be predicted theoretically [33]. Figure 10.3 shows the schematic representation of conducting filler in a single polymer-based composite and binary polymer composite having the same amount of filler.

In addition to selective localization, some other factors will affect the electrical properties of polymer blend nanocomposites, which includes the method of preparation of composites, the morphology of the prepared polymer blend, the interaction

(a)   (b)

■ Polymer A   ■ Polymer B   ⌇ Conducting filler

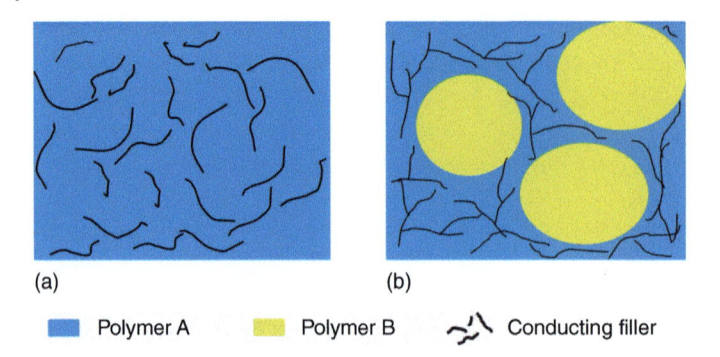

Figure 10.3 Dispersion of the same amount of conducting filler in (a) single polymer composites (conducting network is absent) (b) polymer blend (conducting network is possible due to the selective localization of fillers in the continuous phase).

between the filler and polymer, the dispersion of nanomaterials in the polymers, etc, [1, 47].

The strategy of localization of filler is different in blends having different morphology. Consider the first strategy in the case of blend nanocomposites with domain/droplet/dispersed morphology, if nanofillers are selectively localized in the matrix, the percolation threshold decreases compared with corresponding single polymer-based nanocomposites, due to the phase separation and network formation of fillers at low filler loading. The selective localization of conducting fillers in the continuous phase or matrix phase will lead to the formation of continuous conducting networks and thereby movement of free electrons through the network become possible, hence increasing conductivity due to ohmic conductance and tunneling effect. And other strategies in domain morphology involve the localization of nanofillers either in the dispersed phase or at the interface. In both strategies, the sample is either non-conductive or conductive only at the higher percolation threshold, because only tunneling mechanism is possible in that cases and the localized fillers cannot form a continuous network of conducting fillers. While in the case of blend nanocomposites with co-continuous morphology, the strategy of percolation is different from that of blends with domain morphology. Here nanofillers are selectively localized in any of the polymer components or at the interface and in both case composites may show high conductivity at the lowest percolation threshold. The high conductivity of polymer blend nanocomposite can explain with the double percolation effect [10, 47–50]. All the strategies are represented in the schematic representations given in Figure 10.4. Figure 10.5 represents the cartoon showing the overall conducting mechanism of multiwalled carbon nanotube incorporated poly(trimethylene terephthalate)/polypropylene polymer blend nanocomposites [20].

10.3.3 Conducting Polymers for EMI Shielding Application

As already explained it can be said that today industries are looking for easy processable electrically conducting materials for electronic applications based on

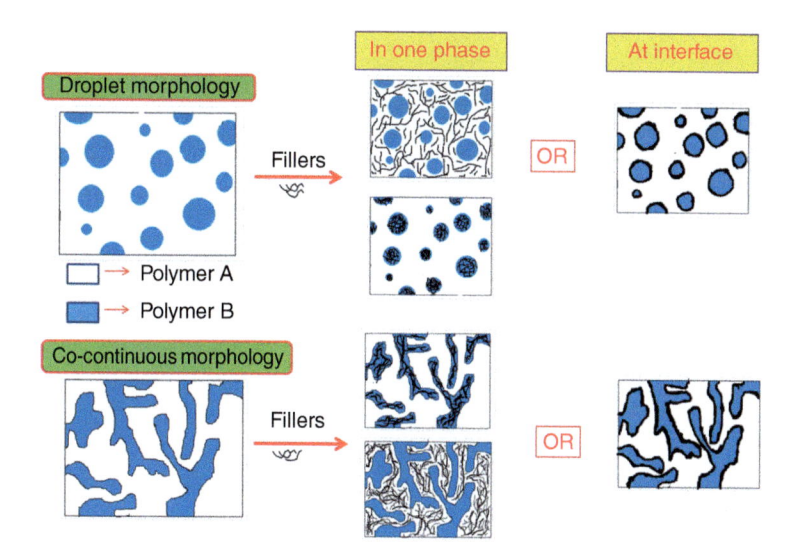

Figure 10.4 Schematic representation showing the selective localization of filler in an immiscible blend [51].

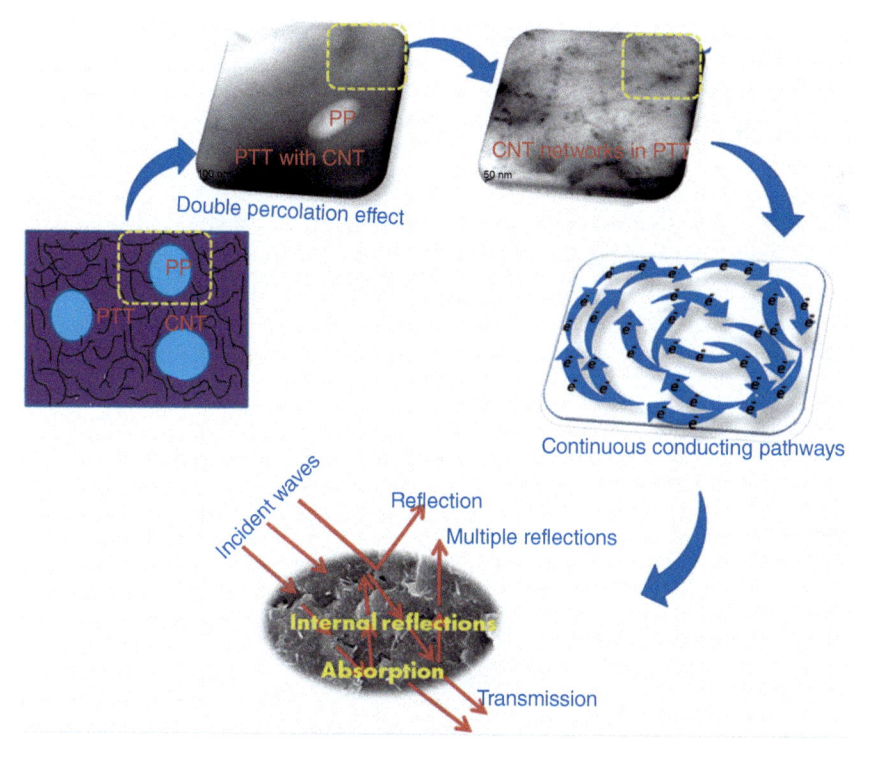

Figure 10.5 Schematic diagram representing EMI shielding mechanism in a PTT/PP/MWCNT system [20].

conducting polymers, conducting fillers in polymer matrices, since it can satisfy the current requirements such as corrosion-resistant, flexible, and lightweight. There are conducting or non-conducting polymers in nature. Intrinsically conducting polymers (CPs) are a novel class of polymers with high conductivity. Its conductivity is mainly due to the conjugation in its polymeric backbone. Using various dopants or conducting fillers, the conductivity of IPCs can be tuned. Conducting polymer composites prepared by mixing conducting fillers with an insulator polymer matrix facing some problems. Carbon-based fillers may lower the mechanical strength of composites due to agglomeration of filler and poor filler–matrix interaction. In addition to this, carbon-based fillers are costly mainly due to the complicated purification or functionalization process. On considering these it can be said that intrinsically conductive polymers (IPCs) have also great attention for several electronic applications due to its low cost, lightweight, easy processability, and high conductivity, in which polyaniline (PANI), polypyrrole (PPy), poly(3,4-ethylene dioxythiophene), etc. have a great interest in many applications. On using conducting polymers highly conducting polymer composites can be prepared. The intrinsic conducting polymers are considered as suitable matrices for the fabrication of conductive composites and it can absorb microwave radiation. Hence, these conducting polymers can be used for EMI shielding applications.

Among various intrinsic conducting polymers, PANI has great importance due to its easy synthesis, good environmental stability, low monomer cost, and reversible doping/de-doping process. From the extensive reports, it can be said that PANI composites prepared by the introduction of conductive fillers and dopants have received great attention for electronic applications, especially as EMI shielding materials. So composites of PANI can be prepared for EMI shielding application by mixing dopants, CNTs with graphite, carbon black, graphene, Fe_2O_3, nanowires, etc. or dispersing CNTs in other polymer matrices such as polystyrene (PS), polyurethane (PU), etc. PPy is another important one in the category of intrinsically conducting polymers that can be used for EMI shielding. Recently researchers also give attention to PPy on considering that it has gained much attention for its numerous advantages such as tunable electrical/dielectric properties, good environmental stability, low density, and simple development. But the EMI shielding effectiveness of pure PPy is poor because of its low electrical conductivity. Several studies are reported based on the enhancement of conductivity of PPy by introducing dopants during polymerization. In addition to intrinsically conducting polymers, magnetic materials and their composites are also considered as potential candidates for EMI shielding applications [52–58].

10.4 Characterization Techniques Used for the Electrical Studies of Polymer Composites

10.4.1 Conductivity Studies of Polymer Composites

As already discussed, polymer composites with high conductivity can be used for EMI shielding applications. The electrical properties of polymer composites can be understood from dielectric studies. The dielectric properties such as dielectric constant (real part of permittivity), dielectric loss (imaginary part of permittivity), and AC conductivities can be measured using the following Eqs. 10.1–10.3, respectively.

$$\varepsilon' = C_p d / \varepsilon_0 A \tag{10.1}$$

$$\varepsilon'' = \varepsilon' \tan \delta \tag{10.2}$$

$$\sigma_{ac} = \omega \varepsilon_0 \varepsilon' \tan \delta \tag{10.3}$$

where ε_0 is permittivity, ε' is the real part of permittivity, ε'' is the imaginary part of permittivity, Cp is parallel capacitance, d is the thickness of the sample, A is the area of the sample, ω is the angular frequency, and $\tan\delta$ is dissipation factor (loss tangent). The real part of impedance (Z') and imaginary part of impedance (Z'') can be measured using following Eqs. 10.4 and 10.5 can be used to calculate the real and imaginary part of the impedance

$$Z' = Z \cos\theta \tag{10.4}$$

$$Z'' = Z \sin\theta \tag{10.5}$$

where Z is the measured impedance and θ is the measured phase angle [48, 59].

Sharika et al. prepared blend nanocomposites of polypropylene (PP) and natural rubber (NR) with different loadings (1,3,5, and 7 wt%) of multiwalled carbon nanotubes (MWCNTs). The dielectric properties of PP/NR/MWCNTs composites are shown in Figure 10.6. Dielectric properties were increased dramatically with the addition of MWCNTs and the blend nanocomposites showed shielding effectiveness of ~29 dB at 3 GHz for 7 wt% of MWCNT loading [60].

In addition to the dielectric properties, the volume resistivity of samples can be measured by two different measurements set up. One is by four-probe method for high conductive samples, and the second is by using two probe methods for low conductive samples. The electrical percolation threshold can be theoretically predicted using the power-law equation (Eq. 10.6) [40, 61–63].

$$\sigma \propto (p - p_c)^\alpha \tag{10.6}$$

where σ is the electrical conductivity, p is the volume fraction of MWCNT, p_c is the electrical percolation threshold value, and α is the critical exponent in conducting region [23, 64, 65].

Figure 10.7 shows the volume conductivity of composites prepared using carbon nanotubes and recycled epoxy by Yuan et al. Here, they reported a facile and effective strategy to build a segregated filler network at epoxy waste particles and converted the system into a high-performance conductive composite. The resultant composites showed and high electrical conductivity at a low percolation threshold and it can be used for EMI shielding application.

An increase in the conductivity with carbon nanotubes and insulator to conductor transition is very clear from the graph. The fitting of conductivity using the

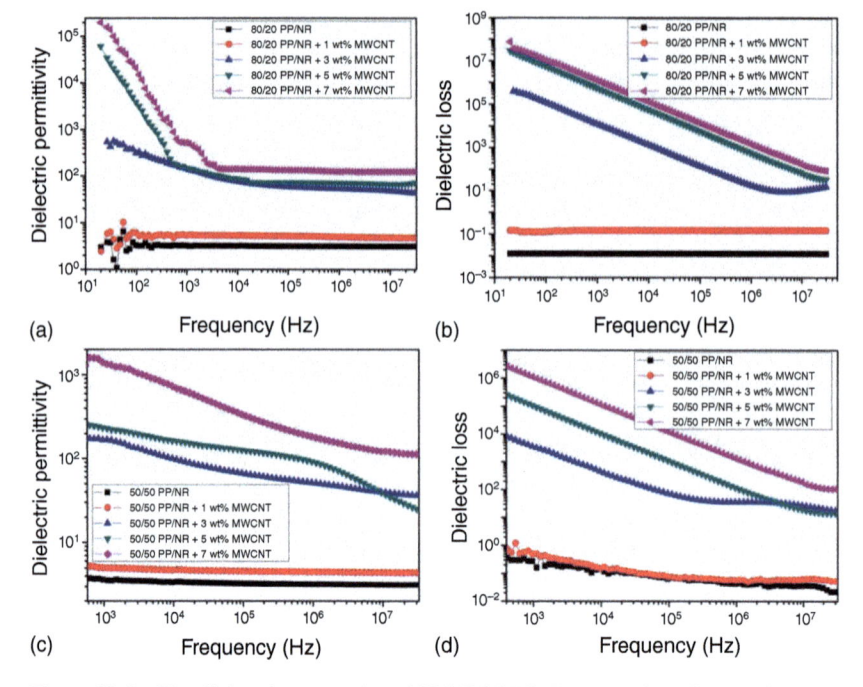

Figure 10.6 The dielectric properties of PP/NR/MWCNTs composites. Source: Sharika et al. [60].

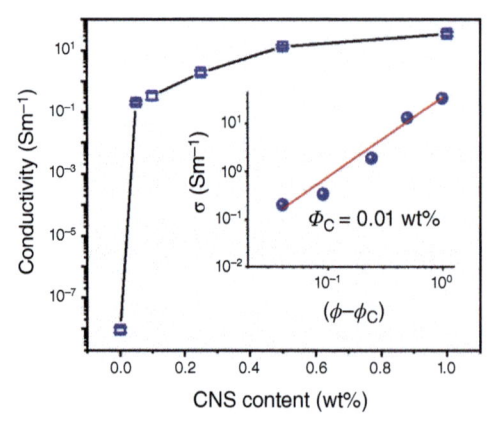

Figure 10.7 Volume conductivity of recycled epoxy composites with carbon nanotubes. Power law fitting is shown in the inset graph. Source: Yuan et al. [66].

power-law equation is given as an insert graph and it reveals that the composite has a percolation threshold of 0.01 wt% [66].

10.4.2 Electromagnetic Interference (EMI) Shielding Studies

The EMI shielding effectiveness (SE) of shielding material means the logarithmic ratio between the incident power (P_I) and the transmitted power (P_T) of an electromagnetic wave and is measured in dB. The total EMI SE includes reflection, absorption, and multiple reflections of EM waves. The total EMI SE depends mainly on the

10.4 Characterization Techniques Used for the Electrical Studies of Polymer Composites

thickness of the material and frequency range of EM waves that we applied; it can be written as

$$SE(dB) = SE_R + SE_A + SE_M \tag{10.7}$$

where SE_A, SE_R, and SE_M corresponds to the shielding effectiveness due to absorption, reflection, and multiple reflections, respectively. Thus main mechanisms of EMI shielding are absorption, reflection, and multiple reflections. Reflection is the primary mechanism and if the $SE_A > 10$ dB or thickness of the sample is greater than the skin depth, multiple reflections can be neglected.

In a vector network analyzer SE can be investigated in terms of scattering parameters (forward reflection coefficient and reverse transmission coefficient) denoted as S11, S12, and S21 [67, 68].

$$SE_R = 10 \log_{10}\left(\frac{1}{(1-[S_{11}]^2)}\right) \tag{10.8}$$

$$SE_A = 10 \log_{10}\left(\frac{1-[S_{11}]^2}{[S_{12}]^2}\right) \tag{10.9}$$

Ajitha et al. [20] studied the applicability of polytrimethylene terephthalate/polypropylene/carbon nanotube (PTT/PP/MWCNTs) blend nanocomposites. They observed that the fabricated composites can be used for the EMI shielding application due to the enhanced conductivity of the PTT/PP polymer blend with the incorporation of MWCNTs and obtained high EMI SE value of ~40 dB for 90PTT/10PP composition with 5 wt% MWCNTs (Figure 10.8).

There are extensive studies reported recently based on the EMI shielding applications of polymer composites. Recently, Liang et al. fabricated lightweight and flexible ternary composites with graphene (Gn), silicon carbide nanowires (SiCnw), and poly(vinylidene fluoride) (PVDF). They measured enhanced dielectric properties, EMI shielding, and thermal conduction for the composites. They reported that these composites have potential applications in EMI shielding and thermal

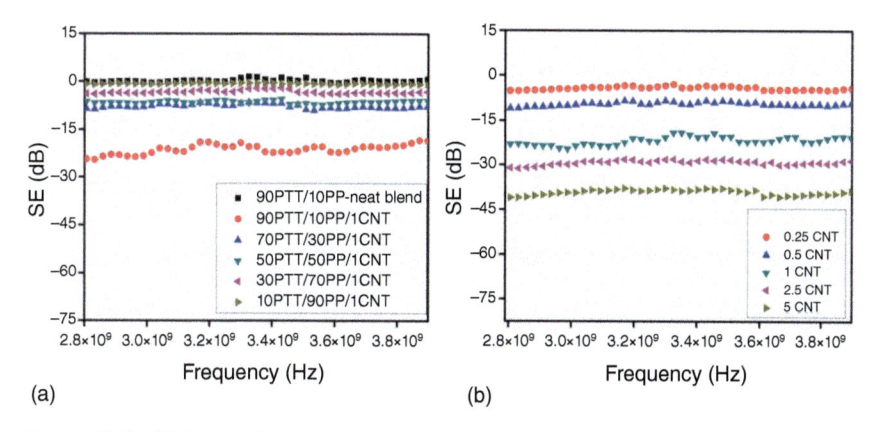

Figure 10.8 EMI shielding effectiveness of (a) PTT/PP blends with 1 wt% CNT, (b) the 90 PTT/10 PP blend with different MWCNTs loadings [20].

management for microelectronics [69]. Gao et al. developed high-performance EMI shielding graphene/PDMS composites at low graphene content. Due to the structural property a 3D conductive network of graphene was formed in the polymer matrix. And they measured an EMI SE value of ~65 dB in the ~0.42 wt% graphene/PDMS composites. The ultralow percolation threshold is achieved for the composites by the structural control [70]. Similarly, Li et al. prepared a lightweight, flexible, and polyacrylonitrile (PAN)@SiO$_2$-Ag composite nanofibrous film with high-performance EMI shielding. They reported that these composite films can be used for electromagnetic shielding and other potential applications, such as in wearable and flexible sensors [71]. Recently Gahlout et al. fabricated composites based on thermoplastic polyurethane [TPU] as matrix and PPy/MWCNT (PCNT) as filler. They synthesized PPy/MWCNT by in situ polymerization of pyrrole monomer by the chemical oxidative method in presence of MWCNT. The fabricated composites showed total shielding effectiveness (SET) up to −46 dB in the frequency range of 8.2–12.4 GHz (X-band). Hence the composites can be used as microwave absorbers [54]. A flexible and flame-retarding thermoplastic polyurethane (TPU)-based EMI-shielding composite was developed by Ji et al. They measured maximum SE of 38.5 dB for the composites with less than 4 wt% CNTs [72]. Zha et al. [3] observed enhanced EMI SE for PVDF/ethylene-α-octene block copolymer (OBC) blend with the incorporation of MWCNT. Otero-Navas et al. [1] reported the applicability of polypropylene/polystyrene (PP/PS)/MWCNTs composites as an EMI shielding material since the fabricated composites can restrict the current leakage; while co-continuous morphology is preferred for applications such as EMI shielding due to the presence of continuous paths for electron conduction. Bizhani et al. [73] reported an SE of 25–29 dB for polycarbonate (PC)/polystyrene-*co*-acrylonitrile (SAN)/MWCNT composites at a low percolation threshold. From the reported extensive studies, it can be said that conductive polymer blend nanocomposites can be used as good EMI shielding materials since it can fulfill the requirements of an effective EMI shielding material.

10.5 Conclusion

The present chapter deals with the applicability of polymer nanocomposites as EMI shielding material. It discusses the outstanding properties of polymer nanocomposites over metals, which are the conventional EMI shielding material. Characterization methods of electrical properties of polymer composites are also discussed in the chapter. Conductive polymer nanocomposites with high conductivity can be used for EMI shielding applications. The theoretical explanation of the conducting nature of single polymer-based nanocomposites and polymer blend nanocomposites are also explained in this chapter. The current research focused on the EMI shielding application of conductive polymer composites, this can be clear from the available literature. An overlook of some recent studies is revealed in this chapter.

References

1 Otero-Navas, I., Arjmand, M., and Sundararaj, U. (2017). Carbon nanotube induced double percolation in polymer blends: morphology, rheology and broadband dielectric properties. *Polymer* 114: 122–134.

2 Zhu, J.-M., Zare, Y., and Rhee, K.Y. (2018). Analysis of the roles of interphase, waviness and agglomeration of CNT in the electrical conductivity and tensile modulus of polymer/CNT nanocomposites by theoretical approaches. *Colloids Surf., A* 539: 29–36.

3 Zha, X.-J., Pu, J.-H., Ma, L.-F. et al. (2018). A particular interfacial strategy in PVDF/OBC/MWCNT nanocomposites for high dielectric performance and electromagnetic interference shielding. *Composites Part A* 105: 118–125.

4 Barrau, S., Demont, P., Peigney, A. et al. (2003). DC and AC conductivity of carbon nanotubes–polyepoxy composites. *Macromolecules* 36 (14): 5187–5194.

5 Fukumaru, T., Fujigaya, T., and Nakashima, N. (2013). Mechanical reinforcement of polybenzoxazole by carbon nanotubes through noncovalent functionalization. *Macromolecules* 46 (10): 4034–4040.

6 Xu, P., Gui, H., Hu, Y. et al. (2014). Dielectric properties of polypropylene-based nanocomposites with ionic liquid-functionalized multiwalled carbon nanotubes. *J. Electron. Mater.* 43 (7): 2754–2758.

7 Oliveira, E.Y., Bode, R., Escárcega-Bobadilla, M.V. et al. (2016). Polymer nanocomposites from self-assembled polystyrene-grafted carbon nanotubes. *New J. Chem.* 40 (5): 4625–4634.

8 Kunjappan, A.M., Poothanari, M.A., Ramachandran, A.A. et al. (2019). High-performance electromagnetic interference shielding material based on an effective mixing protocol. *Polym. Int.* 68 (4): 637–647.

9 Liu, X., Yin, X., Kong, L. et al. (2014). Fabrication and electromagnetic interference shielding effectiveness of carbon nanotube reinforced carbon fiber/pyrolytic carbon composites. *carbon* 68: 501–510.

10 Pawar, S.P., Marathe, D.A., Pattabhi, K., and Bose, S. (2015). Electromagnetic interference shielding through MWNT grafted Fe_3O_4 nanoparticles in PC/SAN blends. *J. Mater. Chem. A* 3 (2): 656–669.

11 Kaur, M., Kakar, S., and Mandal, D. (2011). Electromagnetic interference. In: *3rd International Conference on Electronics Computer Technology*, 1–5. Kanyakumari, India: IEEE. ISBN: CFP1195F-PRT 978-1-4244-8678-6.

12 Kim, S.H., Jang, S.H., Byun, S.W. et al. (2003). Electrical properties and EMI shielding characteristics of polypyrrole–nylon 6 composite fabrics. *J. Appl. Polym. Sci.* 87 (12): 1969–1974.

13 Aswathi, M., Rane, A.V., Ajitha, A. et al. (2018). *Advanced Materials for Electromagnetic Shielding: Fundamentals, Properties, and Applications*, 1–9.

14 Ajitha, A., Surendran, A., Aswathi, M. et al. (2018). *Advanced Carbon Based Foam Materials for EMI Shielding, Advanced Materials for Electromagnetic Shielding: Fundamentals, Properties, and Applications*, vol. 14305, 14305–14325.

15 Joseph, N., Janardhanan, C., and Sebastian, M.T. (2014). Electromagnetic interference shielding properties of butyl rubber-single walled carbon nanotube composites. *Compos. Sci. Technol.* 101: 139–144.

16 Saini, P., Choudhary, V., Singh, B. et al. (2009). Polyaniline–MWCNT nanocomposites for microwave absorption and EMI shielding. *Mater. Chem. Phys.* 113 (2–3): 919–926.

17 Zhan, C., Yu, G., Lu, Y. et al. (2017). Conductive polymer nanocomposites: a critical review of modern advanced devices. *J. Mater. Chem. C* 5 (7): 1569–1585.

18 Sadasivuni, K.K., Ponnamma, D., Kumar, B. et al. (2014). Dielectric properties of modified graphene oxide filled polyurethane nanocomposites and its correlation with rheology. *Compos. Sci. Technol.* 104: 18–25.

19 Gojny, F.H., Wichmann, M.H., Fiedler, B. et al. (2006). Evaluation and identification of electrical and thermal conduction mechanisms in carbon nanotube/epoxy composites. *Polymer* 47 (6): 2036–2045.

20 Mathew, L.P., Kalarikkal, N., Thomas, S., and Volova, T. (2018). An effective EMI shielding material based on poly(trimethylene terephthalate) blend nanocomposites with multiwalled carbon nanotubes. *New J. Chem.* 42 (16): 13915–13926.

21 Ravindren, R., Mondal, S., Nath, K., and Das, N.C. (2019). Investigation of electrical conductivity and electromagnetic interference shielding effectiveness of preferentially distributed conductive filler in highly flexible polymer blends nanocomposites. *Composites Part A* 118: 75–89.

22 Ponnamma, D., Sadasivuni, K.K., Grohens, Y. et al. (2014). Carbon nanotube based elastomer composites–an approach towards multifunctional materials. *J. Mater. Chem. C* 2 (40): 8446–8485.

23 Ram, R., Rahaman, M., and Khastgir, D. (2015). Electrical properties of polyvinylidene fluoride (PVDF)/multi-walled carbon nanotube (MWCNT) semi-transparent composites: modelling of DC conductivity. *Composites Part A* 69: 30–39.

24 Mutiso, R.M. and Winey, K.I. (2015). Electrical properties of polymer nanocomposites containing rod-like nanofillers. *Prog. Polym. Sci.* 40: 63–84.

25 Sandler, J., Kirk, J., Kinloch, I. et al. (2003). Ultra-low electrical percolation threshold in carbon-nanotube-epoxy composites. *Polymer* 44 (19): 5893–5899.

26 Zhao, X., Zhao, J., Cao, J.-P. et al. (1980–2015). Effect of the selective localization of carbon nanotubes in polystyrene/poly(vinylidene fluoride) blends on their dielectric, thermal, and mechanical properties. *Mater. Des.* 56 (2014): 807–815.

27 Lee, S.H., Cho, E., Jeon, S.H., and Youn, J.R. (2007). Rheological and electrical properties of polypropylene composites containing functionalized multi-walled carbon nanotubes and compatibilizers. *Carbon* 45 (14): 2810–2822.

28 Rohini, R. and Bose, S. (2014). Electromagnetic interference shielding materials derived from gelation of multiwall carbon nanotubes in polystyrene/poly(methyl methacrylate) blends. *ACS Appl. Mater. Interfaces* 6 (14): 11302–11310.

29 Bryning, M.B., Islam, M.F., Kikkawa, J.M., and Yodh, A.G. (2005). Very low conductivity threshold in bulk isotropic single-walled carbon nanotube–epoxy composites. *Adv. Mater.* 17 (9): 1186–1191.

30 Du, F., Fischer, J.E., and Winey, K.I. (2003). Coagulation method for preparing single-walled carbon nanotube/poly(methyl methacrylate) composites and their modulus, electrical conductivity, and thermal stability. *J. Polym. Sci., Part B: Polym. Phys.* 41 (24): 3333–3338.

31 Haggenmueller, R., Gommans, H., Rinzler, A. et al. (2000). Aligned single-wall carbon nanotubes in composites by melt processing methods. *Chem. Phys. Lett.* 330 (3–4): 219–225.

32 Li, J., Ma, P.C., Chow, W.S. et al. (2007). Correlations between percolation threshold, dispersion state, and aspect ratio of carbon nanotubes. *Adv. Funct. Mater.* 17 (16): 3207–3215.

33 Mathew, L., Saha, P., Kalarikkal, N. et al. (2018). Tuning of microstructure in engineered poly(trimethylene terephthalate) based blends with nano inclusion as multifunctional additive. *Polym. Test.* 68: 395–404.

34 Thomas, S., Grohens, Y., and Jyotishkumar, P. (2014). *Characterization of Polymer Blends: Miscibility, Morphology and Interfaces*. Hoboken, New Jersey: Wiley.

35 Pawar, S.P., Rzeczkowski, P., Pötschke, P. et al. (2018). Does the processing method resulting in different states of an interconnected network of multiwalled carbon nanotubes in polymeric blend nanocomposites affect EMI shielding properties? *ACS Omega* 3 (5): 5771–5782.

36 Shakir, M.F., Khan, A.N., Khan, R. et al. (2019). EMI shielding properties of polymer blends with inclusion of graphene nano platelets. *Res. Phys.* 14: 102365.

37 Göldel, A., Marmur, A., Kasaliwal, G.R. et al. (2011). Shape-dependent localization of carbon nanotubes and carbon black in an immiscible polymer blend during melt mixing. *Macromolecules* 44 (15): 6094–6102.

38 Ravindren, R., Mondal, S., Nath, K., and Das, N.C. (2019). Synergistic effect of double percolated co-supportive MWCNT-CB conductive network for high-performance EMI shielding application. *Polym. Adv. Technol.* 30 (6): 1506–1517.

39 Ramachandran, A.A., Mathew, L.P., and Thomas, S. (2019). Effect of MA-g-PP compatibilizer on morphology and electrical properties of MWCNT based blend nanocomposites: new strategy to enhance the dispersion of MWCNTs in immiscible poly(trimethylene terephthalate)/polypropylene blends. *Eur. Polym. J.* 118: 595–605.

40 Zhang, K., Yu, H.-O., Shi, Y.-D. et al. (2017). Morphological regulation improved electrical conductivity and electromagnetic interference shielding in poly(L-lactide)/poly(ε-caprolactone)/carbon nanotube nanocomposites via constructing stereocomplex crystallites. *J. Mater. Chem. C* 5 (11): 2807–2817.

41 Pawar, S.P. and Bose, S. (2015). Peculiar morphological transitions induced by nanoparticles in polymeric blends: retarded relaxation or altered interfacial tension? *Phys. Chem. Chem. Phys.* 17 (22): 14470–14478.

42 Lee, T.-W. and Jeong, Y.G. (2014). Enhanced electrical conductivity, mechanical modulus, and thermal stability of immiscible polylactide/polypropylene blends by the selective localization of multi-walled carbon nanotubes. *Compos. Sci. Technol.* 103: 78–84.

43 Bose, S., Cardinaels, R., Özdilek, C. et al. (2014). Effect of multiwall carbon nanotubes on the phase separation of concentrated blends of poly[(α-methyl styrene)-*co*-acrylonitrile] and poly(methyl methacrylate) as studied by melt rheology and conductivity spectroscopy. *Eur. Polym. J.* 53: 253–269.

44 Hoseini, A.H.A., Arjmand, M., Sundararaj, U., and Trifkovic, M. (2017). Tunable electrical conductivity of polystyrene/polyamide-6/carbon nanotube blend nanocomposites via control of morphology and nanofiller localization. *Eur. Polym. J.* 95: 418–429.

45 Abbasi Moud, A., Javadi, A., Nazockdast, H. et al. (2015). Effect of dispersion and selective localization of carbon nanotubes on rheology and electrical conductivity of polyamide 6 (PA 6), Polypropylene (PP), and PA 6/PP nanocomposites. *J. Polym. Sci., Part B: Polym. Phys.* 53 (5): 368–378.

46 Bose, S., Ozdilek, C., Leys, J. et al. (2010). Phase separation as a tool to control dispersion of multiwall carbon nanotubes in polymeric blends. *ACS Appl. Mater. Interfaces* 2 (3): 800–807.

47 Rostami, A., Masoomi, M., Fayazi, M.J., and Vahdati, M. (2015). Role of multiwalled carbon nanotubes (MWCNTs) on rheological, thermal and electrical properties of PC/ABS blend. *RSC Adv.* 5 (41): 32880–32890.

48 Biswas, S., Kar, G.P., and Bose, S. (2016). Simultaneous improvement in structural properties and microwave shielding of polymer blends with carbon nanotubes. *ChemNanoMat* 2 (2): 140–148.

49 Biswas, S., Kar, G.P., and Bose, S. (2015). Microwave absorbers designed from PVDF/SAN blends containing multiwall carbon nanotubes anchored cobalt ferrite via a pyrene derivative. *J. Mater. Chem. A* 3 (23): 12413–12426.

50 Al-Saleh, M.H. (2016). Electrical, EMI shielding and tensile properties of PP/PE blends filled with GNP: CNT hybrid nanofiller. *Synth. Met.* 217: 322–330.

51 Ajitha, A.R., Aswathi, M.K., Reghunadhan, A. et al. (2018). Effect of MWCNTs on Wetting and Thermal Properties of an Immiscible Polymer Blend. *InMacromolecular Symposia* 381 (1): 1800103.

52 Li, P., Du, D., Guo, L. et al. (2016). Stretchable and conductive polymer films for high-performance electromagnetic interference shielding. *J. Mater. Chem. C* 4 (27): 6525–6532.

53 Wang, Y., Gu, F.-Q., Ni, L.-J. et al. (2017). Easily fabricated and lightweight PPy/PDA/AgNW composites for excellent electromagnetic interference shielding. *Nanoscale* 9 (46): 18318–18325.

54 Gahlout, P. and Choudhary, V. (2020). EMI shielding response of polypyrrole–MWCNT/polyurethane composites. *Synth. Met.* 266: 116414.

55 Yang, Z., Zhang, Y., and Wen, B. (2019). Enhanced electromagnetic interference shielding capability in bamboo fiber@polyaniline composites through microwave reflection cavity design. *Compos. Sci. Technol.* 178: 41–49.

56 Fang, F., Li, Y.-Q., Xiao, H.-M. et al. (2016). Layer-structured silver nanowire/polyaniline composite film as a high performance X-band EMI shielding material. *J. Mater. Chem. C* 4 (19): 4193–4203.

57 Bora, P.J., Vinoy, K., Ramamurthy, P.C., and Madras, G. (2017). Electromagnetic interference shielding effectiveness of polyaniline–nickel oxide coated cenosphere composite film. *Compos. Commun.* 4: 37–42.

58 Li, H., Lu, X., Yuan, D. et al. (2017). Lightweight flexible carbon nanotube/polyaniline films with outstanding EMI shielding properties. *J. Mater. Chem. C* 5 (34): 8694–8698.

59 Al-Saleh, M.H. (2016). Carbon nanotube-filled polypropylene/polyethylene blends: compatibilization and electrical properties. *Polym. Bull.* 73 (4): 975–987.

60 Sharika, T., Abraham, J., George, S.C. et al. (2019). Excellent electromagnetic shield derived from MWCNT reinforced NR/PP blend nanocomposites with tailored microstructural properties. *Composites Part B* 173: 106798.

61 Abraham, J., Kailas, L., Kalarikkal, N. et al. (2016). Developing highly conducting and mechanically durable styrene butadiene rubber composites with tailored microstructural properties by a green approach using ionic liquid modified MWCNTs. *RSC Adv.* 6 (39): 32493–32504.

62 Jeong, Y.G. and Jeon, G.W. (2013). Microstructure and performance of multiwalled carbon nanotube/*m*-aramid composite films as electric heating elements. *ACS Appl. Mater. Interfaces* 5 (14): 6527–6534.

63 Thomassin, J.-M., Jerome, C., Pardoen, T. et al. (2013). Polymer/carbon based composites as electromagnetic interference (EMI) shielding materials. *Mater. Sci. Eng. R: Rep.* 74 (7): 211–232.

64 Scarisbrick, R. (1973). Electrically conducting mixtures. *J. Phys. D: Appl. Phys.* 6 (17): 2098.

65 Rahaman, M., Chaki, T., and Khastgir, D. (2012). Modeling of DC conductivity for ethylene vinyl acetate (EVA)/polyaniline conductive composites prepared through insitu polymerization of aniline in EVA matrix. *Compos. Sci. Technol.* 72 (13): 1575–1580.

66 Yuan, D., Guo, H., Ke, K., and Manas-Zloczower, I. (2020). Recyclable conductive epoxy composites with segregated filler network structure for EMI shielding and strain sensing. *Composites Part A* 132: 105837.

67 Kar, G.P., Biswas, S., and Bose, S. (2015). Simultaneous enhancement in mechanical strength, electrical conductivity, and electromagnetic shielding properties in PVDF–ABS blends containing PMMA wrapped multiwall carbon nanotubes. *Phys. Chem. Chem. Phys.* 17 (22): 14856–14865.

68 Poothanari, M.A., Abraham, J., Kalarikkal, N., and Thomas, S. (2018). Excellent electromagnetic interference shielding and high electrical conductivity of compatibilized polycarbonate/polypropylene carbon nanotube blend nanocomposites. *Ind. Eng. Chem. Res.* 57 (12): 4287–4297.

69 Liang, C., Hamidinejad, M., Ma, L. et al. (2020). Lightweight and flexible graphene/SiC-nanowires/poly(vinylidene fluoride) composites for electromagnetic interference shielding and thermal management. *Carbon* 156: 58–66.

70 Gao, W., Zhao, N., Yu, T. et al. (2020). High-efficiency electromagnetic interference shielding realized in nacre-mimetic graphene/polymer composite with extremely low graphene loading. *Carbon* 157: 570–577.

71 Li, T.-T., Wang, Y., Peng, H.-K. et al. (2020). Lightweight, flexible and superhydrophobic composite nanofiber films inspired by nacre for highly electromagnetic interference shielding. *Composites Part A* 128: 105685.

72 Ji, X., Chen, D., Shen, J., and Guo, S. (2019). Flexible and flame-retarding thermoplastic polyurethane-based electromagnetic interference shielding composites. *Chem. Eng. J.* 370: 1341–1349.

73 Bizhani, H., Nayyeri, V., Katbab, A. et al. (2018). Double percolated MWCNTs loaded PC/SAN nanocomposites as an absorbing electromagnetic shield. *Eur. Polym. J.* 100: 209–218.

Index

a
acylhydrazone bonds   198
Ag nanowires   199
aligned carbon nanotubes (ACNTs)   32, 169
allyl-terminated P3HT   161, 162
allyl-terminated poly(3-hexylthiophene-2, 5-diyl) (P3HT)   79, 161
ammonium citrate   196
anisotropic nanofillers   7, 9

b
ball milling   10
barium strontium titanate (BST) nanoparticles   62
BaTiO$_3$ nanoparticles   193
bio-inspired graphite/PDMS polymer composite   139
bio-polar behavior   230
biodegradable polymers, in electronics
  biocompatibility and metabolization   53
  cellulose   62–64
  challenges and prospects   66–67
  chemical structures   54
  chitosan   64–65
  environmental protection   53
  poly(butylene succinate)   54
  poly(ε-caprolactone) (PCL)   54, 58–59
  poly(vinyl alcohol) (PVA)   59–61
  poly(vinyl pyrrolidone) (PVP)   61–62
  polyhydroxyalkanoates   54
  polylactide (PLA)   55–58
  silk fibril (SF)   65–66
  thin film transistor (TFT)   54–55
biodegradation   204, 207
boron nitride (BN)   6, 10, 231
bulk composite   192, 193

c
capacitive pressure sensors   131, 132, 135–138
carbon nanotubes   6, 10, 21, 24, 26, 31–32, 158–161, 169–171, 193, 200, 219, 232, 272, 275–277
carbon rich filaments (CRF)   227, 232, 233
cellulose   31, 35, 54, 62–64, 66–67, 204–207, 233
cellulose-based electro-active paper (EAPap)   62, 63
CeO$_2$/PANI nanocomposites   167
chemical etching   193
chemically-reinforced polymer nanocomposite   144
chitosan   54, 59, 62, 64–67, 204
complementary metal-oxide-semiconductor   77
conducting polymers   7, 160, 162, 267, 269, 270–274
conductive filler   21–25, 27, 29, 31, 34, 39, 40, 41, 61, 202, 247–250, 257, 258, 274
conductive nanofillers   7, 22, 23, 31, 33–35, 37, 40, 41, 57, 267
conductive polymer composites (CPCs)
  conductivity vs. filler loading   249

Polymer Nanocomposite Materials: Applications in Integrated Electronic Devices, First Edition.
Edited by Ye Zhou and Guanglong Ding.
© 2021 WILEY-VCH GmbH. Published 2021 by WILEY-VCH GmbH.

Index

conductive polymer composites (CPCs) (contd.)
  electrical conductivity 21
  fabrication methods
    chemical modification 22
    electrical conductivity 22, 23
    in situ polymerization 27
    melt blending 23–25
    nanofillers and special processing technique 22
    physical blending 22
    solution blending 25–26
  gas sensor 35–38
  nanofillers, selective distribution of
    segregated structure 29, 30
    surface coating 31–32
  percolation threshold 248
  piezoresistive sensor 33–35
  polymer matrix 27–29
  strain sensor 33
  temperature sensor 38–40
conductive polymer nanocomposites 21, 59, 132, 139, 269, 278
conductive polymers (CPs) 21, 85, 90, 161, 164, 165, 167, 171, 269, 274
conventional processing equipment 57
covalent bonding 7, 25, 231
cp(S-PMAT) copolymer 161, 162

d

Diels–Alder reactions 198
dip coating 31, 217
discharge–charge process 161, 162
dispersion-deposition process 159
disulfide bonds 198, 202, 204
2D materials 78, 82, 83, 86, 87, 213, 218, 231–232, 234
double percolation effect 271, 272
double-screw extrusion 10
3D polymer nanocomposite 196, 197
3D porous ternary composites 169, 170
3D printing 57, 105, 106, 206
3D printing method 57
3D printing technique 57, 105

e

electrical conductivity 21–25, 30, 41, 56, 57, 159, 197, 202, 205, 248, 250, 256, 270, 274, 275
electrical double-layer capacitors (EDLCs) 164, 165
electro hydrodynamic (EHD) technique 62, 217, 226, 227, 228, 231
  atomization 231
  method 62
electro-hydrodynamic atomization (EDHA) technique 226
electro-hydrodynamic coating technique 227, 228
electromagnetic interference (EMI) shielding material
  conducting polymers for 272–274
  conductivity studies of 274–276
  materials for 269
  mechanisms of 277
  polymer blend nanocomposites 271–272
  vector network analyzer SE 277
electrospinning 27, 57–59, 63, 194, 195, 203
external quantum efficiency (EQE) 80–82, 88–95, 99, 101–105, 109–112, 114–115, 117–123

f

fabrication techniques 27, 217–218
Fe_3O_4 nanosheets 6
fiber reinforced composites 1
fibre-reinforced plastics (FRPs) 261, 262
fingerprint pattern 35
fluorine doped tin oxide (FTO) 102, 221, 232
fossil fuel consumption 157
functional polymer-based nanocomposite 189
functionalized graphene sheets (FGSs) 159

g

gas sensor 35–38, 41, 64

glass fiber-reinforced plastics 1
gold nanoparticles 1, 57, 65
G/PDMS polymer nanocomposite 149, 151
graphene nano flakes 225, 226
graphene nanoparticles 10
graphene nanoplatelet (GNP) 261, 262
graphene oxide (GO) 27, 59, 63, 152, 159, 193, 224–227, 233, 267

h
hard-disk drive (HDD) 211
healing process 198, 199, 200, 204
heat distortion temperature (HDT) 23, 58, 67
heptane 37, 38
high resolution transmission electron microscope (HRTEM) 170
highest occupied molecular orbit (HOMO) 85, 123
H-PDMS/Ag-PEDOT film 200
human wrist pulse detection 148–149
hydrogel 33, 63–66
hydrogen bonding 31, 64, 172, 173, 198, 200, 202, 224

i
imine bonds 198, 200
in situ polymerization process 9, 22, 23, 27, 40, 59, 160, 162, 167, 194–197, 278
intrinsic systems 198
intrinsically conductive polymers (IPCs) 274
introduced surfactant 40

k
Kapton films 190

l
lithium batteries 158–160
lithium-sulfur batteries 158, 161
lowest unoccupied molecular orbit (LUMO) 85, 97, 98

m
melt blending 23–25, 194, 195
melt spinning method 56
memristor 55, 61, 62, 66, 212, 213, 228
microelectromechanical systems (MEMS) 131, 154
microneedle array (MNA) dermal biosensor 56
micropillar array 35
microstructured composite film 141
mixed filler 254–255, 262
montmorillonite clays (MMT) 6
Moore's law 211, 233
multi-walled carbon nanotubes (MWCNTs) 24–26, 29, 30, 200, 219, 255, 267, 275–278
multifilament elastomer fiber 31

n
nanocomposites 1–11, 53–67, 77–123, 131–154, 157–177, 189–208, 211–234, 247–263, 267–278
nanofiber composite 26, 31, 32, 38
nanofiller aggregations 40
nanofiller incorporated conducting polymer composites 270–271
nanographene platelets (NGPs) 6
nanomaterials 1–3, 5, 6, 9, 78, 132, 153, 154, 192, 194–198, 267, 270–272
nanostructured polyester (PET) 53, 95, 96, 103, 139, 141, 190, 191, 205, 214, 221, 222, 225, 226, 231, 232
natural biodegradable polymers 54, 204
negative temperature coefficient (NTC) effect 38–40, 247, 250, 258, 262
next generation nonvolatile memory (NVM) 212

o
ohmic conductance 272
ohmic conduction 216, 219, 222–224, 226, 229, 230
one-dimensional nanofillers 5–6

organosulfur electrode – cp(S-PMAT) 162, 163
oxide based polymer nanocomposite RSM 218–222

p

pattern-transfer printing process 193
percolation threshold 5, 22, 23, 29, 30, 40, 248, 250, 253, 257, 270–272, 275, 276, 278
phase change material 213
[6,6]-phenyl C$_{61}$ butyric acid methyl ester (PCBM) 79
photoconductor 79, 80–82, 85–88, 99, 102, 103, 115, 118, 122
photodetectors (PD)
　complementary metal-oxide-semiconductor 77
　MEH-PPV-inorganic nanocrystals nanocomposite 112–115
　MEH-PPV-small molecular organic nanocomposite 98–99
　novel semiconductors
　　2D materials 82, 83
　　CQD-based PDs 83
　　highest occupied molecular orbit (HOMO) 85
　　lowest unoccupied molecular orbit (LUMO) 85
　　P3HT and PCBM 87
　　photo-induced excitons 86
　　quantum dots (QDs) 83
　P3HT-inorganic nanocrystals nanocomposite 115–116
　P3HT-small molecular organic nanocomposite 99–100
　photoconductor 80–82
　photodiode (PDi) 80
　polymer-polymer nanocomposite 88–98
　polymer-polymer-small molecular organic nanocomposite 108–109
　polymer-small molecular organic-inorganic nanocrystals nanocomposite 120–123
　polymer-small molecular organic-small molecular organic nanocomposite 110–111
photodiode (PDi) 79–80, 87–99, 101–105, 109–111, 114, 115, 120
photomultiplication 103–105, 110
physical blending 22, 40
piezoresistive pressure sensors 131–132, 137–143, 149, 150, 152
piezoresistive sensor 33–35, 41, 66, 131–132
planar polymer composite film 141
poly (3,4-ethylenedioxythio-phene) poly (styrenesulfonate) (PEDOT:PSS) 28, 61, 111, 214, 218–221, 229–231, 234
poly(ε-caprolactone) (PCL) 54, 58–59, 66–67, 203, 205, 207, 257
poly(3,4-ethylenedioxythiophene) (PEDOT) 29, 167, 169–171, 218
poly(3,4-ethylenedioxythiophene) (Ag-PEDOT) 199–200
poly(3-hexylthiophene-2,5-diyl) (P3HT) 79, 85–87, 91–92, 95, 98–99, 102–107, 110–112, 115–123, 161–162, 232
poly(m-aminothiophenol) (PMAT) 162
poly(anthraquinonyl sulfide) (PAQS) 158–160
poly(L-lactide) (PLLA) 57–58
poly(styrenesulfonate) (PSS) 26, 167, 170, 218
poly(vinyl alcohol) (PVA) 26, 33, 54, 59–61, 64, 66, 167, 175–176, 194–195, 205, 206–207, 214, 218–219, 221, 223–226, 229–231
poly(vinyl pyrrolidone) (PVP) 26, 60–62, 218–219, 221–222, 224–229, 231–233
polyacrylic acid (PAA) 64, 172, 193
polyaniline (PANI) 21, 64–65, 165, 167, 196, 274
polybutylene terephthalate (PBT) nanocomposites 23

Index

polydimethylsiloxane (PDMS) 28, 31–32, 34, 59, 135–137, 140–142, 146, 153, 190, 192–194, 199–200, 205, 227
polydimethylsiloxane-polyurethane (PDMS–PU) polymer 199
polyimide (PI) 27, 53, 159–161, 190, 218
polylactide (PLA) 37, 54–59, 66–67, 205–206
polymer blend 171, 229–230, 257–259, 261–262, 271–272, 277–278
polymer matrix 2–3, 7–8, 21–29, 31, 35, 37, 40–41, 67, 86, 91, 115, 132, 144–145, 154, 159, 176, 190, 193–196, 198, 200–201, 205, 217–218, 232, 247, 249–250, 255, 257, 261, 262, 270, 274, 278
polymer nanocomposite (PoNa) 78–79, 86–88, 95–96, 98–99, 101
polymer nanocomposite (PNC) 140, 141, 145, 163, 213
  advantages 3–5
  ball milling 10
  batteries and supercapacitors 157
  definition 157
  double-screw extrusion 10
  electrode materials
    for batteries 158
    for supercapacitor 164
    polymer–graphene/carbon nanotube 158–161, 165–169
    polymer–inorganic 161–163
    polymer–metal oxide 165
    polymer–metal oxide-graphene/carbon nanotubes 169–171
    polymer–organic salt graphene 163, 164
  electrolytes
    for batteries 171–172
    for supercapacitor 172–173
  flexible and wearable electronic products 159
  heat resistance and biodegradability 2
  in situ synthesis 10–11
  materials 218
  mechanical properties 2
  one-dimensional nanofillers 5–6
  properties of 6–7
  separator
    for batteries 174–175
    for supercapacitors 175–176
  shear mixing 9
  surface chemical and physicochemical properties of 2
  three roller milling 9–10
  three-dimensional nanofillers 6
  two-dimensional nanofillers 6
  ultrasonication-assisted solution mixing 8–9
polymer-graphene nanocomposites 159
polymer–graphene/carbon nanotube 158–161, 165–169
polymer–inorganic 112, 161–163
polymer–metal oxide 165, 169–171
polymer–organic salt graphene 163, 164
polymer–polymer nanocomposite 88–98
polypyrrole (PPy) 21, 165–169, 172–173, 226, 274, 278
polytrimethylene terephthalate/polypropylene/carbon nanotube (PTT/PP/MWCNTs) blend nanocomposites 277
polyurethane (PU) 28, 139, 190, 203, 223, 274
polyvinyl alcohol (PVA) 26, 33, 54, 59–61, 64, 66, 167, 176, 194, 205–207, 218–219, 221, 225, 229, 231
polyvinyl alcohol (PVOH 231
polyvinylpyrrolidone 54, 196, 218
Poole–Frenkel effect 232
porous G/PDMS polymer composite 142–143
positive temperature coefficient (PTC) 247
  conductive chain and thermal expansion model 249
  conductive composites 250
  contacting points 253

positive temperature coefficient (PTC) (*contd.*)
  effects 38, 257
  electrical resistivity 247
  electron tunnelling 250
  explanation and defects 251
  filler dispersion and distribution 253–254
  filler size and shape 250, 253
  in CPCs 248
  mechanism for 249
  mixed filler 254–255
  polymer blend 257–258
  polymer thermal expansion and crystallinity 255–257
  silver coated glass spheres 250
  temperature sensors 259
pressure sensing performance 35–36, 141
pressure sensitivity 131–132, 134, 135, 137, 140, 142–143
pressure sensors
  applications of 154
  capacitive pressure sensors 131, 135–137
  conductive polymer nanocomposites 132
  E-skin application 152–153
  flexible and stretchable pressure sensors 133
  human wrist pulse detection 148–149
  linear relationship 134
  LOD and response speed 134
  microelectromechanical systems (MEMS) 131
  piezoresistive pressure sensors 131, 137–143
  pressure sensitivity 132, 134
  reliability 134
  sensitivity, LOD and response time 153
  subtle human motion detection 149–151
  texture roughness detection 151–152
  triboelectric pressure sensors 143–148

PS/graphene/toluene suspension 24
PVA/MXene nanofibers film 195
PVDF/rGO polymer nanocomposite 152

q

quantum dots (QDs) 6, 55, 62, 83, 114, 218, 227–229, 232
quantum dot based RSM devices 227

r

radio frequency identification (RFID) antenna 259
reduced graphene oxide (rGO) 27, 63, 152–154, 159, 163–164, 167, 169, 193, 227
reliability 64, 66, 134, 189, 197, 202, 204, 208, 224
resistive random access memory (RRAM) based resistive switching 60–61, 211–213, 233
resistive switching memory (RSM)
  alkali-lignin based RSM 223
  cations and anions migration 216
  charge trapping/de-trapping mechanism 232
  conduction mechanisms 216–217
  conductive filaments 214–215
  crossbar arrays 212
  current-voltage (I-V) characteristics curve of 214
  electro-hydrodynamic coating technique 228
  electrons trapping and de-tapping 216
  fabrication techniques 217–218
  graphene based polymer nanocomposite 224–227
  logic-in-memory function 212
  metal based nanoparticles 222–224
  neuromorphic computing 212
  next generation nonvolatile memory 212
  organic polymers and inorganic nanocomposites 213
  oxide based polymer nanocomposite RSM 218–222

performance comparison 243
PNC materials 213, 214
polymer based nanocomposites 229–231
polymer nanocomposite materials 218
quantum dot based RSM devices 227–229
resistance states 212
switching mechanisms 215
two-dimensional (2D) materials 231–232
universal memory 212
working mechanism 212
ring opening polymerization (ROP) 55, 58
roll-to-roll lamination fabrication method 95–96

s

scalable nonvolatile memory techniques 211
scanning electron microscope (SEM) 28, 30–33, 35–36, 136–137, 141–142, 145, 151, 159–163, 165, 167, 169–172, 175–176, 195–196, 203, 206, 219–220, 225, 228, 230
SCLC conduction mechanism 220, 224, 230–231
sea-urchin shaped metal nanoparticles (SSNPs) 139–140
segregated structure 27, 29, 30, 37, 41
self-activating potentiometric sensor 65
self-healing TENG 198–199, 201
self-powered pressure sensors 143
self-powered sensors 147, 190, 200
semicrystalline polycaprolactone (PCL) 203
shape memory polymer (SMP) 201
shape memory PU (SMPU) 203–204
shear mixing 9
silica nanoparticles 10, 172
silk fibril (SF) 65–66, 207
silver coated glass spheres (AgS) 250, 255

single-walled carbon nanotube (SWNT) 31–32, 160–161
sodium alginate film 206
sodium dodecyl benzene sulfonate 196
solution blending 25–26, 194–195
solution blow spinning method 56
solution processing technique 217, 225, 229
space charge limited current (SCLC) mechanism 216–217, 220, 223–225, 229
S-P3HT copolymer 161
S-P3HT sulfur composites 161–162
spin-coating technique 194, 230
static pressure detection 132, 140
strain sensor 33, 41, 59, 65, 198
structural conductive polymers (SCP) 21
subtle human motion detection 149–151, 154
sufficient pressure 146
supercapacitor 157–158, 164–173, 175–176, 197
superhydrophobicity 33
Swagelok cells 161
synthetic biodegradable polymers 54, 59, 205

t

temperature sensor 38–41, 63, 203–204, 248, 259–260, 262
tetrahydrofuran (THF) 37, 38, 57, 59, 230
texture roughness detection 151–152
texture roughness measurement 151
thermoforming-corrosion method 196
thin film transistor (TFT) 55, 60–62, 64, 66, 115
three-dimensional (3D) polymer nanocomposites 196
three roller milling 9
time-resolved photoluminescence method 115
toluene 24, 37–38, 59
transfer-printing technique 95

triboelectric nanogenerators (TENGs) 145
  applications 192
  biodegradation 204–207
  contact electrification and electrostatic induction 189, 190
  electrical output performances 192
  FT mode 192
  functional polymer nanocomposite
    Ce-doped ZnO nanoparticles 196
    conductive sponge 197
    electrospinning 194
    fabrication strategies 194
    in-suit polymerization 195
    melt blending 195
    PCP-TENG 197
    spin-coating 194
    ultrasonic dispersion 196
  grating and rotating structures 194
  grating structure 191
  ingenious structure design 194
  ITO and electrostatic charges 192
  Kapton films 190
  mechanical damage and harsh conditions 189
  monolayer MoS2 193
  nanostructured polyester (PET) and Kapton films 190
  polymer composite 192
  self-healing TENGs
    Ag nanowires and poly(3,4-ethylenedioxythiophene) (Ag-PEDOT) 199
    electrical performance 198
    extrinsic systems 198
    functionalized graphene nanosheets 198
    hydrogen bond-containing elastomeric substrates 198
    intrinsic systems 198
    MWCNTs 200
    non-covalent systems 198
    PDMS-PU 199
    wrinkled graphene 198
  shape memory polymer (SMP) 201–204
  surface functionalization and bulk composite 192
2,4,6-trinitrotoluene (TNT) 57
tristable resistive switching behavior 229
tunneling effect 250, 272

u

ultrasonic dispersion 196
ultrasonication 25, 31, 226
ultrasonication-assisted solution mixing 8–9

v

vinyl hybrid silica nanoparticles (VSNPs) 172–173

w

wet-spinning 27
write once and read many (WORM) 213

z

ZnO nanosheets 6